International Conference on Applied Mechanics, Materials and Civil Engineering (ICAMMCE2020)

Held online due to COVID-19

IOP Conference Series: Materials Science and Engineering
Volume 964

Shanghai, China
20 - 21 September 2020

ISBN: 978-1-7138-3294-2
ISSN: 1757-8981

Printed from e-media with permission by:

Curran Associates, Inc.
57 Morehouse Lane
Red Hook, NY 12571

proceedings.com

Some format issues inherent in the e-media version may also appear in this print version.

Institute of Physics
Dirac House, Temple Back
Bristol BS1 6BE UK

Phone: 44 1 17 929 7481
Fax: 44 1 17 920 0979

techtracking@iop.org

Additional copies of this publication are available from:

Curran Associates, Inc.
57 Morehouse Lane
Red Hook, NY 12571 USA
Phone: 845-758-0400
Fax: 845-758-2633
Email: curran@proceedings.com
Web: www.proceedings.com

TABLE OF CONTENTS

APPLIED MECHANICS AND MECHANICAL ENGINEERING

MATERIALS SCIENCE AND MATERIALS PROCESSING TECHNOLOGY

CIVIL AND STRUCTURAL ENGINEERING

IOP Conf. Series: Materials Science and Engineering **964** (2020) 011001 doi:10.1088/1757-899X/964/1/011001

Preface

Great honor to receive your support of 2020 International Conference on Applied Mechanics, Materials and Civil Engineering (ICAMMCE2020). It will offer an open academic forum for all the participants in the field of Applied Mechanics and Engineering Mechanics, Material Science, Civil Engineering and the other related fields to exchange their latest research findings. On behalf of the organizing committee, we would like to thank for your actively participation and powerfully support!

Under the background of the severe impact of COVID-19 and the travel restrictions, most of the authors can't attend the conference physically. So the organizing committee of ICAMMCE2020 actively responses to the call of the government, reducing the population' s gathering and ensuring the safety and health of all the participants in the conference. Hence, the virtual conference is a better choice for ICAMMCE2020.

It was held virtually via the platform "VooV Meeting" on September 20, 2020. First of all, the organizing committee has opened a test room for all keynote speakers and authors to test the VooV, including video, audio and the screen before the opening. The conference procedures include three parts, including Keynote Speeches, Oral Sessions and Poster Sessions. In the first session of Keynote Speeches, the speakers were allowed 40 minutes to deliver their speeches. In the second session of Oral Sessions, the participants had 10-15 minutes to share their presentation one by one. And the 10-15 minutes Q&A discussion after every presentation was indispensable to achieve a better academic atmosphere.

It is great honor to invite two senior experts at here to share their keynote speeches to us. The first expert is Prof. Jianqin Ma from Chang'an University, Xi'an, China. The second expert is Prof. Igor Ivitsky from National Technical University of Ukraine "Igor Sikorsky Kyiv Polytechnic Institute", Ukraine. Some participants have put forward some interested questions about these keynote speeches, and the experts have clearly explained it. These interaction results in a favorable academic atmosphere. Some excellent oral presenters were awarded for their profound and impressive speeches. All of them has contribute to a huge success of ICAMMCE2020 online conference.

All the submissions in ICAMMCE2020 have gone through a double-blind peer reviewing, the editors and reviewers were responsible to make sure the scientific equality, verification and high standard. There are 108 relevant professors and experts join in the Technical Program Committee for paper reviewing and checking. In view of the main theme and topics, each submission was reviewed by at least two or three experts under a critical reviewing process for scientific paper publication based on the paper contents, topic relevance, structure, language standardization, originality and more aspects. ICAMMCE2020 uses the iThenticate software to detect instances of overlapping and similar text in all submitted papers. Finally, ICAMMCE2020 has received 32 excellent papers, selected from 106 submissions before the final submission deadline. ICAMMCE2020 proceeding contains these following parts:

1. Applied Mechanics and Mechanical Engineering
2. Materials Science and Materials Processing Technology
3. Civil and Structural Engineering

In the end, ICAMMCE2020 want to express the sincerest appreciation to all the active authors with outstanding and valuable academic papers, to the sponsor of Advanced Science and Industry Research Center, Hong Kong, IOP Conference Series: Materials Science and Engineering with effective publication for all excellent paper submissions, all the members and reviewers in ICAMMCE2020 Technical Program Committee with careful reviewing response, for the impressive results of the conference.

IOP Conf. Series: Materials Science and Engineering **964** (2020) 011001 doi:10.1088/1757-899X/964/1/011001

ICAMMCE2020 Organization Committees

IOP Conf. Series: Materials Science and Engineering **964** (2020) 011002 doi:10.1088/1757-899X/964/1/011002

Reviewer List

Prof. Samad Sabbaghi, University of Waterloo, Canada
Prof. Guillermo Escrivá-Escrivá, Universitat Politècnica de València, Spain
Dr. Sanae DAHBI, University Mohammed First Oujda, Morocco
Dr. Andrian Seleznev, Ural Division of Russian Academy of Sciences, Russia
Prof. Igor Ivitsky, National Technical University of Ukraine "Igor Sikorsky Kyiv Polytechnic Institute", Ukraine
Prof. Jianqin Ma, Department of Geotechnical and Tunnel Engineering, School of Highway, Chang'an University, Xi'an, China
Prof. Grzegorz Woroniak, Bialystok University of Technology, Poland
Prof. Octavio Castillo Reyes, Polytechnic University of Catalonia, Spain
Prof. Dan Dobrota, Lucian Blaga University of Sibiu, Romania
Dr. Djamila Bouayad, Bejaia university, Algeria
Dr. Đurđica Kovačić, University of J.J. Strossmayer in Osijek, Croatia
Dr. Nidhi Gupta, Amity University, India
Prof. Valentin Mateev, Technical University of Sofia, Moroccan
Dr. Soumya Mukherjee, Kazi Nazrul University (Public University), India
Dr. R.Murugan, National Institute of Technology Silchar, India
Dr. Karan Veer, Kurukshetra University, India
Prof. Henni Abdellah, Kasdi Merbah University, Algeria
Dr. Mohammad Adil Dar, Indian Institute of Technology - Delhi New Delhi, India
Dr. Chih-Ta Tsai, National Cheng Kung University, Taiwan
Mr. Binod Khadka, Tongji University, China
Prof. Lei Mao, University of Science and Technology of China, China
Prof. Agota Banyai, University of Miskolc, Hungary
Prof. M.Arif Kamal, Aligarh Muslim University, India
Prof. Abdelwahed Barkaoui, International University of Rabat, Morocco
Prof. Yousef Farhaoui, Moulay Ismail University, Morocco
Prof. B. Ramesh, St. Joseph's Institute of Technology, India
Dr. Sumit Kushwaha, Indian Institute of Technology, Kanpur, India
Dr. Jose Miguel Martínez Valle, University of Córdoba, Spain
Prof. Mouattah Kaddour, University Ammar Thlidji of Laghouat, Algeria
Dr. Boutagouga Djamel, University of Tebessa, Algeria
Prof. Sopushynskyy Ivan, Ukrainian National Forestry University,Ukraine
Prof. Miguel Caixinha, University of Beira Interior, Portugal
Dr. Kaviya Somasundaram, Cochin University of Science and Technology,India
Prof. Srinivasa C.V., University Karnataka, India
Dr. Ciortea Elisabeta Mihaela, "1 Decembrie 1918" University of Alba Iulia, Romania
Prof. Arnulfo Luévanos Rojas, Autonomous University of Coahuila, México
Prof. Gholamreza Khalaj, Islamic Azad University, Iran
Prof. Satya Pal Singh, M M M University of Technology, India
Prof. Ali Rahman, Southwest Jiaotong University, China
Prof. Masoud Rahaeifard, Ardakan University, Ardakan, Iran
Prof. Abdelwahed Barkaoui, International University of Rabat, Morocco
Prof. Ana Patricia Pérez Fortes, Universidad Politécnica de Madrid, Spain
Prof. Asselah Amel, University of M'Hamed Bougara-Boumerdes, Algeria
Prof. Khaled Benyounes, University M'Hamed Bougara of Boumerdes, Algeria
Dr. Agnieszka Woszuk, Lublin University of Technology,Poland
Dr. Aydin Shishegaran, Iran University of Science and Technology, Iran
Prof. Shahram Babaie, Islamic Azad University, Iran
Prof. Shehata E Abdel Raheem, Assiut University, Egypt

IOP Conf. Series: Materials Science and Engineering **964** (2020) 011002 doi:10.1088/1757-899X/964/1/011002

IOP Conf. Series: Materials Science and Engineering **964** (2020) 011003 doi:10.1088/1757-899X/964/1/011003

Peer review declaration

All papers published in this volume of *IOP Conference Series: Materials Science and Engineering* have been peer reviewed through processes administered by the Editors. Reviews were conducted by expert referees to the professional and scientific standards expected of a proceedings journal published by IOP Publishing.

- **Type of peer review: Double-blind**
- **Conference submission management system:**
  All the papers were submitted via the conference email: conf@icammce2020.org or icammce2020@sina.com .
- **Number of submissions received: 106**
- **Number of submissions sent for review: 95**
- **Number of submissions accepted: 32**
- **Acceptance Rate (Number of Submissions Accepted / Number of Submissions Received X 100): 30.19%**
- **Average number of reviews per paper: 1. 8**
- **Total number of reviewers involved: 108**
- **Any additional info on review process: ICAMMCE2020 uses the iThenticate software to detect instances of overlapping and similar text in submitted manuscripts.**
- **Contact person for queries: dwfang@yeah.net**

IOP Conf. Series: Materials Science and Engineering **964** (2020) 012001 doi:10.1088/1757-899X/964/1/012001

Seismic Performance Analysis of Irregular Structures under Combined Action of Base Isolation and Energy Dissipation Braces

Ying Wang[1,*], Futing Pan[1], Tan Yang[2] and Chunjiao Dai[1]

[1]Key Laboratory of Building Structure of Anhui Higher Education Institutes, Anhui Xinhua University, Hefei ,China
[2]State Key Laboratory of Fire Science,University of Science and Technology of China, Hefei, China

*Corresponding author email: wangying@axhu.edu.cn

Abstract. Irregular structures are prone to torsional vibration under earthquake action, which aggravates the damage of structures. With the development of energy dissipation technology, LRB and BRB are applied to an irregular steel frame. By comparing and analyzing the seismic response characteristics and laws of the original structure, BRB structure, LRB structure and LRB + BRB structure, the seismic performance and advantages and disadvantages of each control scheme are obtained, especially the influence on torsion effect, which provides technical reference for the practical application of such projects.

1. Introduction
With the development of economy, the application of irregular building structure is more and more common. For the irregular L-shaped frame structure, due to the eccentricity, the lack of lateral resistance and torsional stiffness, the lateral displacement is larger and the damage degree is greater under the earthquake action[1]. At present, the application of energy dissipation technology is increasingly becoming a new way to control structural torsion effect[2]. In this paper, lead rubber bearing (LRB) and buckling restrained energy dissipation brace (BRB) are applied to irregular structures. It is proposed that the lateral stiffness of the upper structure can be changed by reasonable arrangement of BRB to reduce the tensile stress of isolation bearing under earthquake. The seismic response analysis of four different control schemes of an irregular structure, including the original structure, BRB structure, LRB structure, BRB and LRB joint structure, is carried out. The seismic response characteristics and laws of the structures under different control schemes are compared, especially the analysis of the control torsion effect. The research results can be used as reference for similar projects.

2. Buckling-resistant Energy Dissipation Brace
The buckling-resistant energy dissipation brace (BRB) is a new type of brace, usually composed of yield energy dissipation section (L_1, A_1), transition section(L_2, A_2)and connection section (L_3, A_3), as shown in Figure1.The sectional area and stiffness of each section of BRB are different [3]. In practical engineering, the cross-section changes of transition section and connecting section are usually not considered for convenience of calculation. The cross-section area of core material along the whole

length is considered as uniform area A_1 , and the stiffness of BRB is equivalent simplified[4]. The equivalent stiffness of BRB is calculated according to the the equation (1).

$$K_{ef}^{*} = \frac{E_1 A_1}{L_1 + 2L_2 + 2L_3} = \frac{E_1 A_1}{L} \tag{1}$$

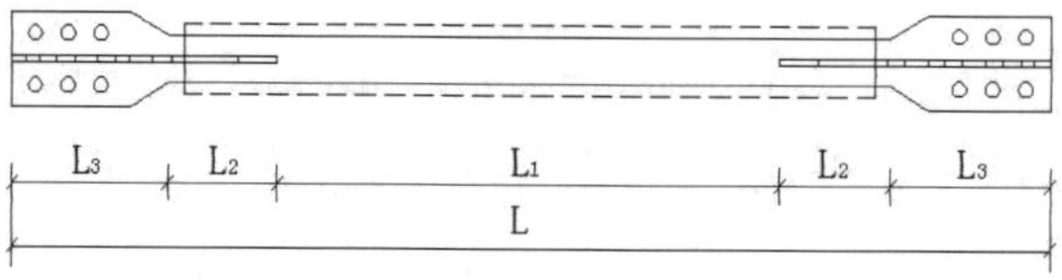

Figure 1. Schematic diagram of buckling restrained brace.

For the BRB structural system, the horizontal seismic effect is first borne by BRB. BRB can significantly improve the ductility and seismic performance of the structural system, and has good energy dissipation capacity, thereby protecting the main structure from damage[5].

3. The Initial Stiffness of BRB and the Lateral Stiffness of the Structure

BRB braces are usually arranged in various ways, such as single oblique direction layout and inverted V-shape layout, as shown in Figure 2.

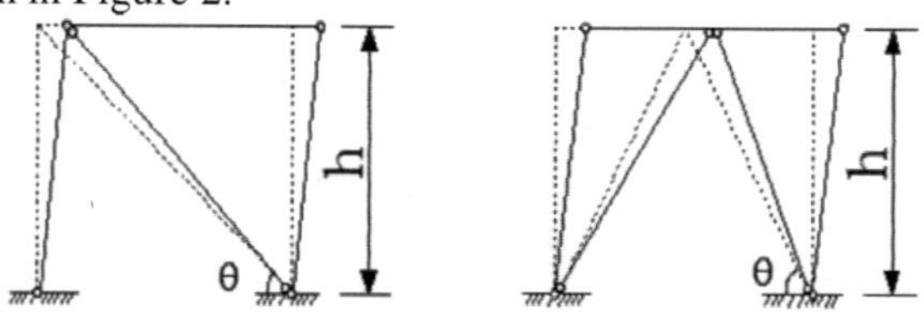

Figure 2. Schematic diagram of BRB single oblique and inverted V layout.

The initial stiffness of BRB in the figure is calculated according to the equation (2) and equation (3).

$$K_B = \frac{(EA_1)\sin\theta\cos^2\theta}{h} \tag{2}$$

$$K_B = \frac{2(EA_1)\sin\theta\cos^2\theta}{h} \tag{3}$$

The lateral stiffness of the structure is K_D, the ratio between the initial stiffness of BRB brace and the lateral stiffness of the structure is $K_F = K_B / K_D$ .Domestic and foreign scholars have conducted qualitative research on the relationship between the lateral stiffness ratio and the main structure, and have achieved corresponding results. For BRB braced frame structure system, K_F needs to be controlled in a reasonable range to effectively reduce the seismic response[3].

4. The Base Seismic Isolation Bearings

The seismic isolation bearing is the main seismic isolation device in the base seismic isolation technology[6,7]. The device not only has the overall automatic reset function, and also can extend the basic period of the building, avoid the main frequency band of the earthquake , significantly reduce the seismic energy input to the structure.

The selection of isolation bearing needs to meet the requirements of minimum size and horizontal yield capacity [8,9]. The vertical compressive stress of rubber isolation bearing of class C building under the representative value of gravity load should not exceed 15MPa. First, calculate the area of the column base bearing according to the formulas $A = F/\sigma$ and $F = 1.0G + 0.5Q$, and then calculate the minimum diameter of the bearing required at the bottom of each column according to the formula $D = 2\sqrt{A/\pi}$. Finally, according to the specification that the total yield force of the seismic isolation layer is not less than 2% of the base reaction force under the action of the representative value of gravity load, the total yield force of the seismic isolation layer is estimated.

5. Engineering Case Analysis

A project is a 12-story steel frame with an L-shaped layout, as shown in Figure 3. The frame is 51m long from north to south and 83m long from east to west . The total height of the building is 43.8m. It belongs to the Class A high-rise building. The height of the bottom floor is 4.2m, and the rest is 3.6m.

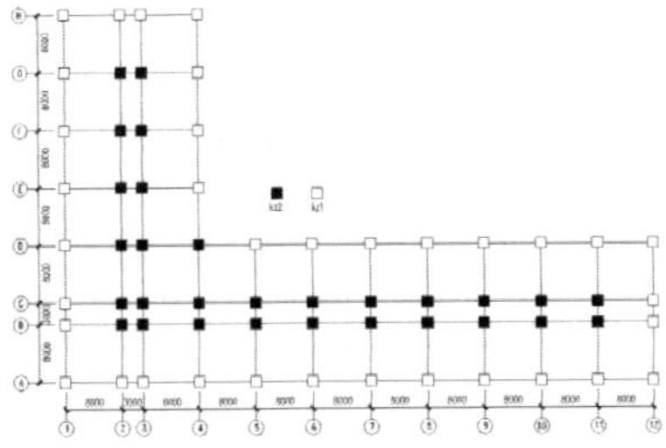

Figure 3. The plan of the structure.

The floor dead load and live load are both 2.0kN/m^2, and the load on the beam is 4.5k N/m. The self-weight of the floor slab is automatically calculated by the program. The seismic fortification intensity of the project is 8 degrees, the design basic earthquake acceleration is 0.3g, the design earthquake group is the first group, the seismic class is three, and the site category is class II . The cross-sectional dimensions of the main structural members are shown in Table 1.

Table 1. Cross-section sizes of the main structural components.

Component name	Section size (unit: mm) 1-6 layers	Section size (unit: mm) 7-12 layers	Material type
Frame column z1	400×400×12	350×350×12	
Frame column z2	500×500×16	450×450×16	Q345
Beam	600×250×12×24		
Floor	Cast-in-place thickness is 100mm		C30

[a]The column is a box-shaped section, and the section size is: side length × side length × wall thickness.
[b] The beam has an I-shaped cross-section, and the dimensions are: height× width × web thickness ×flange thickness.

5.1. Different Seismic Control Solutions

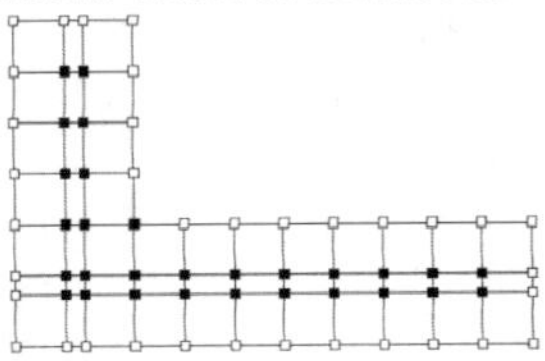

Scheme 1: Original structure

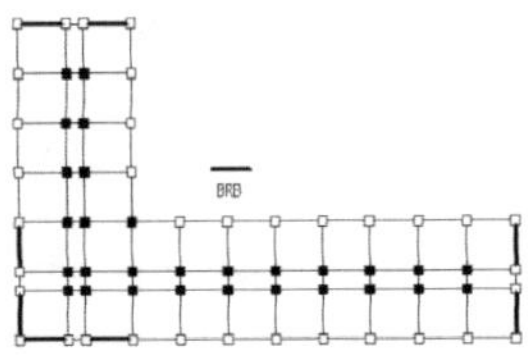

Scheme 2: Structure + BRB

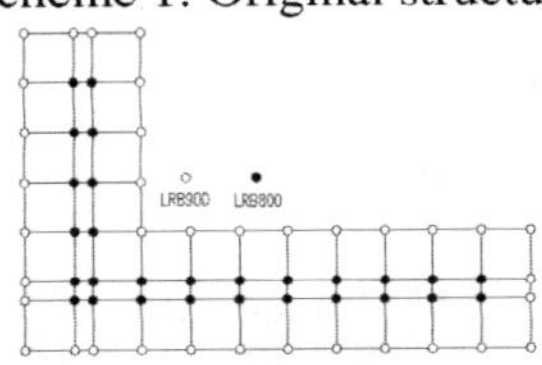

Scheme 3: Structure + LRB

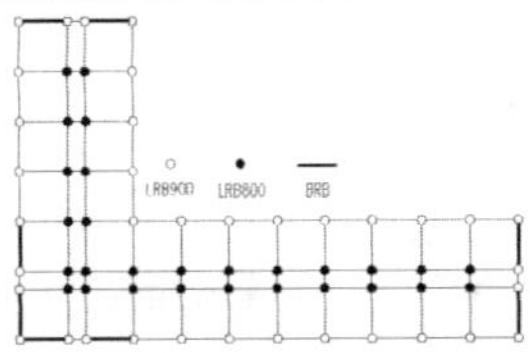

Scheme 4: Structure + BRB + LRB

Figure 4. Four different seismic control schemes.

In order to analyze the impact of BRB, LRB and their combined application on the seismic response of irregular steel frames, seismic control schemes are divided into four types, as shown in Figure 4. Considering comprehensively, the BRB of this project adopts (8-12) m TJ energy dissipation buckling brace (Q235), adopts single oblique layout, and BRB is hinged with beam and column.Two types of Zhentai lead rubber bearingLRB800 and LRB900 are selected for LRB. For the L-shaped structure, when the number of BRB is the same at the limb and corner of the structure, better damping effect can be obtained. Considering the torsional effect, the displacement around the structure may be slightly larger. Considering the overall consideration, LRB900 is arranged along the periphery, and LRB800 is arranged in the middle, as shown in scheme 3.

5.2. Structural Modal Analysis

Midas/gen analysis software is used to analyze the mode of four different control schemes. The maximum number of vibration modes used in the analysis is 18, which meets the requirement of the code.The results are shown in Table 2.

Table 2. Table of vibration mode period and mass participation coefficient of each seismic control scheme.

structure type	Scheme 1			Scheme 2			Scheme 3			Scheme 4		
Mode number	1	2	3	1	2	3	1	2	3	1	2	3
Period T/s	1.983	1.896	1.856	1.446	1.387	1.035	2.523	2.370	2.069	2.159	2.127	1.866
UX(%)	1.780	61.642	18.323	22.872	64.545	0.018	0.829	90.930	2.700	29.293	66.933	0.254
UY(%)	45.468	13.667	21.920	63.663	23.303	0.011	73.090	2.422	18.451	66.663	29.572	0.001
RZ(%)	33.337	6.244	41.656	0.007	0.093	74.971	19.562	1.042	73.002	0.090	0.182	97.913
Period ratio	0.936			0.716			0.820			0.864		

From the above table, the period of the first mode is reduced by 27.08% in scheme 2 compared with scheme 1,Scheme 3 and 4 is 42.68% and 33.02% higher than scheme 2,Scheme 4 is 14.43% lower than scheme 3.This shows that BRB can improve the lateral stiffness of the irregular structure and effectively reduce the natural vibration period of the structure. The LRB can effectively increase the natural vibration period of the structure. Because the lateral stiffness of the upper part is much larger than that of the isolation layer, the effect of scheme 4 on the natural vibration period is not significant. In scheme 1, the first mode is mainly Y-direction translation with large torsion, the second mode is X-direction translation, and the third mode is torsion. The period ratio of the first scheme is 0.936, which is 0.9 higher than that of the code, and the torsion period of the third mode is close to that of the translation period of the first and second modes. However, after setting BRB or LRB, the period ratio of scheme 2, 3 and 4 decreases, among which scheme 2 is the smallest, the participation coefficient of torsional mass of mode 1 is reduced, and the torsional effect of mode 1 and 2 is reduced. It can be seen that reasonable arrangement of BRB can enhance the torsional performance of the structure, adjust the eccentricity of the upper structure and reduce the period ratio of the structure.

5.3. Dynamic Time History Analysis of Each Control Scheme

El-Centro wave, Taft wave and Lanzhou wave are selected to analyze the frequent earthquake response of structures. Due to space limitation, only the analysis results of El-Centro wave are listed.

5.3.1. Interlayer displacement angle. The comparison results are shown in Figure5 and 6.In scheme 1, the inter story displacement angle is relatively large, and the local overrun in Y direction does not meet the seismic code[10]. Compared with scheme 1, scheme 2, 3 and 4 all reduce to varying degrees and meet the requirements. The reduction degree of scheme 2 and 4 is more obvious than that of scheme 3, and scheme 4 is more reduced. This shows that the upper lateral stiffness can be significantly enhanced and the inter story displacement angle can be reduced by reasonably setting BRB for high-rise structures with long natural vibration period. If only LRB is set, the effect is not obvious. Therefore, it is necessary to increase the lateral stiffness of the upper part reasonably, and then set LRB, such as scheme 4, which can significantly reduce the inter story displacement angle, reflecting the advantages of joint setting of BRB and LRB.

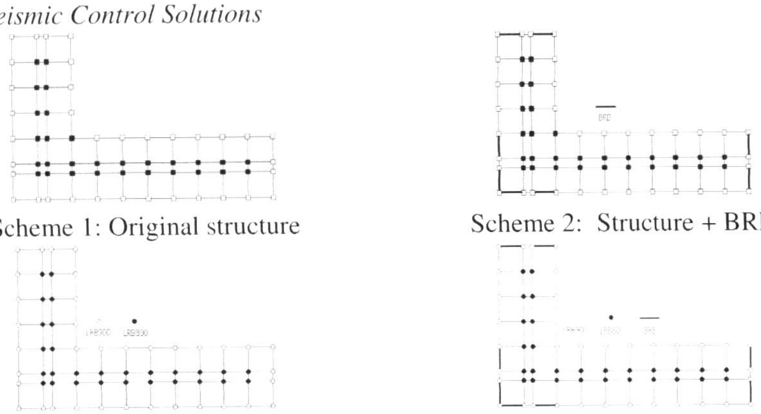

IOP Conf. Series: Materials Science and Engineering **964** (2020) 012001 doi:10.1088/1757-899X/964/1/012001

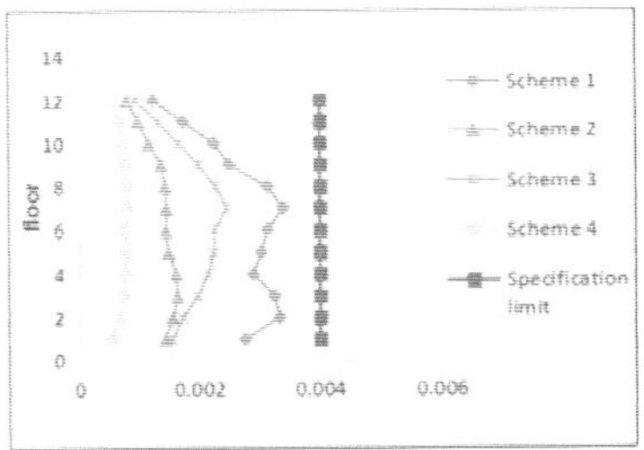

Figure 5. Interlaminar displacement angle in X direction.

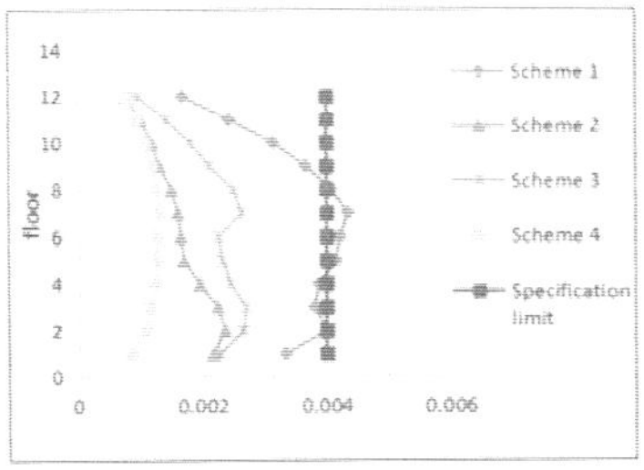

Figure 6. Interlaminar displacement angle in Y direction.

5.3.2. Floor shear. The comparison results are shown in Figure 7 and 8.Compared with scheme 1, the shear forces of each floor in X and Y directions in scheme 2 are increased, while those in scheme 3 and 4 are smaller. By calculating the ratio of story shear of scheme 3 and scheme 1, the maximum horizontal damping coefficient β is 0.395 in X direction and 0.402 in Y direction. After isolation, the structure can be reduced by one degree for design calculation. It is shown that the setting of BRB increases the upper lateral stiffness and increases the shear force of each floor. Setting LRB can weaken the seismic action of the upper part and reduce the shear force of each floor effectively. The joint setting of LRB and BRB can also reduce the shear force of each floor, and make the upper BRB brace more uniform, which is more conducive to BRB energy consumption.

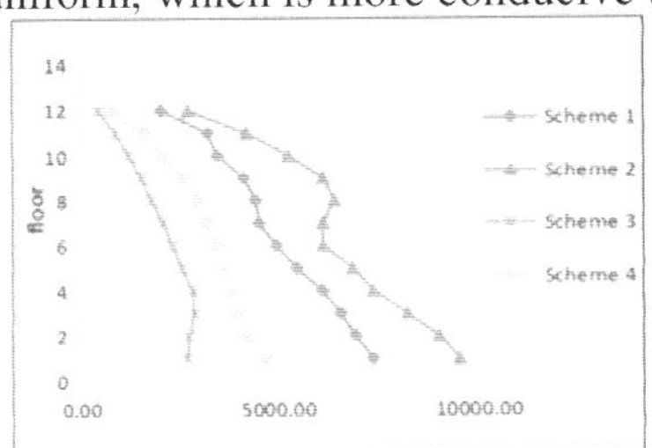

Figure 7. Maximum shear force of each floor in X direction.

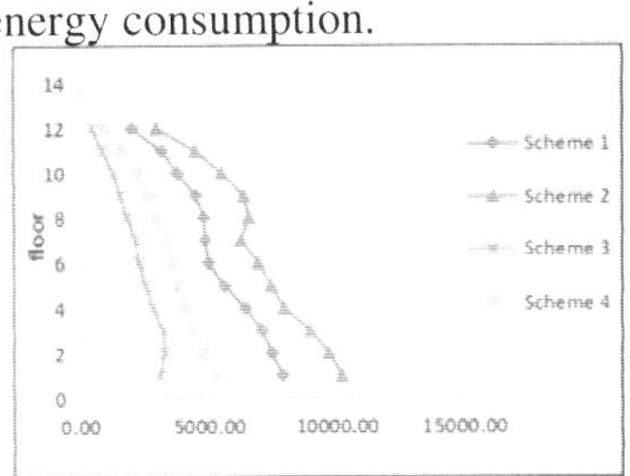

Figure 8. Maximum shear force of each floor in Y direction.

*5.3.3.Maximum horizontal displacement of floor.*The comparison results are shown in Figure 9 and 10. Compared with scheme 1, the maximum horizontal displacement of the top floor in X and Y directions is reduced in scheme 2, 3 and 4. The maximum reduction is 63.8% in scheme 2 and 4, and only 22.7% in scheme 3. It shows that the horizontal displacement can be reduced by setting BRB. If only LRB is set, the maximum horizontal displacement cannot be effectively controlled. If the LRB and BRB are set together and the lateral stiffness of the upper part is enhanced by BRB, the maximum horizontal displacement of the floor can be well controlled, which reflects the advantages of the joint application.

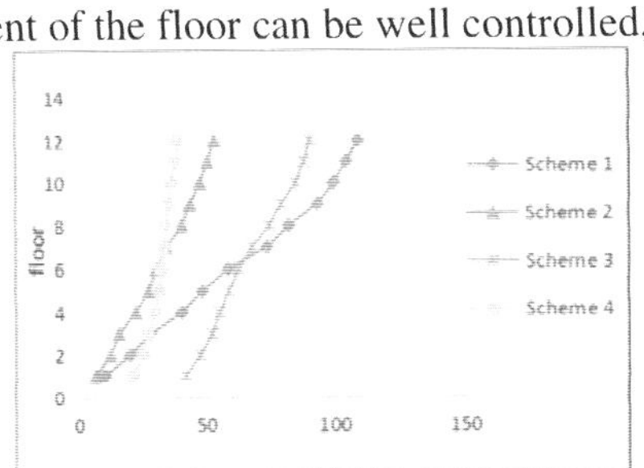

Figure 9. Maximum horizontal displacement of floor in X direction.

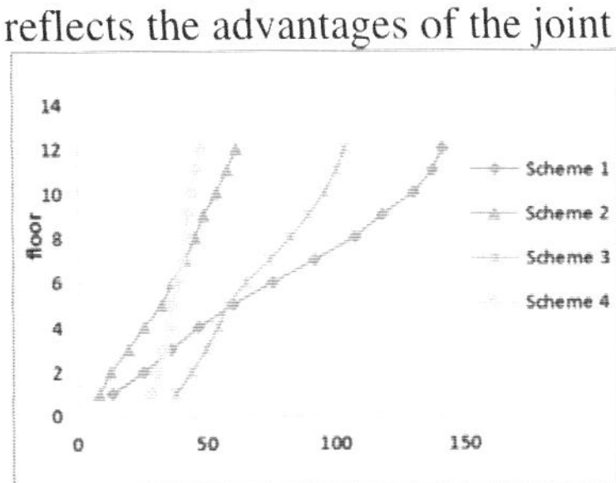

Figure 10. Maximum horizontal displacement of floor in Y direction.

6. Conclusion

By analyzing the seismic performance of an L-shaped irregular high-rise steel frame structure, the lead rubber isolation bearing and the buckling restrained energy dissipation brace are combined in the irregular structure, and the reasonable arrangement of braces in the upper part of the high-rise foundation isolation is proposed to improve the seismic performance of the structure. Through the comparative study of different schemes, the advantages of the combined application are obtained. The results show that:

(1) In the irregular structure, reasonable BRB brace can improve the lateral stiffness of the upper structure, adjust the eccentricity of the upper structure, weaken the torsional effect of the structure, reduce the natural vibration period of the structure, the inter story displacement angle and the horizontal displacement of the floor, but the shear force of each floor will be increased.

(2) Reasonable setting of LRB bearing can weaken the seismic energy transmitted into the superstructure, reduce the inter story displacement angle, horizontal displacement and shear force of each floor, but it will greatly prolong the natural vibration period of the structure.

(3) Adding BRB brace to the upper structure of LRB with base isolation bearing and combining LRB bearing with BRB brace can avoid the shortage of separate application and give full play to the advantages of both. Meanwhile, the structural indexes can meet the requirements of the code, especially for irregular high-rise building structures in high-intensity areas.

Acknowledgments

The research described in this paper was sponsored by the"Natural Science Research Funds for universities of Anhui Province"(KJ2019A0885), the "Provincial and national innovation training programs for College Students" (201912216137,202012216053) and the"Institute of architectural structure design"(yjs201704).

Thanks for the support from the members of the research group and Professor Shen Xiaopu of Anhui Jianzhu University for his careful guidance.

References

[1] Hejal, A.K.Chopra. Earthquake Response of Torsionally Coupled Frame Buildings[J]. Journal of Structural Engineering, 1989, 115(4):834-851.

[2] Fahnestock L A, Sause R, Rides J M . Seismic Response and Performance of Buckling-Restrained Braced Frames[J]. Journal of Structural Engineering, 2007, 9: 1195-1204.

[3] Wang Yonggui. Seismic performance and design method of end strengthened buckling restrained braces [M]. Beijing: China Construction Industry Press, 2015.

[4] JINKOO KIM,YOUNGIL SEO.Seismic disign of low-rise steel frames with buckling-restrained Braces[J].Engineering Structures,2004,(26):543-551.

[5] Cameron,Nicos M, Ian D. Component testing seismic evaluation and characterization of buckling-restrained braces[J].Journal of Structural Engineer(ASCE),2004,130(6):880-894.

[6] Shoaei P., Tahmasebi Orimi H., Zahrai S. M.. Seismic Reliability-based Design of Inelastic Base-isolated Structures with Lead-rubber Bearing Systems[J]. Soil Dynamics and Earthquake Engineering, 2018, 115: 589-605.

[7] Yu B., Becker T. C.. Inelastic Response of Base-isolated Structures Subjected to Impact[J]. Engineering Structures, 2018, 171: 86-93.

[8] Wolff E. D., Ipek C., Constantinou M. C., et al. Torsional Response of Seismically Isolated Structures Revisited[J]. Engineering Structures, 2014, 59(2): 462-468.

[9] Rutman Y. L., Simbort E., Bondarev D. E.. An Analysis of the Dynamics of Seismically Isolated Structures Taking into Account its Torsional Vibrations[J]. Procedia Structural Integrity, 2017, 6:208-215.

[10] Code for seismic design of buildings (GB 50011-2010) [S]. Beijing: China Construction Industry Press, 2016.

IOP Conf. Series: Materials Science and Engineering **964** (2020) 012002 doi:10.1088/1757-899X/964/1/012002

Analysis of Flexural Performance of Steel-timber Composite Cantilever Beam

Futing Pan[1,*], Ying Wang[1], Chunxiao Huang[1] and Li Zhang[2]

[1]Key Laboratory of Building Structure of Anhui Higher Education Institutes,Anhui Xinhua University, Hefei, China
[2]Audit department, Anhui Audit Vocational College, Hefei, China

*Corresponding author email: panfuting@axhu.edu.cn

Abstract. Steel-timber composite member is a new kind of structural member composed of steel plates and planks which are connected by bolts.The finite element analysis results on the flexural behavior of steel-timber composite centilever beam show that the stress of steel plate exceeds the yield strength before the bearing capacity of the composite beam is lost. Under the premise of ensuring the yield of the steel plate,the increase in the thickness of plank has little effect on the in-plane bending stiffness of composite beam,and the lateral bending stiffness of the composite beam is obviously improved which means the plank can effectively restrain the lateral buckling of the core steel plate.Based on the deflection analysis of composite beams, the deflection coefficient of steel-timber composite is derived.

1. Introduction

Considering the national sustainable development strategy, it will be a general trend to select green building materials with less consumption of resources and energy, less impact on the ecological environment and higher recycling efficiency[1,2]. In the summary report on construction and climate change issued by the United Nations Environment Programme in 2012, it was pointed out that the annual greenhouse gas emissions of the construction industry accounted for 30% of the global annual greenhouse gas emissions, while consuming 40% of the global energy. If no rectification measures are taken, the greenhouse gas emissions of the construction industry in the next 20 years will be twice as much as the present.

Steel has the advantages of light weight, high strength, recyclable, less environmental pollution and fast construction speed, but stability has always been a key concern in steel structure design[3].Timber is a kind of natural environmental protection and renewable material, but it has obvious anisotropic characteristics, and the design strength is relatively low, so it is limited in the application of large-span and high-rise buildings, and the failure is always brittle[4,5].The steel-timber composite member is a close and effective combination of steel plate and plank through bolts,which is showed in Fig. 1.Steel with high strength is used to bear external load directly, and wood is used as surface material to provide enough lateral and torsional stiffness for core steel plate, so that the core steel plate will not occur global instability and local buckling before yielding[6,7], so that the design strength can be fully utilized[8].

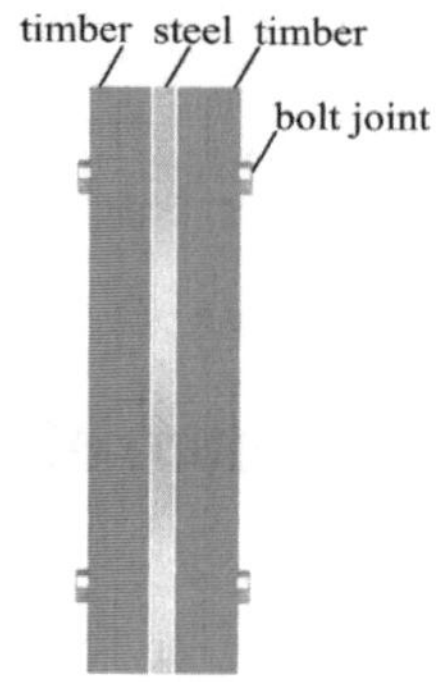

Figure 1. Cross section of steel-timber composite beam.

2. Finite Element Analysis of Steel Wood Composite Cantilever Beam

For comparing and analyzing the bending performance of steel cantilever beam and steel-timber composite cantilever beam, and considering the influence of timber thickness on the bending performance of composite beam[9,10], a set of finite element calculation models of cantilever steel beam and three groups of steel-timber composite cantilever beam are designed.

2.1. Component Design

For the steel-timber composite cantilever beam, both sides of the board provide enough lateral bending stiffness and torsional stiffness for the composite beam, which can effectively suppress the overall buckling instability of the core steel plate. In order to ensure that the local buckling of the steel plate will not occur before yielding, the bolt spacing is reasonably designed by using the thin plate theory. The specific design parameters are shown in Table 1. A transverse concentrated load of 4.5kN is applied on the upper surface of the cantilever end of the steel plate, which ensures that the maximum stress at the fixed end of the steel plate can reach the yield strength.

Table 1. Design parameters of components.

Model number	Dimensions of steel plate (Length×width×thickness/mm)	Dimension of one side board (Length×width×thickness/mm)	Transverse spacing of bolts (mm)	Longitudinal spacing of bolts(mm)
M1	1400×150×6	—	110	200
M2	1400×150×6	1400×150×18	110	200
M3	1400×150×6	1400×150×24	110	200
M4	1400×150×6	1400×150×30	110	200

2.2. Mechanical Properties of Materials

The mechanical properties of steel and timber are tested respectively. The specific results are shown in Table 2.

Table 2. Mechanical properties of materials.

	Steel	Timber
Yield strength(N/mm^2)	280	—
Ultimate tensile strength(N/mm^2)	432	—
Bending strength parrallel to grain(N/mm^2)	—	14.3
Flexural modulus of elasticity($\times 10^5$ N/mm^2)	2.017	0.112

2.3. Results of Finite Element Analysis

According to the shape of the first buckling mode, the initial defect is introduced into the model, and then the large deformation analysis is carried out on the model including the initial defect, and the material nonlinearity is considered at the same time. When the program is calculated to 0.3397 times of the applied load, the calculation does not converge and the calculation is terminated. The vertical and lateral deformation of cantilever end of each load step are shown in Figure 2 and Figure 3.We can see that the steel plate is far from yielding.

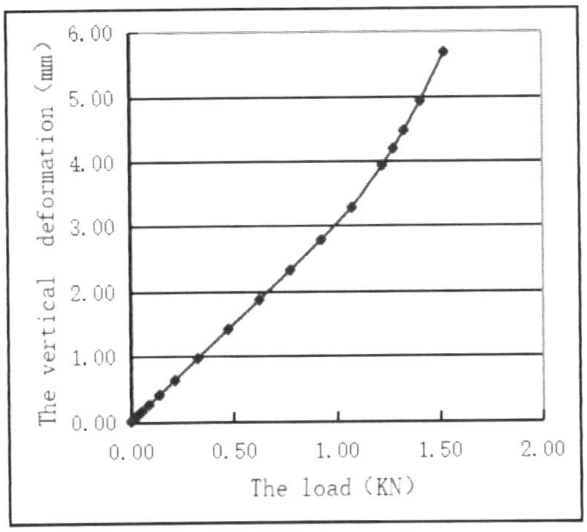

Figure 2. The vertical deformation under the load.

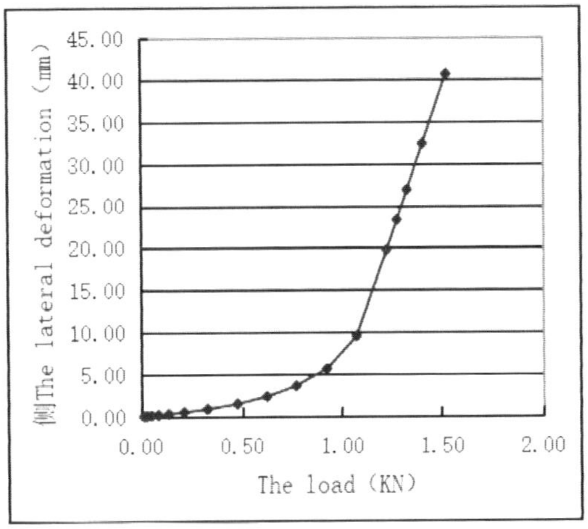

Figure 3. The lateral deformation under the load.

After the static small deformation analysis of three kinds of steel-timber composite cantilever beam, the buckling analysis is carried out, and the buckling mode is obtained. After reading the first buckling mode, the model is loaded with initial defects, and the large deformation analysis is carried out with material nonlinearity considered. The vertical and lateral deformation of cantilever end and the lower surface point of fixed end of each load step are read as shown in Figure 4 and Figure 5.

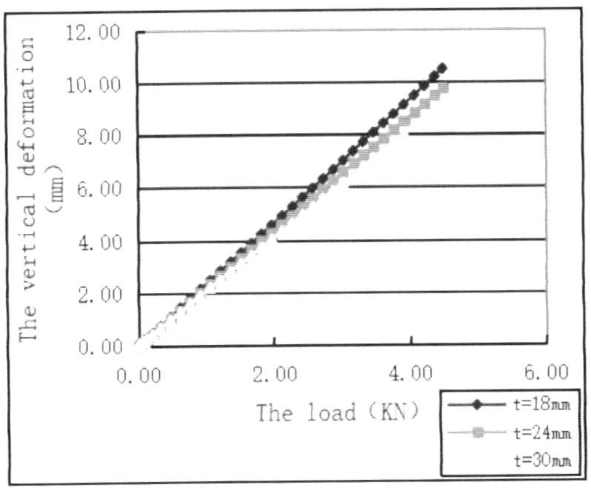

Figure 4. The vertical deformation with various thickness.

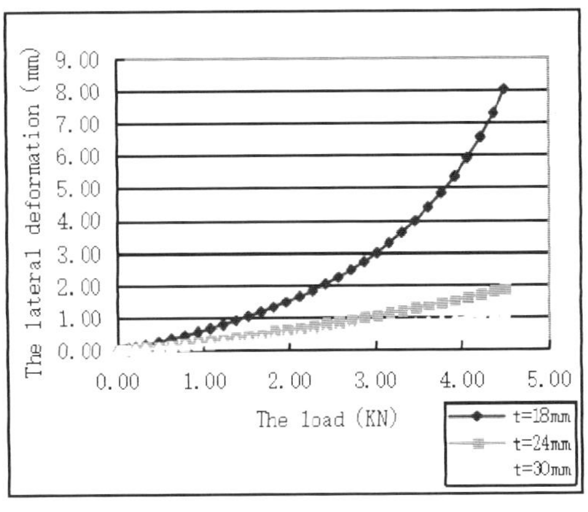

Figure 5. The lateral deformation with various thickness.

3. Theoretical Analysis

3.1. Lateral Bending Stiffness of Steel-timber Composite Cantilever Beam

The results of finite element analysis show that the lateral deformation of steel-timber composite beam is small in elastic stage, and its bending stiffness can be calculated according to Eq. (1). When the thickness of the board is 30mm, the calculation results are shown in Table 3.

$$EI = E_s I_s + E_t I_t \tag{1}$$

Table 3. Flexural rigidity of steel plate,board and composite beam.

$E_s I_s$ ($\times 10^{11} mm^4$)	$E_t I_t$ ($\times 10^{11} mm^4$)	EI ($\times 10^{11} mm^4$)
3.40	1.90	5.30

3.2. Transverse Deformation of Steel-timber Composite Cantilever Beam

Through the analysis of finite element calculation data, the deflection coefficient of steel-timber composite beam is defined as β_{st} .The deflection of the cantilever end of the composite beam is calculated in Eq. (2).

$$f_{\max} = \beta_{st} \frac{Fl^3}{3EI} \tag{2}$$

Taking model 4 as an example, the specific calculation results are shown in Table 4.
The deflection coefficient of steel-timber composite is obtained by the following results, $\beta_{st} = 1.180$ averaged by the values in the table.

Table 4. Calculation results of the deflection coefficient.

F /kN	f / mm	β_{st}
0.78	1.58	1.174
1.23	2.50	1.178
1.83	3.73	1.181
2.13	4.34	1.181
2.58	5.26	1.181
3.03	6.18	1.182
3.48	7.09	1.181
3.93	8.01	1.181
4.23	8.63	1.182

4. Conclusion

(1) Through the finite element analysis and calculation, the steel cantilever beam loses its bearing capacity due to the premature lateral buckling instability. Due to the sufficient lateral stiffness of the composite members provided by the boards on both sides, there is no obvious stiffness mutation of the composite members under the load. The stress of the core steel could reaches the yield strength, and the bearing capacity and ductility of the steel wood composite cantilever beam are significantly improved.

(2) After the steel reaches the yield, the increase of the thickness of the board has less effect on the transverse bending stiffness of the composite beam. Therefore, in the design of steel-timber composite components, the thickness of wood board should be controlled to reduce the weight of the structure and the cost.

(3) The deflection coefficient of steel-timber composite beam is deduced. It can be seen from the calculation results that the deflection coefficient of the composite beam has little difference under different loads, and the calculation result is reliable.

Acknowledgments

The research described in this paper was sponsored by the"Natural Science Research Funds for universities of Anhui Province"(KJ2019A0885) and Key Laboratory of Building Structure of Anhui Higher Education Institutes(Anhui Xinhua University).The supports are gratefully acknowledged.

References

[1] Cheng X,Chen Y Y,Pan L L.Experimental study on steel beam-columns composed of slender H-sections under cyclic bending[J].Journal of Constructional Steel Research, 2013(88) : 279-288.
[2] Koshihara M,Isoda H.Yusa S.The design and installation of a five-story new timber building in Japan [C] // Proceeding of the International Syposium on Timber Structures,Turkey,2009.
[3] Jerzy Jasien´ko,Tomasz P.Nowak.Solid timber beams strengthened with steel plates Experimental studies[J].Construction and Building Materials,63(2014)81-88.
[4] Wolfgang Winter, Kamyar Tavoussi, Tamir Pixner,etc. Timber-steel-hybrid beams for multi-storey buildings [C] //World Conference on Timber Engineering,Auckland,2012.
[5] Gatto K,Uang CM.Effects of loading protocol on cyclic response of wood frame shear walls[J]. Struct Eng,2003,129(10):1384~1393.
[6] Tsai W H, Yang C H, Chang J C, et al. An Activity-Based Costing decision model for life cycle assessment in green building projects[J]. European Journal of Operational Research,2014, 238(2): 607-619.
[7] Ying-Lei Li,Yuan-Qi Li, Zu-Yan Shen. Investigation on flexural strength of cold-formed thin-walled steel beams with built-up box section[J]. Thin-Walled Structures, 2016, 107.
[8] Dinar Camotim, Cilmar Basaglia. On the behaviour, failure and direct strength design of thin-walled steel structural systems[J]. Thin-Walled Structures, 2014, 81.
[9] Yushun Li, Wei Shan, Huangying Shen, Zhen-Wen Zhang, Junzhe Liu. Bending resistance of I-section bamboo-steel composite beams utilizing adhesive bonding[J].Thin-Walled Structures, 2015, 89.
[10] A.Hassanieh,H.R.Valipour,M.A.Bradford.Experimental and analytical behaviour of steel-timber composite connections[J].Construction and Building Materials,2016,118-63-67.

Analytical Solution of the Nonlinear Vibration of the Steel Skeleton Supported Membrane Structure

Mengfei Wang[1], Changjiang Liu[2,3,*], Haibing Xie[1], Su Jiang[1], Mengjia Zhang[2] and Yaozeng Su[2]

[1]College of Environment and Civil Engineering, Chengdu University of Technology, Chengdu, China
[2]School of Civil Engineering, Guangzhou University, Guangzhou, China
[3]Guangdong Engineering Technology Research Center for Complex Steel Structures, Guangzhou University, Guangzhou, 510006, China

*Corresponding author email: cjliu@gzhu.edu.cn

Abstract. Steel skeleton supported membrane structure is the major structural form of membrane structure, accounting for more than 50%. The structural characteristics of membrane structure itself determine that the analysis of its self-vibration characteristics is very important in the design process. According to the von Karman's large deflection theory and the D'Alembert's principle, the governing equations of nonlinear viscous damping vibration of the skeleton supported orthotropic membrane structure were proposed. By applying the Galerkin method and the KBM perturbation means of obtaining the analytical solutions of frequency function, displacement function and mode function of damped nonlinear and linear free vibration of simply supported orthotropic membrane structures on four sides were obtained. In addition, it can be known by example analysis that the transverse stiffness of membrane structure can be improved by increasing prestress and rise span ratio in the practical engineering, and while the influence of geometrically nonlinear, orthotropic, damping and prestress on the dynamic feature of membrane structure must be fully considered. This paper provides some theoretical references for the wind-induced dynamic stability calculation and wind-resistant design of the steel skeleton supported membrane structure.
Keywords: Membrane structure; Orthotropic; Geometric nonlinearity; Linear; Free vibration.

1. Introduction

Matched with the traditional rigid structure, the structural characteristics of the membrane structure determine that it is essential to analyze the shape-finding and self-vibration characteristics during the design process[1]. A.H.Sofiyev[2], Li[3]and other scholars have studied the nonlinear vibration of membrane. However, in the vibration analysis of steel skeleton supported membrane structure, there are few studies that consider both the geometric nonlinearity of membrane surface deformation and the orthotropic characteristics of material. L.V.Swpanova[4], Ali.A.Yazdi[5] and other scholars proved the effectiveness of the perturbation method and the Galerkin method in solving nonlinear problems. In this paper, the KBM perturbation method and the Galerkin method were used to solve the partial differential governing equations of nonlinear free vibration of supported membrane structures. The nonlinear and linear vibration frequencies of skeleton supported membrane structure under different parameter variables are analyzed and calculated through calculation examples. The results of this paper provide a handy analytical way for calculating the natural frequency and transverse displacement of the skeleton supported orthotropic membrane structure with viscous damping.

2. Structural Models and Governing Equations

The structural model studied in this paper is cylindrical shell-shaped steel framework supported membrane structure, it's shown in figure 1. The membrane material is orthotropic, and the boundary condition is four edges clamped, x and y are two orthogonal directions. a and b denote the lengths of the membranes in the x and y directions; N_{ox} and N_{oy} represent the pretension in x and y directions; f_2 is the rise span ratio in x direction. The coordinate system is xyz, point C is the center of the horizontal projection of the membrane surface to the plane xoy.

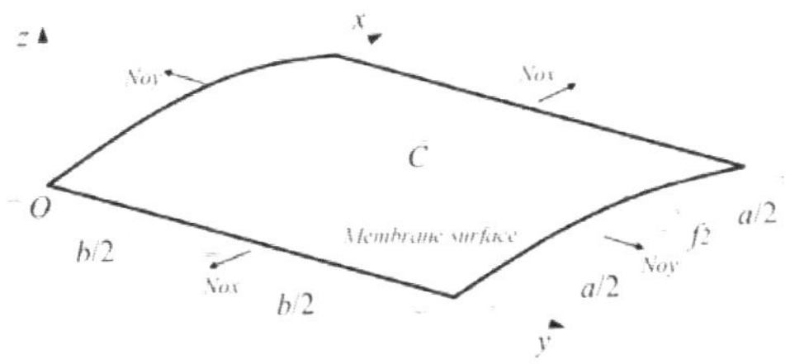

Figure 1. Skeleton supported membrane structural model.

2.1. Governing Equations and Boundary Conditions

According to the governing equation derived by Liu[6]et al., and introduce stress function $\varphi = \varphi(x, y, t)$ [7]. The motion equation and compatible equation of the skeleton supported membrane structure can be derived as follows:

$$\rho_0 \frac{\partial^2 w}{\partial t^2} + c\frac{\partial w}{\partial t} - (h\frac{\partial^2 \varphi}{\partial y^2} + N_{0x})\frac{\partial^2 w}{\partial x^2} - k_{0x} h\frac{\partial^2 \varphi}{\partial y^2} - \left(h\frac{\partial^2 \varphi}{\partial x^2} + N_{0y}\right)\frac{\partial^2 w}{\partial y^2} = 0 \tag{1}$$

$$\frac{1}{E_1}\frac{\partial^4 \varphi}{\partial y^4} + \frac{1}{E_2}\frac{\partial^4 \varphi}{\partial x^4} = \left(\frac{\partial^2 w}{\partial x \partial y}\right)^2 - \frac{\partial^2 w}{\partial x^2}\frac{\partial^2 w}{\partial y^2} - k_{0x}\frac{\partial^2 w}{\partial y^2} \tag{2}$$

Where: ρ_0 represents the membrane material surface density; c represents the damping coefficient; w represents deflection $w(x,y,t)$; h represents the membrane material thickness; E_1 and E_2 represent Young's modulus in x and y direction; μ_1 and μ_2 denote Poisson's ratio in x and y directions; G represents shear modulus; k_{0x} denotes initial principal curvature in the x direction.

The corresponding displacement and stress boundary conditions are shown as follows:

$$\begin{cases} w(0,y,t)=0, \dfrac{\partial^2 w}{\partial x^2}(0,y,t)=0 \\ w(a,y,t)=0, \dfrac{\partial^2 w}{\partial x^2}(a,y,t)=0 \end{cases}, \begin{cases} w(x,0,t)=0, \dfrac{\partial^2 w}{\partial y^2}(x,0,t)=0 \\ w(x,b,t)=0, \dfrac{\partial^2 w}{\partial y^2}(x,b,t)=0 \end{cases}, \begin{cases} \dfrac{\partial^2 \varphi}{\partial x^2}(0,y,t)=0 \\ \dfrac{\partial^2 \varphi}{\partial x^2}(a,y,t)=0 \end{cases}, \begin{cases} \dfrac{\partial^2 \varphi}{\partial y^2}(x,0,t)=0 \\ \dfrac{\partial^2 \varphi}{\partial y^2}(x,b,t)=0 \end{cases} \tag{3}$$

2.2. Solution of Governing Equation

The functions satisfying the boundary condition equation (3) are separated as follows[8,9]:

$$w(x,y,t) = T_{mn}(t) \cdot W_{mn}(x,y) = T(t) \cdot \sin\frac{m\pi x}{a}\sin\frac{n\pi y}{b} \tag{4}$$

Where, $W_{mn}(x,y)$ is the given deformation function; $T_{mn}(t)$ is a function of time; m and n represent the order of vibration mode in x and y directions respectively.

The substitution of equation (4) into equation (3) yields, assuming that the stress $\varphi(x,y,t)$ satisfying the stress boundary condition equation (3) is:

$$\varphi(x,y,t)=T^2\left(\alpha\cos\frac{2m\pi x}{a}+\beta\cos\frac{2n\pi y}{b}+\delta_1 x^2+\delta_2 y^2\right)+T(t)\sin\frac{m\pi x}{a}\sin\frac{n\pi y}{b}\xi \tag{5}$$

$$\alpha=\frac{E_2 n^2 a^2}{32m^2 b^2},\beta=\frac{E_1 m^2 b^2}{32n^2 a^2},\ \xi=\frac{k_{0x}(n\pi/b)^2}{\left((n\pi/b)^4/E_1\right)+\left((m\pi/a)^4/E_2\right)},\ \delta_1=\frac{\pi^2 E_2 n^2}{16b^2},\ \delta_2=\frac{\pi^2 E_1 m^2}{16a^2}$$

Equation (3) and equation (5) are substituted into Equation (1), and the Bubnov-Galerkin method [10] is used to transform it into homogeneous differential equation. Then, by using KBM perturbation method[11] to solve the equation, the approximate expression of nonlinear free vibration frequency can be obtained:

$$\omega_n=\omega_0-\frac{3\varepsilon\alpha_1 F^2 e^{\varepsilon\alpha_3 t}}{8\omega_0} \tag{6}$$

$$A_0=\frac{ab}{4}\rho_0, B_0=\frac{ab}{4}c, E_0=h\pi^4\frac{3E_1 m^4 b^4+3E_2 n^4 a^4}{64a^3 b^3}, C_0=\frac{m^2\pi^2 b^2 N_{0x}+n^2\pi^2 a^2\left(N_{0y}+k_{0x}h\xi\right)}{4ab},$$

$$\varepsilon=\frac{h^2}{ab}<<1, \omega_0^2=\frac{C_0}{A_0}, \alpha_1=-\frac{E_0}{\varepsilon A_0},\ \alpha_3=-\frac{B_0}{\varepsilon A_0},$$

Where, F represents the initial displacement.

It can be seen from the above equation that the frequency of nonlinear free vibration is about the amplitude. In the case that only linear vibration is considered, the high-order trace are omitted. The approximate expression of linear free vibration frequency can be obtained as follows:

$$\omega_l=\omega_0=\sqrt{\frac{C_0}{A_0}} \tag{7}$$

The analytic expressions of nonlinear and linear $T(t)$ functions can be obtained and substituted into equation (3) respectively, the displacement functions of damped nonlinear and linear free vibration of the skeleton supported prestressed membrane structure are obtained:

$$w_n(x,y,t)=\sum_{m=1}^{\infty}\sum_{n=1}^{\infty}\sin\frac{m\pi x}{a}\sin\frac{n\pi y}{b}\left(Fe^{\frac{1}{2}\alpha_3\varepsilon t}\cdot\cos\left(\left(\omega_0-\frac{3\alpha_1\varepsilon F^2 e^{\alpha_3\varepsilon t}}{8\omega_0}\right)t\right)\right) \tag{8}$$

$$w_l(x,y,t)=\sum_{m=1}^{\infty}\sum_{n=1}^{\infty}\sin\frac{m\pi x}{a}\sin\frac{n\pi y}{b}\left(Fe^{\frac{1}{2}\alpha_3\varepsilon t}\cdot\cos(\omega_0 t)\right) \tag{9}$$

By superposing the initial surface function of the skeleton supported membrane with its displacement equation (8) and equation (9), the modes S_n and S_l of the skeleton supported membrane can be obtained:

$$S_n=-\frac{f_2\left(x-(a/2)\right)^2}{(a/2)^2}+\sum_{m=1}^{\infty}\sum_{n=1}^{\infty}\sin\frac{m\pi x}{a}\sin\frac{n\pi y}{b}\left(Fe^{\frac{1}{2}\alpha_3\varepsilon t}\cdot\cos\left(\left(\omega_0-\frac{3\alpha_1\varepsilon F^2 e^{\alpha_3\varepsilon t}}{8\omega_0}\right)t\right)\right) \tag{10}$$

$$S_l=-\frac{f_2\left(x-(a/2)\right)^2}{(a/2)^2}+\sum_{m=1}^{\infty}\sum_{n=1}^{\infty}\sin\frac{m\pi x}{a}\sin\frac{n\pi y}{b}\left(Fe^{\frac{1}{2}\alpha_3\varepsilon t}\cdot\cos(\omega_0 t)\right) \tag{11}$$

3. Analysis of Example

Take the orthotropic membrane materials commonly used in engineering practice as an case: E_1=1.4×10^{6}kN/m^2, E_2=0.9×10^{6}kN/m^2[6], the membrane density ρ_0=1.7kg/m^2, the damping c=120Ns/m, the thickness of the membrane h=1.0m, the length a=1m and width b=1m, the prestress $N_{0x}=N_{0y}$=1000 N/m. According to equation (6) and (7), assuming that F=0.05m and t=0.01s. The first three frequencies of different parameters (initial displacement, prestress, damping, aspect span ratio, etc.) of nonlinear and linear free vibration at the rise span ratio 1/10 and 1/12 were respectively calculated, the calculation results are shown in figures 2-6 and table 1 (1NR10 represents f_2=1/10, 1NR12 represents f_2=1/12 . According to equation (10), the first three modes of the first three orders are plotted at t=0s and t=0.0035s, as shown in figure 7.

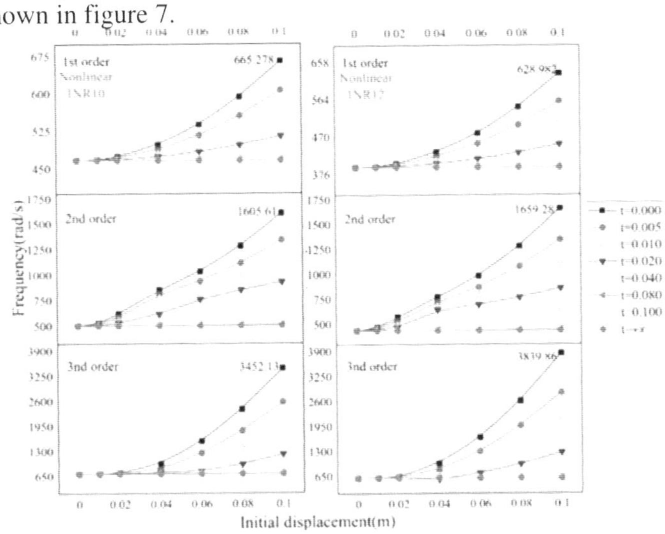

Figure 2. Nonlinear vibration frequencies of each order under different initial displcements.

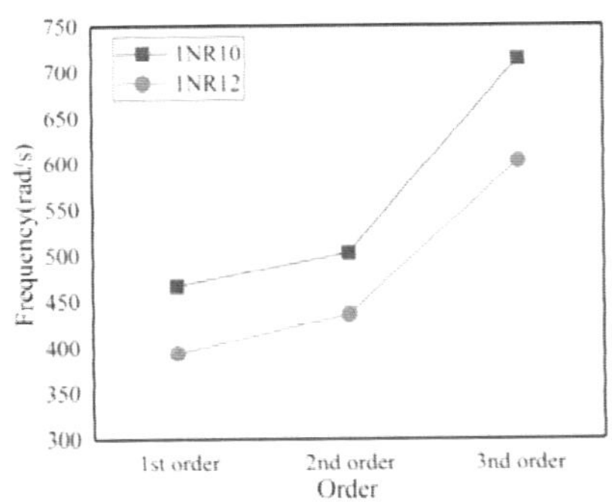

Figure 3. Linear vibration frequencies of each order .

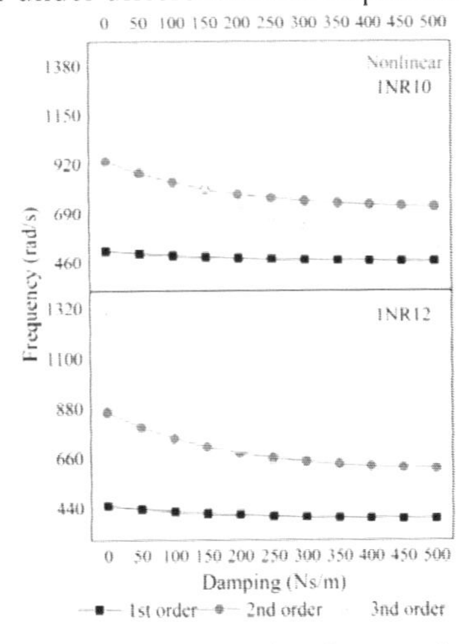

Figure 4. Nonlinear vibration frequencies of each order under different damping.

According to equation (7), it can be seen that the value of linear vibration frequency is independent of the initial displacement and time. Analysis of figures 2-4 and figure 6 show that: When considering the damping and geometric nonlinearity of membrane vibration, the frequency of free vibration increases as initial displacement and order increases, decreases as damping enhances. The frequency of nonlinear and linear vibration increase with the increase of f_2, prestress and order. As can be seen from figure 5: When $E_1>E_2$ and value of E_2 is fixed, the frequencies of nonlinear and linear free vibration decrease as the elastic modulus ratio decreases, and increases with the order aggrandize. When $E_1<E_2$ and value of E_1 is fixed, the law is opposite. As can be seen from table 1: If the a and b values of the two orthogonal directions is interchanged or the rise span ratio is changed, the frequency of linear free vibration is smaller than that of nonlinear free vibration. From the results of figure 7: It can be seen that equation (10) can be used to calculate the vibration modal equation of each order conveniently.

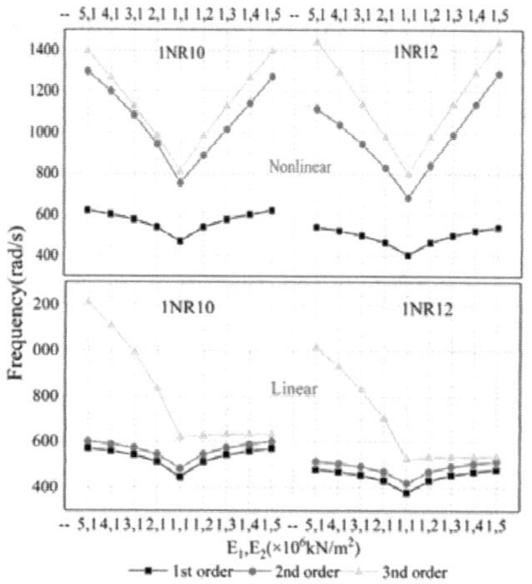

Figure 5. Nonlinear and linear free vibration frequencies under different E_1:E_2.

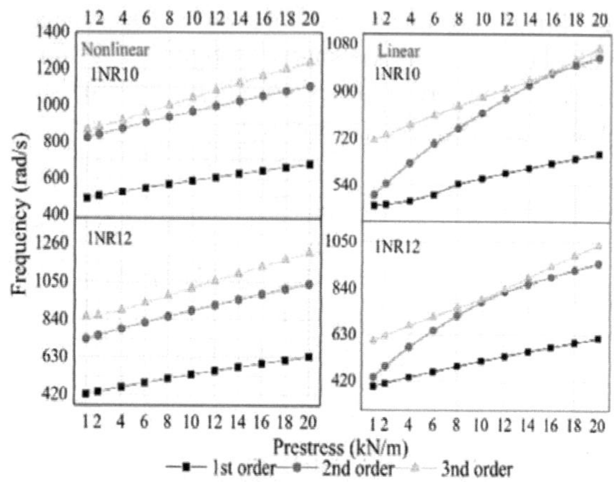

Figure 6. Nonlinear and linear free vibration frequencies under different prestress.

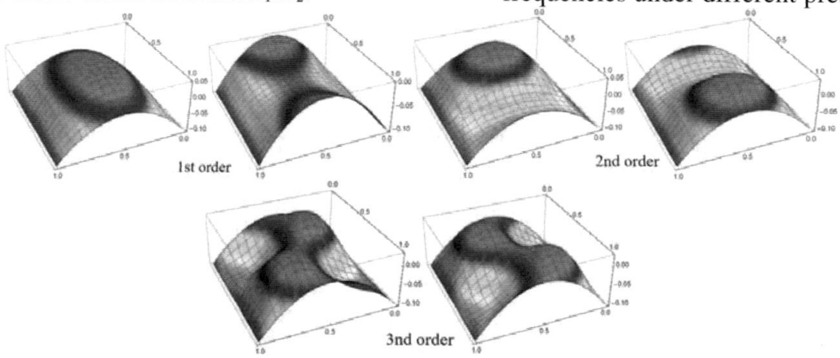

Figure 7. The first three modes of nonlinear free vibration.(m).

Table 1. The nonlinear and linear free vibration frequency(rad/s) of each order with different aspect span ratios under different rise span ratios.

	f_2	a,b (m)	1,1	1,2	2,1	1,3	3,1	1,4	4,1	1,5	5,1
		1st order	491.25	209.75	218.51	171.11	153.55	167.18	140.31	167.02	137.03
	1/10	2nd order	823.79	491.25	540.25	290.47	585.64	209.75	607.44	180.83	615.30
		3nd order	866.69	743.56	566.30	821.74	586.17	853.79	604.16	866.67	612.14
Nonlinear		1st order	422.55	195.77	196.94	169.24	148.05	167.08	138.80	167.10	136.51
	1/12	2nd order	732.64	422.55	548.58	257.99	594.00	195.77	611.05	175.41	616.92
		3nd order	855.67	765.40	575.11	832.22	593.57	857.97	607.48	868.53	613.67
		1st order	466.75	166.18	193.08	102.96	113.29	86.54	90.64	81.11	82.94
	1/10	2nd order	502.68	222.23	243.00	173.07	179.41	161.23	163.46	157.14	158.09
Linear		3nd order	713.71	466.75	239.62	260.33	174.26	166.18	160.14	123.56	155.88
		1st order	393.49	146.27	167.65	96.61	104.33	84.17	87.12	80.09	81.38
	1/12	2nd order	435.51	207.77	223.33	169.37	173.89	159.97	161.54	156.61	157.28
		3nd order	602.17	393.49	217.74	222.77	168.47	146.27	158.16	112.52	155.05

4. Conclusion

(1) The rise span ratio, prestress and aspect span ratio have an effect on the linear and nonlinear free vibration of membrane, while the nonlinear free vibration of membrane is also affected by the initial displacement and damping parameters. (2) With the same initial displacement, the frequency of nonlinear free vibration decreases with the aggrandize of time and viscous damping, and increases with the aggrandize of prestress. The frequency of linear free vibration increases with the aggrandize of prestress and order. In the practical engineering, the transverse stiffness of membrane structure can be improved by increasing prestress. (3) In consideration of the geometrical nonlinearity and linearity of the membrane, the values of nonlinear and linear vibration frequencies increase as the f_2 increase. Therefore, the transverse stiffness of membrane structure can be improved by increasing the rise span ratio. (4) When the aspect span ratio of the orthogonal direction changes, the frequency values of the nonlinear and linear free oscillations will change. (5) The research results of this paper provide a theoretical reference for the subsequent research on the wind and rain resistance dynamic response of the skeleton supported membrane structure.

Acknowledgments

This work was supported by the National Natural Science Foundation of China (Project Numbers 516 08060), Guangdong Basic and Applied Basic Research Foundation (Project Number 2019A1515063).

References

[1] Liu, X., Cai, G.P., Peng, F.J., Zhang, H. (2018) Active control of large-amplitude vibration of a membrane structure. Nonlinear Dynamics, 93(2):629-642.

[2] Sofiyev, A.H., Hui, D., Haciyev, V.C., Erdem, H., Yuan, G. Q., Schnack, E., Guldal, V. (2017) The nonlinear vibration of orthotropic functionally graded cylindrical shells surrounded by an elastic foundation within first order shear deformation theory. Composites Part B: Engineering, 116:170-185.

[3] Li, D., Zheng, Z.L., Todd, M.D. (2018) Nonlinear Vibration of Orthotropic Rectangular Membrane Structures Including Modal Coupling. Journal Of Applied Mechanics-Transactions Of the Asme, 85(6): 061004.

[4] Stepanova, L.V., Igonin S.A. (2014) Perturbation method for solving the nonlinear eigenvalue problem arising from fatigue crack growth problem in a damaged medium. Applied Mathematical Modelling, 38(14): 3436-3455.

[5] Yazdi, A.A. (2016) Assessment of Homotopy Perturbation Method for Study the Forced Nonlinear Vibration of Orthotropic Circular Plate on Elastic Foundation. Latin American Journal of Solids and Structures, 13(2): 243-256.

[6] Liu, C.J., Wang, F., Liu, J., Deng, X.W., Zhang, Z.L., Xie, H.B. (2019) Theoretical and Numerical Studies on Damped Nonlinear Vibration of Orthotropic Saddle Membrane Structures Excited by Hailstone Impact Load. Shock and Vibration, https://doi.org/10.1155/2019/9234832.

[7] Liu, C.J., Deng, X.W., Zheng, Z.L. (2017) Nonlinear wind-induced aerodynamic stability of orthotropic saddle membrane structures. Journal of Wind Engineering and Industrial Aerodynamics, 164:119-127.

[8] Liu, C.J., Zheng, Z.L., Jun, L., Guo, J.J., Wu, K. (2013) Dynamic Analysis for Nonlinear Vibration Of Prestressed Orthotropic Membranes with Viscous Damping. International Journal of Structural Stability and Dynamics, 13(02):1350018.

[9] Zheng, Z.L., Song, W.J., Liu, C.J., He, X.T., Sun, J.Y., Xu, Y.P. (2012) Study on Dynamic Response of Rectangular Orthotropic Membranes Under Impact Loading. Journal of Adhesion Science and Teehnol, 26(10-11):1467-1479.

[10] Tang, Y.Q., Li, T.J. (2017) Equivalent-force density method as a shape-finding tool for cable-membrane structures. Engineering Structures, 151:11-19.

[11] Zheng, Z.L., Guo, J.J., Song, W.J., He, X.T., Lu, F., Xie, C.X., Sun, J.Y. (2014) Nonlinear Free Vibration Analysis of Axisymmetric Polar Orthotropic Circular Membranes under the Fixed Boundary Condition. Mathematical Problems in Engineering, http://dx.doi.org/10.1155/2014/65 1356.

IOP Conf. Series: Materials Science and Engineering **964** (2020) 012004 doi:10.1088/1757-899X/964/1/012004

Analytical Solutions of Two-body Spring Vibrator Problem

Ping Zhu[1,2,*]

[1]School of Science and Technology, Puer University, China
[2]Open Key Laboratory of Mechanics of university in Yunnan Province, China

*Corresponding author email: zhuupp@163.com

Abstract. In two-body problems, the problem of the two-body spring vibrator is a more typical and interesting mechanics problem. But the relative motion of the two-body spring vibrator involves elliptical integrals, so the problem is very complex and had not yet been solved analytically in the past time. In this paper, using the elliptical integral theory, we obtain the analytic solutions of the two-body spring vibrator motion law, so that we can discuss further the motion characteristics.

1. Introduction

In dynamical system, the two-body central force problem represents the most fundamental kind of interaction between two bodies[1-3], such as the gravitational interaction problem in astrophysics[4-6], the electromagnetic interaction problem in atomic physics[7-9], and the strong interaction problem in nuclear physics. Of course, there are other possibilities, too. For instance, in two-body problems, the problem of the two-body spring vibrator is a more typical and interesting mechanics problem. Here we shall investigate the problem of the two-body spring vibrator which a pair of particles with masses m_1 and m_2 are connected by a mechanical spring whose yield strength coefficient is k. Because the dynamical problem of the two-body spring vibrator involves elliptical integrals, it is difficult for us to solve the problem. The analytical solution of the two-body spring vibrator problem is not given in the past time. Just for this reason, it is very significant to solve and to discuss the problem.

In this paper, using the method of the theory of elliptic integrals and the mathematic trick, the two-body spring vibrator problem is investigated, and the analytical solutions of the two-body spring vibrator problem are given.

This paper is organized as follows: in section II, making use of the method of analytical mechanics, we make the theoretical analysis of the two-body spring vibrator problem. In Sec. III, the analytical solutions of the two-body spring vibrator problem for different conditions are shown. Finally, summaries and conclusions of the results conclude the paper.

2. Theoretical Analysis

Two-body spring vibrator problem: The two material points the mass of which is respectively m_1 and m_2 are connected with a elastic bar, natural length of which is l, the elastic constant is k_0, and the mass of the elastic bar may be ignored. When we take $t=0$, the system possesses an initial state. After this time, After that, it moves under the action of the force $\vec{F}(t)$. We discuss the motion law of the system.

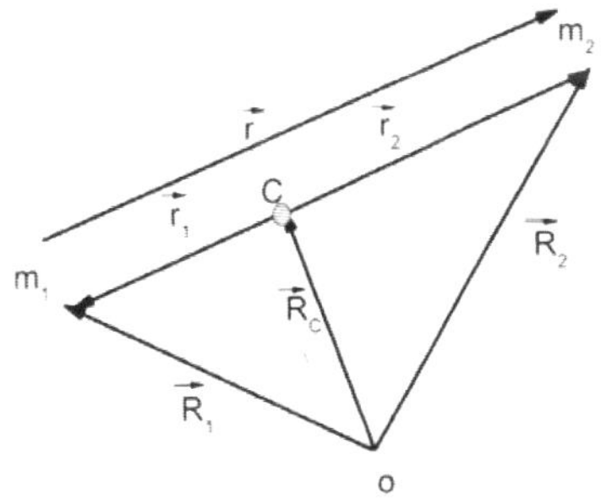

Figure 1. The relative displacements of the two-body spring vibrator.

First, one may decompose the motion of the two- body spring vibrator into the motion of the mass center and the motion of the vibrator relative to the mass center.

In Figure 1, the point o is the fixed coordinate origin, C is the mass center of the system, $\vec{r}$ is the displacement of m_2 relative to m_1. From Figure 1, we get

$$\vec{r_1} = -\frac{m_2}{m_1+m_2}\vec{r}, \quad \vec{r_2} = \frac{m_1}{m_1+m_2}\vec{r}. \tag{1}$$

From Eq.(1), we see that if the relative displacement $\vec{r}$ is determinated, the positions of m_1 and m_2 are yet determinated.

According to the motion law of the mass center, we have

$$\overrightarrow{R_c} = \overrightarrow{R_{c0}} + \int_0^t \overrightarrow{v_{c0}}dt + \int_0^t\int_0^t \vec{f}(t')dt'dt, \tag{2}$$

where

$$\vec{f}(t) = \frac{\vec{F}(t)}{m_1+m_2}, \tag{3}$$

which is the force per unit mass. When $t=0$, $\vec{r_1} = \overrightarrow{r_{10}}$, $\vec{r_1} = \overrightarrow{r_{20}}$, and $\overrightarrow{v_c} = \overrightarrow{v_{c0}}$, we obtain

$$\overrightarrow{R_c} = \frac{m_1\overrightarrow{r_{10}} + m_2\overrightarrow{r_{20}}}{m_1+m_2} + \overrightarrow{v_{c0}}\, t + \frac{1}{2}\vec{f}(t)t^2. \tag{4}$$

Lagrange function of the system is given by

$$L = L_1(\dot{\overrightarrow{R_c}}, \overrightarrow{R_c}) + L_2(\dot{\vec{r}}, \vec{r})\cdot, \tag{5}$$

$$L_1 = \frac{1}{2}(m_1+m_2)\dot{\overrightarrow{R_c}}^2 - U^e(\overrightarrow{R_c}), \quad L_2 = \frac{1}{2}\mu\dot{\vec{r}}^2 \text{ - } U'(\vec{r}). \tag{6}$$

where $U^e(R_c)$ is the potential function of the system in theforce field $\vec{F}(t)$, $\mu = m_1m_2/(m_1+m_2)$ is convert mass, and $U'(\vec{r})$ is the interaction potential function of the system. Since the interaction fore of the two-body spring vibrator is the central force, the relative motion trajectories are on a fixed plan, and we take this plane as a planc of polar coordinates. The relative Lagrange function is given by

$$L_2 = \frac{1}{2}\mu(\dot{r}^2 + r^2\dot{\theta}^2) + \frac{1}{2}k_0(r-l)^2 . \tag{7}$$

Obviously, the relative Lagrange function L_2 does not contain the time, the mechanical energy is a conserved quantity, namely,

$$E = \frac{1}{2}\mu(\dot{r}^2 + r^2\dot{\theta}^2) + \frac{1}{2}k_0(r-l)^2 = \text{constant} . \tag{8}$$

L_2 does not obvisouly contain the polar angle, the angular momentum is a conserved quantity, namely,

$$J = \mu r^2\dot{\theta} = \text{constant} . \tag{9}$$

From Eqs.(12) and (13), we can obtain the orbital equation

$$\theta = \frac{J}{\sqrt{\mu k_0}}\int_{r_0}^{r}\frac{dr}{r\sqrt{-[r^4 - 2lr^3 - (2E - k_0 l^2)r^2/k_0 + J^2/\mu k_0]}} . \tag{10}$$

Equation (10) is an elliptical integral. We define

$$f(r) = r^4 - 2lr^3 - \frac{(2E - k_0 l^2)r^2}{k_0} + \frac{J^2}{\mu k_0} . \tag{11}$$

To obtain the solution of Eq.(10), we must discuss the roots of the equation

$$f(r) = 0 . \tag{12}$$

Defining

$$r = z + l/2,\quad q = \frac{k_0 l^2 - 2E}{k_0} - \frac{3l^2}{2},\quad w = \frac{l(k_0 l^2 - 2E)}{k_0} - l^3,\quad s = \frac{J^2}{k_0\mu} + \frac{(k_0 l^2 - 2E)l^2}{4k_0} - \frac{3l^4}{16} .$$

Then we get

$$r^4 - 2lr^3 - \frac{(2E - k_0 l^2)r^2}{k_0} + \frac{J^2}{\mu k_0} = z^4 + qz^2 + wz + s = (z^2 + \beta z + h)(z^2 + \beta z + p) = 0 , \tag{13}$$

where $2h = q + \beta^2 - \frac{w}{\beta}$, $2p = q + \beta^2 + \frac{w}{\beta}$, $hp = s$.

By eliminating h and p, we get β satisfying equation

$$\beta^6 + 2q\beta^4 + (q^2 - 4s)\beta^2 - w^2 = 0 . \tag{14}$$

By solving Eq.(14), we get the roots β_1^2, β_2^2, and β_3^2 of the equation, and take the positive square roots β_1, β_2, and β_3. Then, the roots of Eq. (10) are given by

$$\begin{aligned} r_1 &= \frac{1}{2}(\beta_1 + \beta_2 + \beta_3) + \frac{l}{2}, & r_2 &= \frac{1}{2}(\beta_1 - \beta_2 - \beta_3) + \frac{l}{2} \\ r_3 &= \frac{1}{2}(-\beta_1 + \beta_2 - \beta_3) + \frac{l}{2}, & r_4 &= \frac{1}{2}(-\beta_1 - \beta_2 + \beta_3) + \frac{l}{2} \end{aligned} . \tag{15}$$

The discriminant of the equation is

$$\Delta = 4\left[\frac{(k_0 l^2-2E)^2}{3k_0^2}+\frac{4J^2}{k_0\mu}\right]^3-\left[\frac{(k_0 l^2-2E)^3}{27k_0^3}+\frac{4l^2J^2}{k_0\mu}-\frac{8(k_0 l^2-2E)J^2}{3k_0^2\mu}\right]^2. \tag{16}$$

3. The Analytical Solutions of the Two-body Spring Vibrator

3.1. The Discriminant $\Delta > 0$

When $\Delta > 0$, Eq. $f(r)=0$ have four different real roots r_1, r_2, r_3, and r_4, and we assume $r_1 < r_2 < r_3 < r_4$. Then we have $f(r)=(r-r_1)(r-r_2)(r-r_3)(r-r_4)$. If we require that $\sqrt{-f(r)}$ is significant, r should satisfy $-f(r)>0$. Hence the relations of the displacement r with the roots only have two cases: $r_1 < r < r_2 < r_3 < r_4$, and $r_1 < r_2 < r_3 < r < r_4$.

3.1.1. $r_1 < r < r_2 < r_3 < r_4$. From Eq. (10), we have

$$\theta = \frac{J}{\sqrt{\mu k_0}}\int_{r_0}^{r_2}\frac{dr}{r\sqrt{(r-r_1)(r-r_2)(r-r_3)(r_4-r)}}-\frac{J}{\sqrt{\mu k_0}}\int_{r}^{r_2}\frac{dr}{r\sqrt{(r-r_1)(r-r_2)(r-r_3)(r_4-r)}}, \tag{17}$$

where $r \ge r_0$. By the elliptical integral, we have[11]

$$\theta = \frac{Jg'}{r_2\alpha_1^2\sqrt{\mu k_0}}(\alpha_1^2-\alpha^2)\left\{\left[\prod(\varphi_0,\alpha_1^2,k)-\prod(\varphi,\alpha_1^2,k)\right]+\alpha^2\left[F(\varphi_0,k)-F(\varphi,k)\right]\right\}, \tag{18}$$

Where $sn^2u=[(r_3-r_1)(r_2-r)]/[(r_2-r_1)(r_3-r)]$, $k^2=[(r_4-r_3)(r_2-r_1)]/[(r_4-r_2)(r_3-r_1)]$, $k^2<\alpha^2=(r_2-r_1)/(r_3-r_1)<1$, $\varphi=\sin^{-1}\sqrt{[(r_3-r_1)(r_2-r)]/[(r_2-r_1)(r_3-r)]}$, $snu_1=\sin\varphi$, $\alpha_1^2=(r_3/r_2)\alpha^2$, $F(\varphi,k)$ is the elliptic integral function of the first kind, and $\prod(\varphi,\alpha,k)$ is the elliptic integral function of the third kind.

3.1.2. $r_1 < r_2 < r_3 < r < r_4$. From Eq. (10), we have[11]

$$\theta = \frac{Jg'}{r_4\sqrt{\mu k_0}}\left\{(1-\frac{r_4}{r_1})\left[\prod(\varphi_0,\frac{r_1}{r_4}\alpha^2,k)-\prod(\varphi,\frac{r_1}{r_4}\alpha^2,k)\right]+\frac{r_4}{r_1}[F(\varphi_0,k)-F(\varphi,k)]\right\}, \tag{19}$$

Where $sn^2u=[(r_3-r_1)(r_4-r)]/[(r_4-r_3)(r_3-r_1)]$, $k^2=[(r_4-r_3)(r_2-r_1)]/[(r_4-r_2)(r_3-r_1)]$, $g'=2/\sqrt{(r_4-r_2)(r_3-r_1)}$, $\alpha^2=(r_3-r_4)/(r_3-r_1)<0$, and $\varphi=\sin^{-1}\sqrt{[(r_3-r_1)(r_4-r)]/[(r_4-r_3)(r-r_1)]}$.

3.2. The Discriminant $\Delta = 0$

When the discriminant $\Delta=0$, the equation $f(r)=0$ possesses two groups of equal real roots or one group of equal real roots. However, two groups of equal real roots make $\sqrt{-f(r)}$ cannot hold water; only when the equation $f(r)=0$ possesses one group of equal real roots, the quadratic radical $\sqrt{-f(r)}$ can be significant, and Eq. (10) becomes a general integral in this case.

We suppose that the one group of the real roots of Eq. $f(r)=0$ are $r_1=r_2, r_3\ne r_4\ne r_1$, and $r_3<r_4$.

3.2.1. $r_1 = r_2 < r_3 < r_4$. If we require that $\sqrt{-f(r)}$ is significant, r have to satisfy $r_3 < r < r_4$. In this case, we get

$$\theta = -\frac{J}{r_1\sqrt{\mu k_0 r_3 r_4}}[\sin^{-1}\frac{(r_3+r_4)r-2r_3r_4}{r(r_4-r_3)} - \sin^{-1}\frac{(r_3+r_4)r_0-2r_3r_4}{r_0(r_4-r_3)}] + \frac{J}{r_1\sqrt{\mu k_0(r_3-r_1)(r_4-r_1)}}\times$$

$$\left\{\sin^{-1}\frac{(r_3-2r_1+r_4)(r-r_1)-2(r_3-r_1)(r_4-r_1)}{(r-r_1)(r_4-r_3)} - \sin^{-1}\frac{(r_3-2r_1+r_4)(r_0-r_1)-2(r_3-r_1)(r_4-r_1)}{(r_0-r_1)(r_4-r_3)}\right\} \quad . \tag{20}$$

3.2.2. $r_3 < r_1 = r_2 < r < r_4$. When $r_3 < r_1 = r_2 < r < r_4$, If we require that $\sqrt{-f(r)}$ is significant, r have to satisfy $r_3 < r < r_1 = r_2 < r_4$ or $r_3 < r_1 = r_2 < r < r_4$.

3.2.2.1. $r_3 < r < r_1 = r_2 < r_4$. Similarly, we have

$$\vartheta = \frac{J}{r_1\sqrt{\mu k_0 r_3 r_4}}[\sin^{-1}\frac{(r_3+r_4)r-2r_3r_4}{r(r_4-r_3)} - \sin^{-1}\frac{(r_3+r_4)r_0-2r_3r_4}{r_0(r_4-r_3)}] + \frac{J}{r_1\sqrt{\mu k_0(r_4-r_1)(r_1-r_3)}}\times$$

$$\left\{\ln[\frac{\sqrt{-r^2+(r_3+r_4)r-r_3r_4}+\sqrt{(r_4-r_1)(r_1-r_3)}}{(r_1-r)} + \frac{r_3+r_4-2r_1}{2\sqrt{(r_4-r_1)(r_1-r_3)}} - \ln[\frac{\sqrt{-r_0^2+(r_3+r_4)r_0-r_3r_4)}+\sqrt{(r_4-r_1)(r_1-r_3)}}{(r_1-r_0)} + \frac{r_3+r_4-2r_1}{2\sqrt{(r_4-r_1)(r_1-r_3)}}\right\}. \tag{21}$$

3.2.2.2. $r_3 < r_1 = r_2 < r < r_4$. Similarly, we have

$$\theta = \frac{J}{r_1\sqrt{\mu k_0 r_3 r_4}}[\sin^{-1}\frac{(r_3+r_4)r-2r_3r_4}{r(r_4-r_3)} - \sin^{-1}\frac{(r_3+r_4)r_0-2r_3r_4}{r_0(r_4-r_3)}] - \frac{J}{r_1\sqrt{\mu k_0(r_4-r_1)(r_1-r_3)}}$$

$$\times\left\{\ln[\frac{\sqrt{-r^2+(r_3+r_4)r-r_3r_4}+\sqrt{(r_4-r_1)(r_1-r_3)}}{(r-r_1)} + \frac{2r_1-r_3-r_4}{2\sqrt{(r_4-r_1)(r_1-r_3)}} - \ln[\frac{\sqrt{-r_0^2+(r_3+r_4)r_0-r_3r_4}+\sqrt{(r_4-r_1)(r_1-r_3)}}{(r_0-r_1)} + \frac{2r_1-r_3-r_4}{2\sqrt{(r_4-r_1)(r_1-r_3)}}\right\} \tag{22}$$

3.2.3. $r_3 < r_4 < r_1 = r_2$. When $r_3 < r_4 < r_1 = r_2$, If we require that $\sqrt{-f(r)}$ is significant, r have to satisfy $r_3 < r < r_4 < r_1 = r_2$. In this case, we get

$$\theta = \frac{J}{r_1\sqrt{\mu k_0 r_3 r_4}}[\sin^{-1}\frac{(r_3+r_4)r-2r_3r_4}{r(r_4-r_3)} - \sin^{-1}\frac{(r_3+r_4)r_0-2r_3r_4}{r_0(r_4-r_3)}] - \frac{J}{r_1\sqrt{\mu k_0(r_1-r_3)(r_1-r_4)}}\times$$

$$\left\{\sin^{-1}[\frac{(2r_1-r_3-r_4)(r_1-r)-2(r_1-r_3)(r_1-r_4)}{(r_1-r)(r_4-r_3)} - \sin^{-1}[\frac{(2r_1-r_3-r_4)(r_1-r_0)-2(r_1-r_3)(r_1-r_4)}{(r_1-r_0)(r_4-r_3)}\right\} \tag{23}$$

3.3. The Discriminant $\Delta < 0$

When the discriminant $\Delta < 0$, Eq. $f(r) = 0$ possesses conjugate complex roots. If Eq. $f(r) = 0$ possesses two groups of conjugate complex roots, and we suppose $r_1 = \overline{r_2}$, and $r_3 = \overline{r_4}$.

Because the displacement r is a real number, we have $(r - r_1)(r - r_2) = (r - r_1)\overline{(r - r_1)} > 0$, and $(r - r_3)(r - r_4) = (r - r_3)\overline{(r - r_3)} > 0$. In this case, the quadratic radical $\sqrt{-f(r)} =$

$\sqrt{-(r-r_1)(r-r_2)(r-r_3)(r-r_4)}$ cannot hold water.

If Eq. $f(r)=0$ possesses only one couple of conjugate complex roots, and we suppose $r_1=\overline{r_2}$, and $r_3<r_4$.

When only the displacement r satisfies $r_3<r<r_4$, the quadratic radical $\sqrt{-f(r)}$ is significant. Then we have

$$\theta=-\frac{J}{\sqrt{\mu k_0}}\int_{r_3}^{r_0}\frac{dr}{r\sqrt{(r-r_1)(r-\overline{r_1})(r-r_3)(r_4-r)}}+-\frac{J}{\sqrt{\mu k_0}}\int_{r_3}^{r}\frac{dr}{r\sqrt{(r-r_1)(r-\overline{r_1})(r-r_3)(r_4-r)}} \tag{24}$$

By the elliptic integral, we obtain [12]

$$\theta=-\frac{J(A+B)g'}{\sqrt{\mu k_0(Ar_3-Br_4)}}\times\left\{\alpha_1F(\varphi_0,k)+\frac{\alpha-\alpha_1}{1-\alpha^2}[\prod(\varphi_0,\frac{\alpha^2}{\alpha^2-1},k)-\alpha f_1]\right\}+$$
$$\frac{J(A+B)g'}{\sqrt{\mu k_0(Ar_3-Br_4)}}\times\left\{\alpha_1F(\varphi,k)+\frac{\alpha-\alpha_1}{1-\alpha^2}[\prod(\varphi,\frac{\alpha^2}{\alpha^2-1},k)-\alpha f_1]\right\}, \tag{25}$$

Where $\alpha_1^2=-(r_1-r_2)/4$, $cnu=[(r_4-r)B-(r-r_3)A]/[(r_4-r)B+(r-r_3)A]$, $A^2=(r_4-(r_1+r_2)/2)^2+(r_1+r_2)^2/4$, $B^2=(r_3-(r_1+r_2)/2)^2+(r_1+r_2)^2/4$, $g'=1/\sqrt{AB}$, $\varphi=amu_1=\cos^{-1}\{[(r_4-r)B-(r-r_3)A]/[(r_4-r)B+(r-r_3)A]\}$, $\alpha=(r_3A-r_4B)/(r_3A+r_4B)$, $\alpha=(A-B)/(A+B)$, $f_1=\sqrt{(1-\alpha^2)/(k^2-k'^2\alpha^2)}tg^{-1}[\sqrt{(k^2+k'^2\alpha^2)/(1-\alpha^2)}sdu]$, and $k'=\sqrt{1-k^2}$.

Now, we have given all analytic solutions of the two-body spring vibrator.

4. Summaries and Conclusions

Giving the specific values of E,l,m_1,m_2,k_0, and J, we may solve the equation $f(r)=0$ and get the relevant roots, so that we can attain the specific orbit equation of the system, and discuss the motion laws of the two-body spring vibrator.

Considering an ideal case for $l=0$, from Eq.(15) we can get the four different real roots of the equation $f(r)=0$ is given by

$$r_1=-[\frac{E+\sqrt{E^2-k_0J^2/\mu}}{k_0}]^{\frac{1}{2}},\quad r_2=-[\frac{E-\sqrt{E^2-k_0J^2/\mu}}{k_0}]^{\frac{1}{2}},$$
$$r_3=[\frac{E-\sqrt{E^2-k_0J^2/\mu}}{k_0}]^{\frac{1}{2}},\quad r_4=[\frac{E+\sqrt{E^2-k_0J^2/\mu}}{k_0}]^{\frac{1}{2}}. \tag{26}$$

Obviously, only when the relative displacement r satisfies $-\sqrt{(E+\sqrt{E^2-k_0J^2/\mu})/k_0}<r<\sqrt{(E+\sqrt{E^2-k_0J^2/\mu})/k_0}$, can the motion equation of the two-body spring vibrator Eq. (10) get real solutions. Therefore the relative displacement of the two-body spring vibrator cannot exist in the region $-\sqrt{(E+\sqrt{E^2-k_0J^2/\mu})/k_0}<r<\sqrt{(E-\sqrt{E^2-k_0J^2/\mu})/k_0}$. The region contains the area of $0 < r < r_3$, so one may infer that even if the natural length is zero, two spring vibrators cannot be infinitely close as well.

Utilizing numerical simulation we can give motion diagrams of the two-body spring vibrator, and can analyze further the motion characteristics. On the other hand, the interaction force of the two-body spring vibrator is similar to the molecular interaction force, so the above discussions may aid us to understand further properties of the molecular force and the influence of the molecular force on the molecular radiation.

References

[1] Moscardo F and San-Fabian E1991 Phys. Rev. A Vol 44(3)p 1549.
[2] Calvo M P and J. M. Sanz-serna Siam J M 1993 J. Sci. Comput. Vol14(4)p 916.
[3] Damour T Jaranowski P and G. Schafer G 2000 Phys. Rev. D Vol 62 p 021501(R)4962.
[4] Hestens D1999 New Founctios for Classical Mechanic (New York : Kluwer Academic Publishers).
[5] Gutzwiller M C 1998 *Rev. Modern Phys.* Vol 70 p 589.
[6] SCHEERES D J 2002 *Celestial Mechanics and Dynamical Astronomy*Vol 83p 155.
[7] Schild A 1961 *Physical Review* Vol131p 2762.
[8] Angelov V G 1990 *Journal of Mathematical Analysis and Applications* Vol151p 448.
[9] LucaF J D2007 *Journal of Mathematical Physics* Vol48(1)p 012901.
[10] Wang Z X and Guo D R 1987*Intruoduction Special Functtion*(Beijing : China Science Press)p 729-733.
[11] Byrd P F and FredmanM D1970 *Handbook of Elliptic Integrals for Engineers and Scientists* (Berlin: Springer-Verlag)pp109-205.
[12] Byrd P F and FredmanM D1970 *Handbook of Elliptic Integrals for Engineers and Scientists* (Berlin: Springer-Verlag)pp133-206.

Effect of Fractional Parameters on Plane Waves of Generalized Electromagneto–thermoelastic Diffusion in a Rotating Half-space

Jianzhong Sun* and Yaping Wu

School of Civil Engineering, Lanzhou Jiaotong University, Lanzhou 730070, Gansu, China

*Corresponding author email: sunjzh@mail.lzjtu.cn

Abstract. This paper studies the effect of score parameters on the dynamic response of a thermally conductive half-space at a constant angular velocity and with variable material properties. The problem is formulated in the context of fractional order generalized thermoelastic diffusion theory. The exact expressions for the dimensionless results are derived and graphed.
Keywords: Fractional calculus; Rotating half-space; Diffusion relaxation time.

1. Introduction

Thermal diffusion in a deformed body is due to the temperature field, mass diffusion field, and strain field. Nowaski proposed the theory of thermoelastic diffusion. Sherief [1]generalized this theory to predict the propagation of thermoelastic and diffusive waves. Most materials such as amorphous media and porous materials, and physical conditions such as transient loading, have brought serious challenges to classical generalized thermoelastic theory[2]. To solve such problems, Ignaczak and Ostoja-Starzewski [3] introduced time-fractional thermoelasticity theory.

This paper uses generalized electromagnetic–thermoelastic theory to investigate the effect of fractional parameters on two-dimensional electromagneto–thermoelastic diffusion in a conducting half-space that has temperature-dependent properties and rotates at constant angular velocity.

2. Formulation of the Problem

In Figure 1, we consider the thermoelastic problem of an isotropic, homogeneous, and semi-infinite perfect conductor occupying the Cartesian half-space $x > 0$. The simplified equations of electrodynamics for the elastic medium are

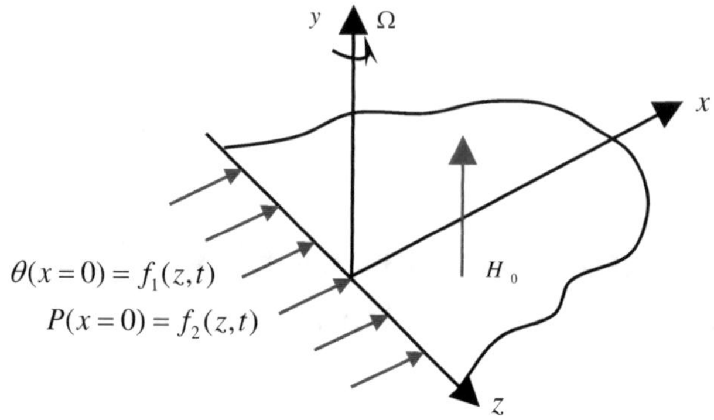

Figure 1. Model of the problem.

$$\partial_j \varepsilon_{ijk} h_k = J_i + D_{i,t} \tag{1}$$

$$\partial_j E_k \varepsilon_{ijk} = -B_{i,t} \tag{2}$$

$$\mathrm{B}_{i,i} = 0, D_{i,i} = \rho_e \tag{3}$$

$$B_i = \mu_0 (H_i + h_i), \quad D_i = \varepsilon_0 E_i \tag{4}$$

$$J_i = \sigma_0 (E_i + \mu_0 \varepsilon_{ijk} u_{k,t} H_j) \tag{5}$$

The motion equation is

$$\rho[u_{i,tt} + \Omega_p \Omega_q u_m \varepsilon_{pqs} \varepsilon_{smi} + 2\varepsilon_{ijk} u_{i,t} \Omega_j] = \sigma_{ij,j} + F_i \tag{6}$$

where ρ and σ_{ij} are the density and the components of the stress tensor, and the Lorentz force F_i is given by

$$F_i = \mu_0 \varepsilon_{ijk} J_k H_j \tag{7}$$

The constitutive equations are

$$\sigma_{ij} = 2\mu e_{ij} + \delta_{ij} \left(\lambda_0 e_{kk} - \gamma_1 \theta - \gamma_2 P \right) \tag{8}$$

$$C = nP + d\theta + \gamma_2 e_{kk} \tag{9}$$

$$\rho S = \gamma_1 e_{kk} + l_1 \theta + dP \tag{10}$$

The strain as a function of displacement is

$$e_{ij} = \frac{1}{2}\left(u_{i,j} + u_{j,i}\right) \tag{11}$$

The generalized heat transfer model is

$$[\kappa_{ij} T_{,j}]_{,i} = (1 + \frac{\tau_0^\alpha}{\alpha!} \frac{\partial^\alpha}{\partial t^\alpha})[l_1 T_0 \theta_{,t} + dT_0 P_{,t} + \gamma_1 T_0 e_{,t}], \quad 0 < \alpha \le 1 \tag{12}$$

Generalized mass diffusion equation :

$$[D_{ij} P_{,j}]_{,i} = (1 + \frac{\tau^\alpha}{\alpha!} \frac{\partial^\alpha}{\partial t^\alpha})[nP_{,t} + d\theta_{,t} + \gamma_2 e_{,t}], \quad 0 < \alpha \le 1 \tag{13}$$

and

$$\theta = T - T_0, \lambda_0 = \lambda - \frac{\beta_2^2}{b}, \gamma_1 = \beta_1 + \frac{a}{b}\beta_2, \gamma_2 = \frac{\beta_2}{b}, l_1 = \frac{\rho C_E}{T_0} + \frac{a^2}{b}, d = \frac{a}{b}, n = \frac{1}{b},$$
$$\beta_1 = (3\lambda + 2\mu)\alpha_t, \quad \beta_2 = (3\lambda + 2\mu)\alpha_c.$$

Meanwhile, the material properties are assumed to depend on temperature:

$$\begin{aligned} &\lambda = \lambda' f(T), \quad \mu = \mu' f(T), \quad \lambda_0 = \lambda_0' f(T), \quad \kappa = \kappa' f(T), \quad D = D' f(T) \\ &\beta_1 = \beta_1' f(T), \quad \beta_2 = \beta_2' f(T), \quad \gamma_1 = \gamma_1' f(T), \quad \gamma_2 = \gamma_2' f(T) \end{aligned} \tag{14}$$

where $\lambda', \mu', \kappa', D', \beta_1', \beta_2', \gamma_1'$, and γ_2' are known constants;

$$f(T) = 1 - \alpha^* T \tag{15}$$

where α^* is an empirical parameter. Using Eqs. (14) and (15) in Eqs. (6), (8), (9), (12), and (13), one has

$$\begin{aligned} &\rho[\ddot{u}_i + \{\Omega \times (\Omega \times u)\}_i + (2\Omega \times \dot{u})_i] \\ &= [2\mu e_{ij} + \delta_{ij}(\lambda_0 e_{kk} - \gamma_1\theta - \gamma_2 P)]_{,j} f(T) + \\ &[2\mu e_{ij} + \delta_{ij}(\lambda_0 e_{kk} - \gamma_1\theta - \gamma_2 P)] f(T)_{,j} + \mu_0 (J \times H)_i \end{aligned} \tag{16}$$

$$[\kappa_{ij} T_{,j} f(T)]_{,i} = (1 + \frac{\tau_0^\alpha}{\alpha!}\frac{\partial^\alpha}{\partial t^\alpha})[l_1 T_0 \theta_{,t} + d T_0 P_{,t} + \gamma_1 T_0 e_{,t}], \quad 0 < \alpha \le 1 \tag{17}$$

$$[D_{ij} P_{,j} f(T)]_{,i} = (1 + \frac{\tau^\alpha}{\alpha!}\frac{\partial^\alpha}{\partial t^\alpha})[n P_{,t} + d\theta_{,t} + \gamma_2 e_{,t}], \quad 0 < \alpha \le 1 \tag{18}$$

$$\sigma_{ij} = f(T)[2\mu e_{ij} + \delta_{ij}(\lambda_0 e_{kk} - \gamma_1\theta - \gamma_2 P)], \quad C = nP + d\theta + f(T)\gamma_2 e_{kk} \tag{19}$$

The displacement components are

$$u_x = u(x,z,t), \quad u_y = 0, \quad u_z = w(x,z,t) \tag{20}$$

$$e_{xy} = e_{yy} = e_{yz} = 0 \tag{21}$$

The components of magnetic intensity are

$$H_x = 0, \quad H_y = H_0 + h(x,z,t), \quad H_z = 0 \tag{22}$$

The following relations can be obtained:

$$E_x = E_1, \quad E_y = 0, \quad E_z = E_3 \tag{23}$$

$$J_1 = J_x = \sigma_0(E_1 - \mu_0 H_0 w_{,t}), \quad J_2 = J_y = 0, \quad J_3 = J_z = \sigma_0(E_3 + \mu_0 H_0 u_{,t}) \tag{24}$$

Equations (1), (4), and (24) give the two equations

$$h_{,z} = -\sigma_0(E_1 - \mu_e H_0 w_{,t}) - \varepsilon_0 E_{1,t} \tag{25a}$$

$$h_{,x} = \sigma_0(E_3 + \mu_e H_0 u_{,t}) + \varepsilon_0 E_{3,t} \tag{25b}$$

From Eqs. (2) and (4), one can derive

$$E_{3,x} - E_{1,z} = \mu_0 h_{,t} \tag{26}$$

From Eqs. (7) and (24), we obtain

$$F_x = -\mu_0\sigma_0 H_0(E_3 + \mu_0 H_0 u_{,t}) \tag{27a}$$

$$F_z = 0 \tag{27b}$$

$$F_z = \mu_0\sigma_0 H_0(E_1 - \mu_0 H_0 w_{,t}) \tag{27c}$$

In the context of linear theory, one has

$$f(T) \approx 1 - \alpha^* T_0 = M \tag{28}$$

Substituting Eq. (28) into Eqs. (16)–(19), we get

$$\begin{aligned}\rho(u_{,tt} - \Omega^2 u + 2\Omega w_{,t}) = M(\lambda_0 + 2\mu)u_{,xx} + M\mu u_{,zz} + M(\lambda_0 + \mu)w_{,xz}\\ - M\gamma_1\theta_{,x} - M\gamma_2 P_{,x} - \sigma_0\mu_0 H_0(E_3 + \mu_0 H_0 u_{,t})\end{aligned} \tag{29}$$

$$\begin{aligned}\rho(w_{,tt} - \Omega^2 w - 2\Omega u_{,t}) = M(\lambda_0 + 2\mu)w_{,zz} + M\mu w_{,xx} + M(\lambda_0 + \mu)u_{,xz}\\ - M\gamma_1\theta_{,z} - M\gamma_2 P_{,z} + \sigma_0\mu_0 H_0(E_1 - \mu_0 H_0 w_{,t})\end{aligned} \tag{30}$$

$$\kappa M\nabla^2\theta = (1 + \frac{\tau_0^\alpha}{\alpha!}\frac{\partial^\alpha}{\partial t^\alpha})[T_0 l_1\theta_{,t} + T_0 d P_{,t} + M T_0\gamma_1 e_{,t}] \tag{31}$$

$$DM\nabla^2 P = (1 + \frac{\tau^\alpha}{\alpha!}\frac{\partial^\alpha}{\partial t^\alpha})[nP_{,t} + d\theta_{,t} + M\gamma_2 e_{,t}] \tag{32}$$

Then the constitutive equations are

$$\sigma_{xx} = M(2\mu u_{,x} + \lambda_0 e - \gamma_1\theta - \gamma_2 P) \tag{33a}$$

$$\sigma_{zz} = M(2\mu w_{,z} + \lambda_0 e - \gamma_1\theta - \gamma_2 P) \tag{33b}$$

$$\sigma_{xz} = M\mu(u_{,z} + w_{,x}) \tag{33c}$$

$$C = nP + d\theta + M\gamma_2 e_{kk} \tag{33d}$$

The cubical dilatation can be expressed as

$$e = u_{,x} + w_{,z} \tag{34}$$

The following dimensionless variables are introduced:

$$\begin{gathered}x^* = c_1\eta x,\quad z^* = c_1\eta z,\quad u^* = c_1\eta u,\quad w^* = c_1\eta w,\quad t^* = c_1^2\eta t,\quad \tau^* = c_1^2\eta\tau\\ \tau_0^* = c_1^2\eta\tau_0,\quad \theta^* = \frac{\beta_1\theta}{\lambda + 2\mu},\quad C^* = \frac{\beta_2 C}{\lambda + 2\mu},\quad \sigma_{ij}^* = \frac{\sigma_{ij}}{\lambda + 2\mu},\quad P^* = \frac{P}{\beta_2}\\ E_i^* = \frac{\eta E_i}{\sigma_0\mu_0^2 H_0 c_1},\quad h^* = \frac{\eta h}{\mu_0\sigma_0 H_0},\quad \eta = \frac{\rho C_E}{\kappa},\quad c_1^2 = \frac{\lambda + 2\mu}{\rho},\quad i,j = 1,3\end{gathered} \tag{35}$$

Dropping the asterisks for simplicity, the governing equations are normalized in terms of Eq. (35) as:

$$\begin{aligned}\beta_M^2[u_{,tt} - \Omega^2 u + 2\Omega w_{,t}] = \beta'^2 u_{,xx} + u_{,zz} + (\beta' - 1)w_{,xz} - \alpha_1\beta^2\theta_{,x} - \alpha_2 P_{,x}\\ - \beta^2\alpha\gamma(\gamma E_3 + u_{,t})\end{aligned} \tag{36}$$

$$\beta_M^2[w_{,tt}-\Omega^2 w-2\Omega u_{,t}]=\beta'^2 w_{,zz}+w_{,xx}+(\beta'-1)u_{,xz}-\alpha_1\beta^2\theta_{,z}-\alpha_2 P_{,z} + \beta^2\alpha\gamma(\gamma E_1-w_{,t}) \quad (37)$$

$$\nabla^2\theta=\xi_1(1+\frac{\tau_0^\alpha}{\alpha!}\frac{\partial^\alpha}{\partial t^\alpha})\theta_{,t}+\xi_2(1+\frac{\tau_0^\alpha}{\alpha!}\frac{\partial^\alpha}{\partial t^\alpha})e_{,t}+\xi_3(1+\frac{\tau_0^\alpha}{\alpha!}\frac{\partial^\alpha}{\partial t^\alpha})p_{,t} \quad (38)$$

$$\nabla^2 P=\eta_1(1+\frac{\tau_0^\alpha}{\alpha!}\frac{\partial^\alpha}{\partial t^\alpha})\theta_{,t}+\eta_2(1+\frac{\tau_0^\alpha}{\alpha!}\frac{\partial^\alpha}{\partial t^\alpha})e_{,t}+\eta_3(1+\frac{\tau_0^\alpha}{\alpha!}\frac{\partial^\alpha}{\partial t^\alpha})p_{,t} \quad (39)$$

$$h_{,z}=-\gamma E_1-V^2 E_{1,t}+w_{,t} \quad (40)$$

$$h_{,x}=\gamma E_3+V^2 E_{3,t}+u_{,t} \quad (41)$$

$$h_{,t}=E_{3,x}-E_{1,z} \quad (42)$$

where

$$\gamma=\frac{\sigma_0\mu_0}{\eta},\beta^2=\frac{\lambda+2\mu}{\mu},\beta'^2=\frac{\lambda_0+2\mu}{\mu_0},\beta_M^2=\frac{\lambda+2\mu}{M\mu}$$

$$\alpha=\frac{\mu_0 H_0^{\ 2}}{\lambda+2\mu}\alpha_1=\frac{\gamma_1}{\beta_1},\alpha_2=\frac{\gamma_2\beta_2}{\mu},\xi_1=\frac{T_0 l_1}{M\kappa\eta}$$

$$\xi_2=\frac{T_0\beta_1\gamma_1}{\kappa\eta(\lambda+2\mu)},\xi_3=\frac{T_0 d\beta_1\beta_2}{M\kappa\eta(\lambda+2\mu)},\eta_1=\frac{d\left(\lambda+2\mu\right)}{MD\eta\beta_1\beta_2}$$

$$\eta_2=\frac{\gamma_2}{D\eta\beta_2},\eta_3=\frac{n}{MD\eta},c^2=\frac{1}{\mu_0\varepsilon_0},V^2=\frac{c_1^2}{c^2}$$

Then one has

$$\sigma_{xx}=\frac{M}{\beta^2}(2u_{,x}+\frac{\lambda_0}{\mu}e-\alpha_1\beta^2\theta-\alpha_2 P) \quad (43a)$$

$$\sigma_{zz}=\frac{M}{\beta^2}(2w_{,z}+\frac{\lambda_0}{\mu}e-\alpha_1\beta^2\theta-\alpha_2 P) \quad (43b)$$

$$\sigma_{xz}=\frac{M}{\beta^2}(u_{,z}+w_{,x}) \quad (43c)$$

$$C=\frac{\alpha_2}{\beta^2}(P+Me)+\frac{d\beta_2}{\beta_1}\theta \quad (43d)$$

Deal with equations 36 and 37,we arrive at

$$(\nabla^2+\varepsilon_M\Omega^2-\varepsilon\alpha\gamma\frac{\partial}{\partial t}-\varepsilon_M\frac{\partial^2}{\partial t^2})e-\varepsilon\alpha_1\nabla^2\theta-\varepsilon_1\nabla^2 P-\varepsilon\alpha\gamma^2 h_{,t}+2\Omega\varepsilon_M\psi_{,t}=0 \quad (44)$$

where ψ is

$$\psi=u_{,z}-w_{,x} \quad (45)$$

Deal with equations 36 and 37,we arrive at

$$(\nabla^2 + \beta_M^2\Omega^2 - \beta^2\alpha\gamma\frac{\partial}{\partial t} - \beta_M^2\frac{\partial^2}{\partial t^2})\psi = -2\Omega\beta_M^2 e_{,t} - \beta^2\alpha\gamma^2(E_{1,x} + E_{3,z}) \tag{46}$$

Now we deal with 40 and 41,we arrive at

$$(\gamma + V^2\frac{\partial}{\partial t})(E_{1,x} + E_{3,z}) = -\psi_{,t} \tag{47}$$

Applying Eq. (47) to Eq. (46) gives

$$[(\nabla^2 + \beta_M^2\Omega^2 - \beta^2\alpha\gamma\frac{\partial}{\partial t} - \beta_M^2\frac{\partial^2}{\partial t^2})(\gamma + V^2\frac{\partial}{\partial t}) + \beta^2\alpha\gamma^2\frac{\partial}{\partial t}]\psi = 2\Omega\beta_M^2(\gamma + V^2\frac{\partial}{\partial t})e_{,t} \tag{48}$$

where

$$\varepsilon = \frac{\beta^2}{\beta'^2},\ \varepsilon_1 = \frac{\alpha_2}{\beta'^2},\ \varepsilon_M = \frac{\beta_M^2}{\beta'^2}$$

Deal with 40 and 41,we arrive at

$$\nabla^2 h = e_{,t} + \gamma h_{,t} + V^2 h_{,tt} \tag{49}$$

3. Normal Mode Analysis

The solutions are decomposed as

$$\begin{aligned}&\left[u, v, e, \psi, \theta, h, E_i, P, C, \sigma_{ij}\right](x, y, t) = \\ &\left[u^*(x), v^*(x), e^*(x), \psi^*(x), \theta^*(x), h^*(x), E_i^*(x), P^*(x), C^*(x), \sigma_{ij}^*(x)\right]\exp(\omega t + isz)\end{aligned} \tag{50}$$

4. Numerical Results and Discussion

The material copper is chosen for numerical evaluations [9]. The results for different physical variables are presented for x = 10 at t =0.1, t=0.2, and t=0.3. The fractional parameter α is taken into account with the values 0.5, 0.8, and 1.0, with the temperature-dependent modulus $M = 1 - \alpha^* T_0 = 2$ and the rotation $\Omega = 0.01$. The real parts of the non-dimensional results are shown in Figures 2–7.

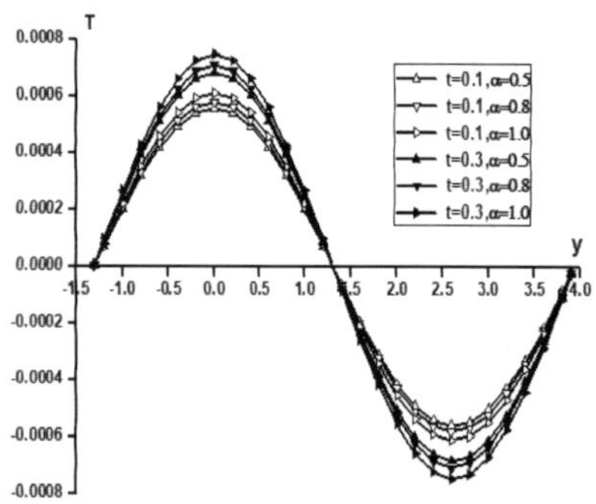

Figure 2. Temperature distribution (x=10).

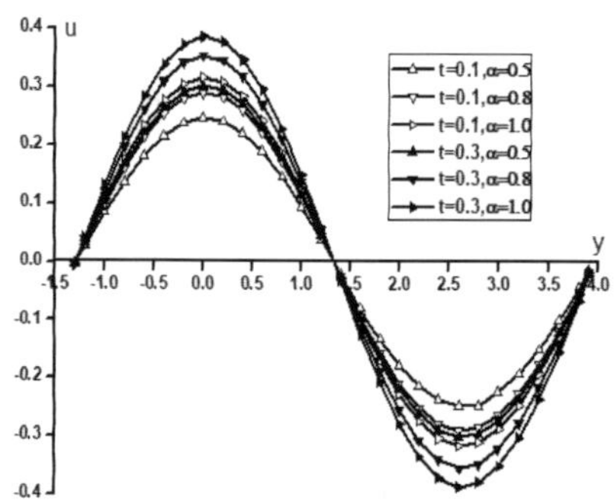

Figure 3 Horizontal displacement distribution (x=10).

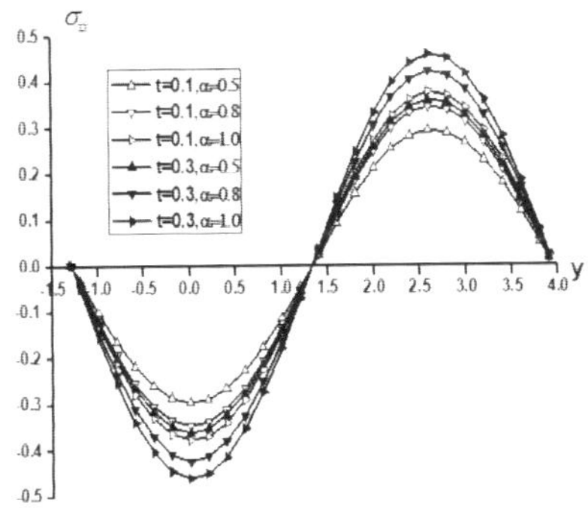

Figure 4. Distribution of σ_{xx} (x=10)

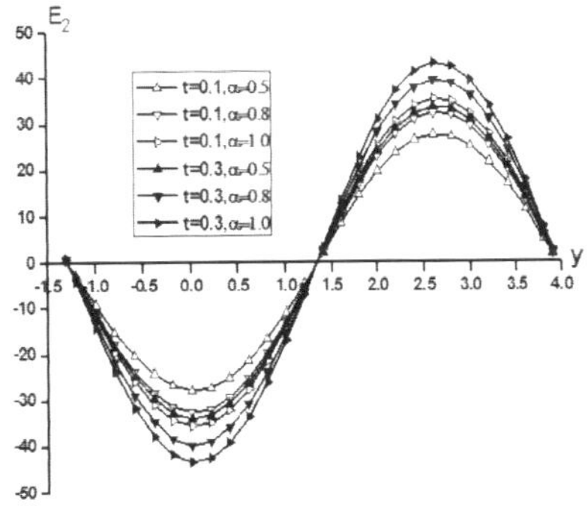

Figure 5. Distribution of E_2 (x=10)

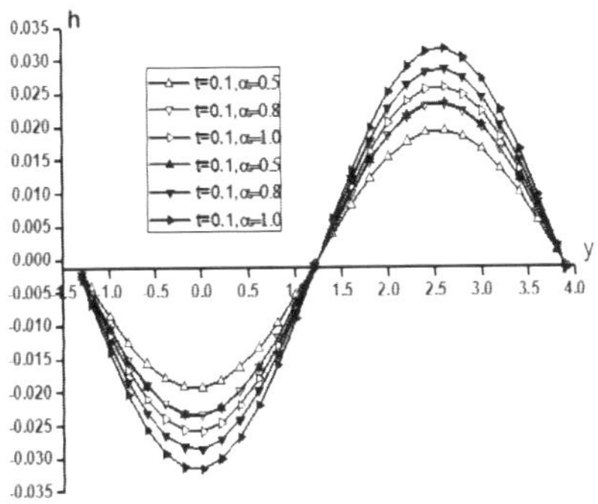

Figure 6. Distribution of h (x=10)

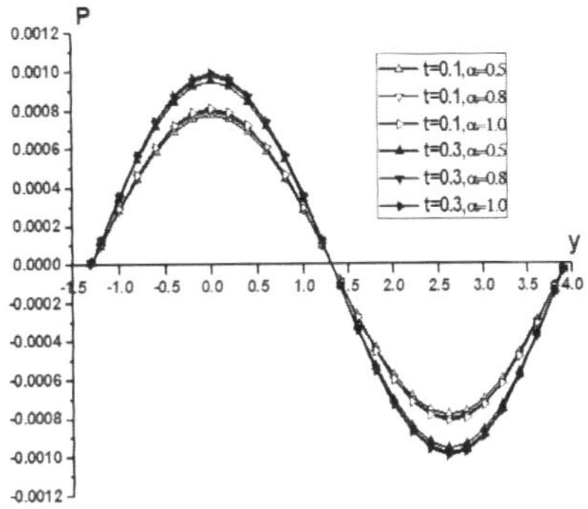

Figure 7. Distribution of P (x=10)

5. Conclusion

This paper has studied fractional electromagneto–thermoelastic diffusion with variable material properties. The following conclusions can be drawn from the above calculations:

(1) The problem of fractional electromagnetic heat diffusion in a heat-conducting and electrically conducting rotating half space can be solved using the regular mode method. This obtains the tenth-order characteristic equations and exact solutions of the physical variables.

(2) When the fractional parameter increases in a given time, the magnitudes of the temperature, displacement, stress, induced magnetic field, electric field, and concentration increase continuously. However, the influence on the chemical potential distribution is very small. The solution graphs show that different fractional parameters are associated with different problems.

Acknowledgments

This work was supported by the National Natural Science Foundation of China (grant no. 51236003). Mark Kurban from Liwen Bianji, Edanz Editing China (www.liwenbianji.cn/ac), edited a draft of this manuscript.

References

[1] O.P. Niraula, N. Noda, Derivation of material constants in non-linear electromagneto-thermo-elasticity, J. Therm. Stress. 33 (2010) 1011-1034.

[2] W.Q. Chen, K.Y. Lee, H.J. Ding, General solution for transversely isotropic magneto-electro-thermo-elasticity and the potential theory method, Int. J. Eng. Sci. 42 (2004) 1361-1379.

[3] I. Ignaczak, M. Ostoja-Starzewski, Thermoelasticity with Finite Wave Speeds, Oxford University (2010).

The Effects of the Size and Location of the Recirculation Windows of Combustion Chambers in a Coke Oven on Emissions of NO and Unburnt Fuels by CFD Numerical Simulation

Yinping Cao[1], Zhengang Du[1], Zhen Hu[2], Shuangling Jin[2], Wangzhao Lu[2] and Rui Zhang[2,*]

[1]Baoshan Iron&Steel Co., Ltd., 655 Fujin Road, Shanghai 201900, P. R. China
[2]School of Materials Sciences, Shanghai Institute of Technology, 100 Haiquan Road, Shanghai 201418, P. R. China

*Corresponding author email: zhangrui@sit.edu.cn

Abstract. NOx emission from combustion chambers of coke ovens is receiving renewed attention in recent years in China. Combustion process optimization and post combustion treatments are used to reduce NOx emission. In this paper, the effects of size and location of recirculation windows in the heating flues of combustion chambers in a coke oven from Baoshan Iron & Steel Co., Ltd. on emissions of NO and unburnt fuels were investigated by CFD numerical simulation, targeting for redesign of recirculation windows. Results indicate that with increasing the window size, the amount of recycled flue gases increases, maximum flame temperature decreases and temperature uniformity is improved, leading to a reduction of NO emission. The emission of unburnt fuels increases with the window size at higher heat flux, but this is less obvious at lower heat flux. With moving the windows towards inlets of fuels and air, the average temperatures increase slightly, maximum flame temperature and temperature uniformity keep unchanged, the emission of unburnt fuels decreases apparently, the NO emission increases slightly at lower heat flux, but keeps unchanged at higher heat flux. The unburnt fuel emission is dominated by the average temperature while NO emission by both the average temperature and temperature uniformity. The effect of the window size on NO emission is more apparent than that of the location of the windows while the latter is more obvious than the former to reduce the unburnt fuel emission.
Keywords: CFD; NOx emission; Flue gas; Combustion.

1. Introduction

Anthropogenic NOx emissions have negative impacts on environment and human health such as acid rain, stratospheric ozone depletion and particulate matter formation. NOx emitted from coke ovens in China accounts for around 7% of total, ranking the fourth largest NOx source. Ministry of Environmental Protection of the People's Republic of China has stipulated in 2014 that NOx content in flue gas from coke ovens in cities has to be less than 150 mg/L, and 500 mg/L for old ovens and 200 mg/L for newly erected coke ovens in rural areas. Selective catalytic reduction is used to reduce NOx contents in flue gas from old coke ovens in cities, which has a high operation cost. For newly erected coke ovens, methods such as the staging combustion and external recirculation of combustion gases have been used to control NOx emissions. Although an internal flue gas circulation is applied in most of the coke ovens in steel industries such as the Baoshan Iron & Steel Co., Ltd. in China, it is not as efficient as the external recirculation and the NOx content in flue gas in Baoshan Iron & Steel Co. Ltd. is still higher than 400 mg/L. The catalyst used in selective catalytic reduction with ammonia to convert NOx to N_2 in most of the iron and steel companies in China has a carcinogenic V_2O_5 component, which pose a threat to human health. And ammonia with an unpleasant smell is emitted in this method, causing another environment problem.

In a coke oven battery, the thermal energy generated by burning gaseous fuels in combustion chambers heats up adjacent coking chambers, where the coal-to-coke conversion is completed. The temperature required for completing the coal-to-coke conversion in the coking chambers is over 1273 K. Therefore, the temperature in combustion chambers has to be higher than 1273 K so that suitable conditions for generating large amounts of harmful NOx are guaranteed. The formation of NOx in combustion chambers of coke ovens is complex, depending on temperature, concentration, flow fields of gas, unmixedness of fuel and air, NOx reburning and postflame reactions. Numerical simulations based on computational fluid dynamics is the most powerful and appropriate method to investigate heat and mass transfer, flow patterns and chemical reactions in combustion chambers, which can be traced back to as early as 1990s[1-3]. With the development of computation software and the improvement of computation power, the NOx formation in combustion chambers has been simulated in great details in recent years and much progress has been achieved in understanding the complex process and NOx formation mechanism, which is helpful for NOx emission control in combustion chambers.

Many efforts have been made to develop the models and methods in numerical simulation of coking and combustion chambers to describe and understand the complex coking and combustion process, some of which are focused on NOx emission control. Heat transfer occurs between coking chambers and combustion chambers. So the simulation strategy has two types, the coupled simulation of both combustion and coking chambers, and the decoupled simulation on combustion chamber alone. Weiss et al carried out a decoupled simulation on combustion chamber, focusing on NOx and CO emission control using heat flux into coking chamber as the boundary condition[4]. They found that the air excess and the geometry of gas supply channels under a nonpremixed combustion mode had an influence on the NO formation and that the double-stage combustion produced lower NO emissions than the single-stage combustion, due to the uniform temperature distribution for the former. A. Q. Zhang, et al simulated the thermal processes in combustion and coking chamber of a coke oven by two decoupled methods, a parallel decoupled method using the heat flux obtained by the energy balance method as the boundary condition for the heating walls, and a serial decoupled method with no need of heat flux of the heating walls as the boundary condition, but instead using the simulation results of the steady heat transfer in the combustion chamber as the inputs of the coking chamber[5]. They found that both methods were in good agreement with the measured temperature evolution in coking chamber, but the parallel method was time-saving. Heat flux averaged in space and in one-hour periods in literature was defined as the unsteady boundary conditions for the heating walls of the combustion chamber by J. Smolka et al[6] to perform a decoupled simulation on temperature variation on the external flue wall of the combustion chamber in a coke oven using an accurate heating flue geometry in a real unit. These authors investigated further the coupled CFD simulation on the combustion and coking chambers[7], which reflected the variation of boundary condition with coking time and along the height of heating walls. Jin K, et al proposed a three dimensional transient model for coupled coking and combustion and focused on the influence of air-staging on the NOx emissions[8]. They found that the staging combustion played an important role in improving temperature uniformity and reducing NOx concentration of exhaust. The efficiency of the air-staging method for NOx emission reduction was also verified by B. Staiger in solid fuel-fired boilers[9]. Adams BR and Wang DH found that reburning leads to reduced final NOx emissions[10]. Hermann W, et al.[2] carried out an investigation on the influence of entrance angle of air inlet on the flow patterns and NO formation in the combustion chamber of a coke oven with a three-staged air supply and found that an angle of 30° in the third stage intensified a good mixing behavior between fuel gas and air by vortex, and as a consequence, leading to a reduced formation of NO. Wessipe K, et al. indicated that a very uniform flow pattern and temperature distribution along the flue height was obtained by placing honeycomb bodies in heating flues, resulting in a decrease of NOx emission[11]. The effect of flue gas recirculation was investigated by J. Baltasar et al. in a gas-fired lab furnace and concluded that a marked decrease of NOx emissions with this method without significant effects on flame stability, overall combustion efficiency and emissions of CO and unburnt hydrocarbons[12]. The similar conclusions were drawn by F. Qian et al., who found that flue gas recirculation allowed NOx to enter the combustor with the chance of being reduced and that flue gas contained more CO_2 to suppress NOx formation[13]. Gamrat et al. investigated the influence of external flue gas recirculation on NOx formation in the combustion chamber of a coke oven[14]. They used averaged heating flux of 2000, 3000, 4000 and 5000 W/m^2 as the boundary conditions of the heating walls and found that the NOx concentration dropped from near 1000 to around 200 ppm under an external recirculation ratio from 0 to 0.3, indicating that the external recirculation was a crucial method to reduce NOx emissions in the coke industry. Compared with the external flue gas recirculation, however, the importance of the internal flue gas recirculation in combustion chambers of coking ovens was focused much less in literatures.

In this paper, internal flue gas recirculation in combustion chambers of coking ovens was investigated, focusing on emission reduction for both NO and unburnt fuels. For this purpose, a real coking oven in Baoshan Iron & Steel Co. Ltd. was used as the starting point, and the size and location of the recirculation windows were varied and their effects on emissions of NO and unburnt fuels were investigated by a decoupled simulation method using different heat flux averaged along height of heating walls as the boundary conditions.

2. Numerical Models and Simulation Method

2.1 Physical Model

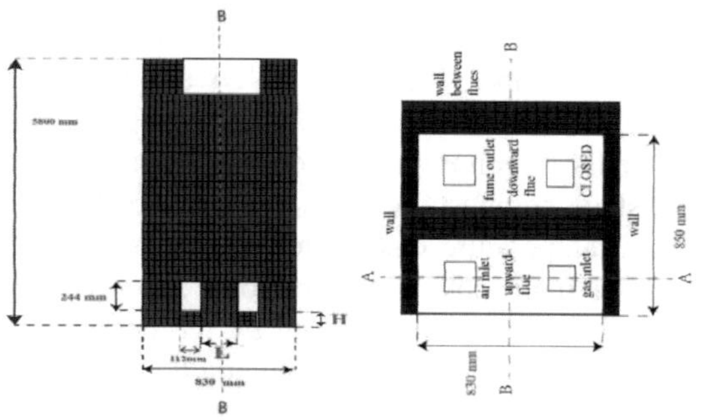

Figure 1. A schematic of the combustion chamber (the front view in the left and the top view in the right). The H in the left graph means the distance of the low limit of the recirculation windows to the inlets of air and fuels.

In a coke oven battery, combustion chambers and coking chambers are alternately arrayed. The physical model includes a combustion chamber only in the decoupled simulation. As shown in Figure 1, there are a upward heating flue, where air and fuel are supplied, a downward heating flue, where the flue gas is emitted, a bridge window in the upper part, where the gases change their flow direction, and two symmetrical windows in the lower part, where internal recirculation of flue gas occurs in each combustion chamber. The gaseous fuels, air inlets and the flue gas outlet are located on the bottom walls. In each flue, two parallel vertical walls, noted as heating walls, are in contact with the two coke oven chambers while the other two, named as dividing walls, are connected to the neighboring flues. Gaseous fuels and air are preheated to 1373K through the regenerative chamber before entering the combustion chamber. The combustion inside the combustion chamber lasts for approximately 30 min. Then, the flow direction is reversed, i.e., the air inlet becomes the flue gas outlet and vice versa. The time required for the coal-to-coke conversion in the coking chamber in one coking cycle is approximately 20 h. So, the coking and combustion are unsteady. However, the combustion can be divided into a finite number of steady state process, so the combustion is in a steady state in a certain coking moment. During the coking cycle, heat is transferred from the combustion chamber to the coking chamber through the heating walls. The up wall, bottom walls and dividing walls are considered to be adiabatic while heat flux from the combustion chamber to the coking chamber occurs through the heating walls, which varies from charging coal (0 h) to discharging coke (20 h) and along the height direction of the heating walls.

2.2. Governing Equations

The general conservation equations of combustion in combustion chamber are described as follows:

2.2.1. Energy equation.

$$\frac{\partial(\rho H)}{\partial t}+\frac{\partial(\rho u_j H)}{\partial x_i}=\frac{\partial}{\partial x_j}\left[\frac{\lambda}{C_p}\left(\frac{\partial H}{\partial x_i}+\frac{\partial H}{\partial x_j}\right)\right] \quad (1)$$

Where λ is fluid thermal conductivity (($W \cdot m^{-1} \cdot K^{-1}$), Cp is fluid specific heat ($J \cdot kg^{-1} \cdot K^{-1}$) and H is fluid enthalpy ($J \cdot kg^{-1}$).

2.2.2. Momentum equation.

$$\frac{\partial(\rho u_i u_j)}{\partial t}+\frac{\partial(\rho u_i u_j)}{\partial x_j}=-\frac{\partial p}{\partial x_i}+\frac{\partial}{\partial x_j}\left[\mu\left(\frac{\partial u_i}{\partial x_j}+\frac{\partial u_j}{\partial x_i}\right)\right] \quad (2)$$

Where u_j is fluid velocity in j direction ($m \cdot s^{-1}$), μ is viscosity (Pa·s) and p is pressure (Pa).

2.2.3. Continuity equation.

$$\frac{\partial \rho}{\partial t}+\frac{\partial(\rho u_i)}{\partial x_i}=0 \quad (3)$$

Where ρ is fluid density ($kg \cdot m^{-3}$) and u_i is fluid velocity in i direction ($m \cdot s^{-1}$).

2.2.4. Two-equation k–ε realizable turbulence.

$$\frac{\partial(\rho k)}{\partial t}+\frac{\partial(\rho k u_i)}{\partial x_i}=\frac{\partial}{\partial x_j}\left[\left(\mu+\frac{\rho C_\mu k^2}{\varepsilon\sigma_k}\right)\frac{\partial k}{\partial x_j}\right]+G_k+G_b-\rho\varepsilon-Y_M \quad (4)$$

$$\frac{\partial(\rho\varepsilon)}{\partial t}+\frac{\partial(\rho\varepsilon u_i)}{\partial x_i}=\frac{\partial}{\partial x_j}\left[\left(\mu+\frac{\rho C_\mu k^2}{\varepsilon\sigma_\varepsilon}\right)\frac{\partial \varepsilon}{\partial x_j}\right]+C_{1\varepsilon}\frac{\varepsilon}{k}(G_k+C_{3\varepsilon}G_b)-C_{2\varepsilon}\rho\frac{\varepsilon^2}{k} \quad (5)$$

where k is turbulence kinetic energy ($m^2 \cdot s^{-1}$), ε is turbulence kinetic energy rate of dissipation ($m^2 \cdot s^{-1}$), G_k represents the generation of turbulence kinetic energy due to the mean velocity gradient (kg $\cdot m^{-1} \cdot s^{-1}$), G_b is the generation of turbulence kinetic energy due to buoyancy (kg $\cdot m^{-1} \cdot s^{-1}$), Y_M represents the contribution of the fluctuating dilatation in compressible turbulence to the overall dissipation rate ((kg $\cdot m^{-1} \cdot s^{-3}$), $C_{1\varepsilon}$, $C_{2\varepsilon}$, $C_{3\varepsilon}$ are constants, $C_{1\varepsilon} = 1.44$, $C_{2\varepsilon} = 1.92$,$C_{3\varepsilon} = 0.09$, and σ_k and σ_ε are the turbulent Prandtl numbers for k and ε, $\sigma_k = 1.0$, $\sigma_\varepsilon = 1.3$.

2.3. Combustion Model

2.3.1. Reactions.

$$CH_4 + 2O_2 = 2H_2O + CO$$

$$CO + 0.5O_2 = CO_2$$

$$H_2 + 0.5O_2 = H_2O$$

$$C_2H_4 + 2O_2 = 2H_2O + 2CO$$

2.3.2. Equations.

$$\frac{\partial(\rho\bar{f})}{\partial t}+\frac{\partial(\rho u_i\bar{f})}{\partial x_i}=\frac{\partial}{\partial x_i}\left(\frac{\rho C_\mu K^2}{\varepsilon\sigma_t}\frac{\partial\bar{f}}{\partial x_i}\right)+\iiint B\rho f_f f_{ox}\times exp\left(-\frac{E}{RT}\right)p(f_f,f_{ox},T)df\,df_{ox}\,dT \quad (6)$$

Where E is the activation energy, $\bar{f}$ is mixture fraction (%), σ_t is constant. R is the Molar gas constant, the subscript f is the fuel, the subscript ox is the oxidizing agent.

2.4. Radiation Model

$$q_r = -\frac{1}{3(\alpha+\sigma_s)-C\cdot\sigma_s}\cdot\nabla G \quad (7)$$

$$\nabla\left(\frac{1}{3(\alpha+\sigma_s)-C\cdot\sigma_s}\cdot\nabla G\right) - \alpha G + 4\alpha\sigma T^4 = S_G \quad (8)$$

Where q_r is radiation flux (W), α is absorption coefficient (m^{-1}), $\sigma_s$is scattering coefficient (m^{-1}), G is incident radiation, C is linear phase singularity phase function coefficient, σ is Stephen-Boltzmann Constant, W $\cdot m^{-2}\cdot K^{-4}$ and S_G is radiation source term.

2.5. NO Formation Model (Thermal NO Model)
NOx formation reactions:

$$O + N_2 \rightarrow N + NO$$

$$N + O_2 \rightarrow O + NO$$

$$N + OH \rightarrow H + NO$$

The thermal NOx mechanism assumes that the rate of formation and consumption of N atoms are equal, so that the NOx formation can be defined as:

$$\frac{d[NO]}{dt} = 2k_{f,1}[O][N_2]\frac{\left(1-\frac{k_{r,1}k_{r,2}[NO]^2}{k_{f,1}[N_2]k_{f,2}}\right)}{\left(1+\frac{k_{r,1}[NO]}{k_{f,2}[O_2]+k_{f,3}[OH]}\right)} \quad (9)$$

Where $k_{f,1}$, $k_{f,2}$ and $k_{f,3}$ are the rate constants for the forward reactions, and $k_{r,1}$ and $k_{r,2}$ are the rate constants for the reverse reactions. To solve this equation, the concentrations of O and OH must be determined. In this study, the O and OH radicals were modelled using a partial-equilibrium model.

2.6. Boundary Conditions and Parameters

2.6.1. Boundary conditions.
Gas inlet: $m_{gas} = m_0, T_{gas} = T_0$;
Air inlet: $m_{air} = m_1, T_{air} = T_1$;
Flue gas outlet: $P_{out} = P_0$;
The top and bottom surfaces as well as the dividing walls are assumed to be adiabatic. The heat flux through heating walls of the upward and downward flues is varied from 2000 to 5000 W/m^2 with reference to the heat flux in real coke ovens.

2.6.2. Component of gases. The fuel gas is a mixture gas of coke oven gas (COG) and blast furnace gas (BFG) with a volume ratio of 9.39%, whose composition is listed in Table 1. The flow rates and inlet temperatures of air and fuel gas are tabulated in Table 2.

Table 1. Composition of fuel gases.

Components	H_2	O_2	N_2	CH_4	CO	CO_2	C_2H_4
Volume,%	8.44	0.09	46.11	2.22	21.07	21.84	0.26

Table 2. The inlet boundary condition of the combustion chamber.

Compositions	Temperature (K)	Mass flow rate (kg.s^{-1})	k (m^2.s^{-2})	ε (m^2.s^{-3})
Air	1373	0.02214(α = 1.1)	0.38	0.39
Fuel gas	1373	0.02155	0.43	0.44

The above models were solved numerically with the FLUENT package. The spatial domain was divided into 300,000 finite elements with structured non-uniform grids in Cartesian coordinates. The discretization of governing equations employed the central difference scheme for diffusion terms and the second-order difference scheme for advective terms. The SIMPLE series algorithm was applied to handle the pressure-velocity coupling. The calculation precision was less than 10^{-5} for relative error.

In this investigation, the size of the recirculation windows was varied by decreasing the L value from 200 mm for the original windows in Baoshan Iron&Steel Co. Ltd. to 0 mm, where two windows are merged into one. The location of the windows was varied by decreasing the H value from 150 mm (the initial location corresponds to 122 mm) to 0 mm, where the recirculating flue gas is in the same level as the inlets of air and fuels. The effects of the size and location of the recirculation windows on the average and maximum temperature of the heating flue, concentrations of NO and unburnt fuels were investigated under different heat flux from 2000 to 5000 W/m^2.

3. Results and Discussion

3.1. The Volume-averaged Temperatures and Maximum Flame Temperatures

Figure 2 shows the changing trends of the volume-averaged temperatures and maximum flame temperatures in the combustion chamber with the locations of the recirculation windows under different recirculation window sizes and heat flux. It is found that with the moving of the recirculation windows away from the bottom of the combustion chamber (increasing H values from 0 to 150 mm), the volume-averaged temperatures decrease and level off for all window sizes (L0, L50, L100, L150 and L200) under all heat flux from 2000 to 5000 W/m^2. The average temperatures decease with increasing heat flux under the same window size and location as expected. The average temperature is lower for a smaller window size (L200) than for a larger window size (L0) when the H value is 0 mm under all heat flux. But this trend is gradually reversed when the H value is increasing to 150 mm under the higher heat flux of 4000 and 5000 W/m^2. With the moving of the windows towards the bottom of the combustion chamber, the mixing between fuel gases, air and flue gases becomes well, which increases the combustion temperature as the temperature of flue gas is higher than that of the fuel gases and air in the inlets. As a result, the combustion reactions become fast and the conversion of fuels to H_2O and CO_2 becomes high, which account for the increased average temperatures with moving the windows towards the bottom of the combustion chamber. With increasing window sizes (L decreases from 200 to 0 mm), the amount of flue gas recirculated increases. As a result, the combustion temperature increases, but fuel gases and air are diluted. Under well-mixed cases where the H values are small, the former is dominated, and as a result the average temperature is higher for a larger window size. Under a high H value of 150 mm where the mixing of flue gas, air and fuels is the worst, the latter is dominated gradually, this phenomena is more obvious at higher heat flux.

The changing trend of maximum flame temperature with the location (H values) of the reciuculation windows is not obvious, fluctuates and remains unchanged roughly under all heat flux and window sizes. The higher is the window size, the lower is the maximum flame temperature. This could be explained by the fact that at a larger window size, the amount of flue gases recirculated is higher, which causes a dilution effect for fuel gases and air, and as a result the combustion temperature is lower. The higher is the heat flux, the lower is the maximum flame temperature under the same window size as expected.

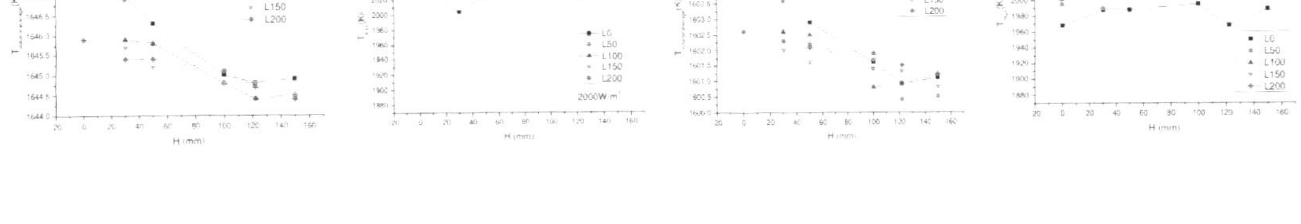

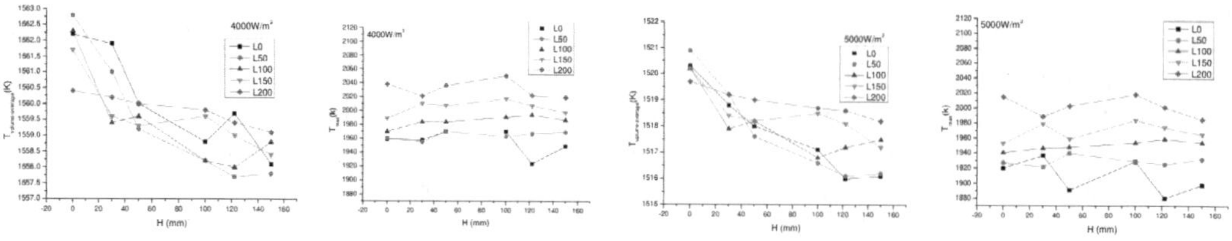

Figure 2. The volume-averaged temperatures (left column) and maximum flame temperatures (right column) of the combustion chamber with H values (locations of the recirculation windows) at different L values (the recirculation window sizes) under different heat flux.

Figure 3 shows the differences between the maximum and average temperatures in the combustion chamber versus H values (locations of the recirculation windows) at different window sizes and heat flux.

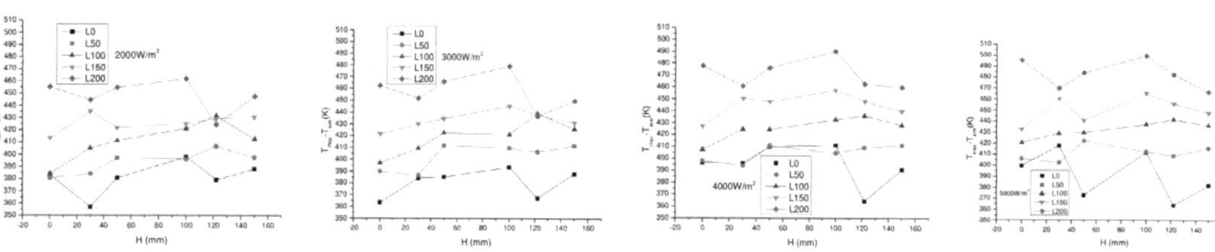

Figure 3. The differences between the maximum and average temperatures in the combustion chamber with H values (locations of the recirculation windows) at different L values (the recirculation window sizes) under different heat flux.

The changing trend is quite similar to that of maximum flame temperatures. As Figure 2 shown, the average temperature changes are smaller with H values and window sizes (within 5 K), but the maximum flame temperature changes are larger at different window sizes. The differences between the maximum and average temperatures, which reflect temperature uniformity, depend on the changing trend of maximum flame temperature with window sizes. This indicates that the temperature in the combustion chamber become uniformity with increasing window size under all window locations and heat flux. The differences between the maximum and average temperatures increase slightly with increasing heat flux under the same location and size of the windows, indicating that the lower is the heat flux, the more is the uniformity of the temperature in the combustion chamber.

3.2. The NO Concentrations of Flue Gases in the Outlet of Combustion Chamber

Figure 4 shows the variations of the NO concentrations in the outlet of the combustion chamber with locations of the recirculation windows at different window sizes and heat flux. It is found that at the heat flux of 2000, 3000 and 4000 W/m^2 the NO concentrations decrease slightly with increasing H values at the smaller window sizes of L150 and L200, keep almost unchanged at the larger window sizes of L 0, L 50 and L100. At the highest heat flux of 5000 W/m^2 the NO concentrations keep almost unchanged with H values for all window sizes. With increasing H values, the average temperatures decrease, causing a decreasing of NO formation rate and therefore a decrease of the NO concentrations. This is true at heat flux of 2000, 3000 and 4000 W/m^2 for the smaller window sizes of L150 and L200 as the combustion chamber temperatures are less uniform at smaller window sizes, which is favorable for NO formation. For the larger window sizes of L 0, L 50 and L100, the combustion chamber temperatures are more uniform, which is unfavorable for NO formation. As a compromise result, the NO concentrations keep unchanged for the cases of larger window sizes of L 0, L 50 and L100. At the highest heat flux of 5000 W/m^2, both the average and maximum flame temperatures are lowest, and NO formation rates are lowest, indicating NO formation rate is least sensitive to temperatures. As the variation of average temperatures with H values is quite small (less than 5 K) and the maximum flame temperatures keep mostly unchanged with H values, as a result, the NO concentrations keep unchanged with H values for all window sizes under the highest heat flux of 5000 W/m^2.

The NO concentrations decrease with increasing heat flux at the same location of the recirculation windows for all window sizes. This is reasonable in that both the average and maximum flame temperatures decrease with increasing heat flux and therefore the NO formation rates follow the same changing trend. The NO concentrations decrease with increasing window size under the same location of the recirculation windows for all heat flux. With increasing window size, the amount of the recirculated flue gases increases, leading to a decrease of maximum flame temperatures and therefore the NO formation rates with increasing window size under all heat flux. Although the average temperatures are also affected by the window size, this effect is not dominant as the variation of average temperatures is quite small (within 5 K).

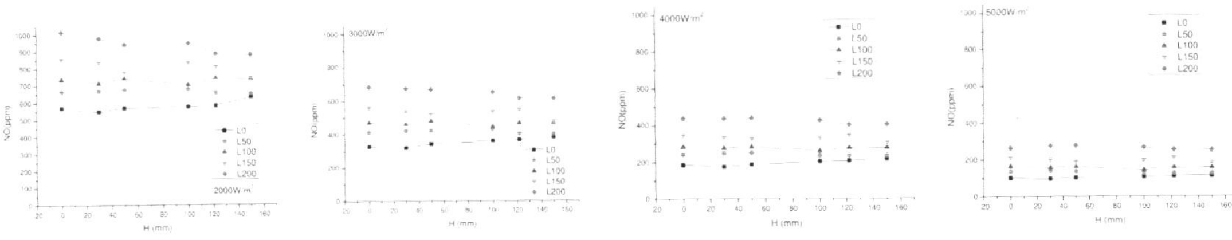

Figure 4. The NO concentrations in the outlet of the combustion chamber with H values (locations of the recirculation windows) at different L values (the recirculation window sizes) under different heat flux.

3.3. The Concentrations of Unburnt Hydrocarbons (CH_4, C_2H_4, Denoted as C_nH_m in Figure 5), CO and H_2 of the Flue Gases in the Outlet of Combustion Chamber

Figure 5, 6 and 7 show, respectively, the concentrations of unburnt hydrocarbons, CO and H2 in the outlet of the combustion chamber versus the locations of the recirculation windows at different window sizes and heat flux. It can be seen from Figure 5 that the concentrations of unburnt hydrocarbons increase and level off with increasing H values at all window sizes and heat flux, indicating that the increased mixing extent at low H values increases the completeness of hydrocarbon combustion. This is understandable in that the average temperatures decrease and level off with increasing H value at all window sizes and heat flux. As can be seen in Figure 3, the temperature uniformity increases with increasing window size. But the concentration of unburnt hydrocarbons does not as shown in Figure 5. The temperature uniformity has little effect on the combustion reactions of fuel gases. This indicates that the concentrations of unburnt hydrocarbons are dominated by the average temperatures. However, as discussed above, the NO concentration is dependent on both the average temperature and temperature uniformity. This could be accounted for by the fact that the activation energies for the formation of NO are higher than those for the combustion of hydrocarbons.

The concentrations of unburnt hydrocarbons increase with increasing heat flux under the same window sizes and locations since the average temperatures are lower at higher heat flux under otherwise identical conditions.

At a high heat flux of 5000 W/m^2, the concentrations of unburnt hydrocarbons increase with increasing window sizes at the same location of the windows (the same H value). With decreasing the heat flux, this changing trend becomes less obvious. This could be ascribed to the fact that at lower heat flux, the average temperatures are higher, and therefore its effect for the extent of combustion of hydrocarbons become less significant.

The changing trends of the concentrations of unburnt CO (Figure 6) and H_2 (Figure 7) with the locations of windows are quite the same as that of unburnt hydrocarbons, indicating that a dominant role of the average temperature for the combustion reactions of CH_4, C_2H_4, CO and H_2. The concentration of unburnt CO is highest, followed by H_2 and hydrocarbons under the same heat flux, window size and location, which has the same concentration order as the inlet fuel gases.

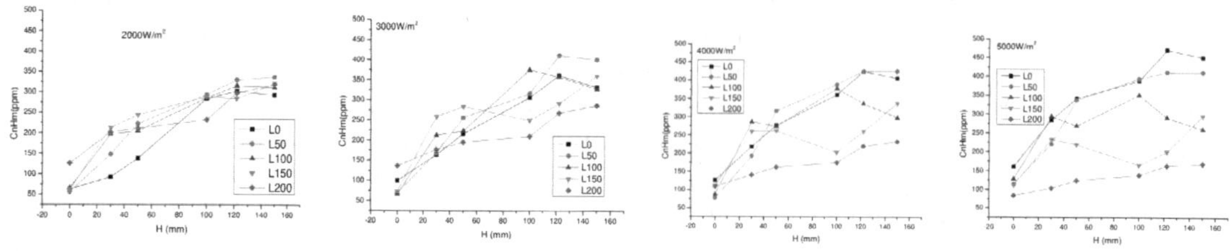

Figure 5. The concentrations of unburnt hydrocarbons in the outlet of the combustion chamber with H values (locations of the recirculation windows) at different L values (the recirculation window sizes) under different heat flux.

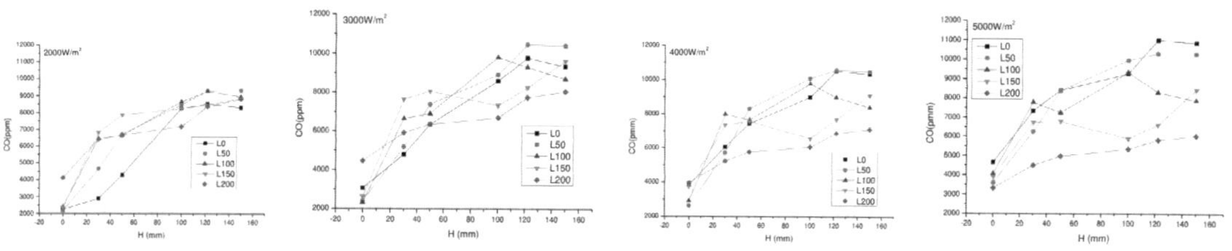

Figure 6. The concentrations of CO in the outlet of the combustion chamber with H values (locations of the recirculation windows) at different L values (the recirculation window sizes) under different heat flux.

4. Conclusion

Starting from the original size (L200) and location (H122) of the recirculation windows, we change their values to see how they affect the emissions of NO and fuel gases by CFD simulation with the FLUENT software package. It is found that both the size and location of the recirculation windows have effect on the volume-averaged temperature and maximum temperatures. The emissions of NO and unburnt fuel gases (CH_4, C_2H_4, CO and H_2) are determined in turn by the average temperature and temperature uniformity in the combustion chamber. The concentration of NO is determined both by the average temperature, maximum flame temperature and temperature uniformity in the combustion chamber. But the emissions of the unburnt fuel gases are mainly dependent on the average temperature. Increasing the window size leads to an increase of amount of flue gases recirculated and an improved temperature uniformity, by which NO emission is alleviated. The emissions of fuel gases at higher heat flux increase with the window size, but this trend is less obvious at lower heat flux. Moving the windows towards the bottom of the combustion chamber increases the average temperature, by which NO emission increases a little bit at lower heat flux, but remains unchanged at higher heat flux. The emissions of fuel gases decrease with moving the windows towards the bottom of the combustion chamber. To optimize the emissions of NOx and unburnt fuels, the layouts of the recirculation windows should be redesigned.

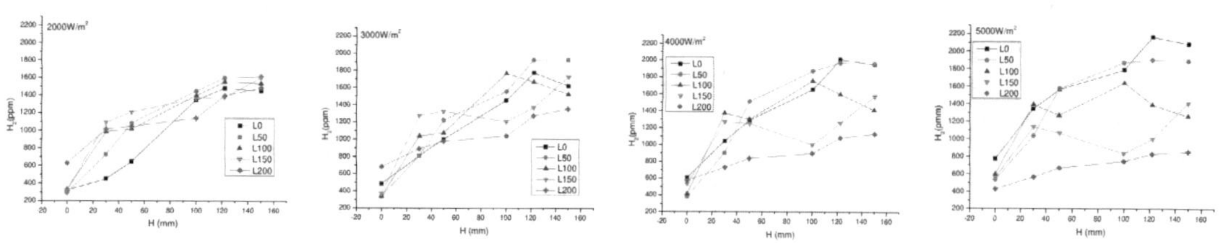

Figure 7. The concentrations of unburnt hydrogen in the outlet of the combustion chamber with H values (locations of the recirculation windows) at different L values (the recirculation window sizes) under different heat flux.

Acknowledgment

This work was supported by the National Nature Science Foundation of China (U1710252).

References

[1] Eisenhut W, Friedrich F and Reinke M 1990 Coking plant environment in West Germany. *Cokemaking Int.* 1:74-79.

[2] Hermann W, Hass W, Kuhn P and Sucker D 1992 Flow patterns and NO formation of the Kaiserstuhl III heating system: investigation on a model twin flue and predictions. *Cokemaking Int.* 4:71-74.

[3] Correa SM 1993 Turbulence-chemistry interactions in the intermediate regime of premixed combustion. *Combustion and Flame* 93:41–60.

[4] Weiss C, Rieger J, Rummer B 2012 Formation and Control of Nitrogen Oxide in the Heating System of a Coke Oven. *Environmental Engineering Science* 29: 555-562.

[5] Zhang AQ, Feng YZ, Zhang XX, Wang MD, Yang JF and Yan X 2013 Decoupling simulation of thermal processes in coupled combustion and coking chambers of a coke oven. *ISIJ International* 53:995-1001.

[6] Smolka J, Fic A, Nowak AJ, Kosyrczyk L 2014 3-D periodic CFD model of the heating system in a coke oven battery. *International Journal of Numerical Methods for Heat & Fluid Flow* 24:891-906.

[7] Smolka J, Slupik L, Fic A, Nowak AJ, Kosyrczyk L 2016 3-D Coupled CFD Model of a Periodic Operation of a Heating Flue and Coke Ovens in a Coke Oven Battery. *Fuel* 165: 94-104.

[8] Jin K, Feng YH, Zhang XX, Wang MD, Yang JF, Ma XB 2013 Simulation of transport phenomena in coke oven with staging combustion. *Applied Thermal Engineering* 58:354-362.

[9] Staiger B, Unterberger S, Berger R, Hein KRG (2005). Development of an air staging technology to reduce NOx emissions in grate fired boilers. *Energy* 30:1429-1438.

[10] Adams BR and Wang DH 2001 NOx emissions in a steel reheat furnace firing by product fuels. *Am. Soc. Mech. Eng., Heat Transf. Div.* 369:21-26.

[11] Wessipe K, Baer H and Rohde W 1997 Optimization of combustion and reduction of NOx formation at coke chambers by application of honeycomb bodies in heating flues. *Cokemaking Int.* 9:42-46.

[12] Baltasar J, Carvalho MG, Coelho P, Costa M 1997 Flue gas recirculation in a gas-fired laboratory furnace: measurements and modelling. *Fuel* 76: 919–929.

[13] Qian F, Chyang C, Yeh C, Tso J 2013 Effect of operating conditions on NOx and CO emissions in a pilot-scale vortexing fluidized-bed combustor with flue gas recirculation. *Chem. Eng. Technol.* 36:268-276.

[14] Gamrat S, Poraj J, Bodys J, Smolka J, Adamczyk W 2016 Influence of external flue gas recirculation on gas combustion in a coke oven heating system. *Fuel Processing Technology* 152:430-437.

Testing the Micro Climate in Sports Shoes by a New Manikins Measurement System

Wei Zhao*, Pengcheng Yuan and Linghua Kong

Laboratory of Internet of Things, College of Mechanical and Automotive Engineering, Fu Jian University of Technology, Fuzhou, Fujian, 350118, China

*Corresponding author email: zhaowei@fjut.edu.cn

Abstract. To measure the temperature and humidity of micro climate in the shoes is a necessary step to design and manufacture new kind of ones. The temperature and humidity in moving human foot are important factors of the shoes' thermal-humidity comfort. A new moving manikins test system for measuring temperature and humidity in sports shoes is introduced. Experiments and calculated results by the traditional method and by the moving manikins method shows that the new moving manikins test system can measure the temperature and humidity of sports shoes correctly, repeatably and normatively.

1. Introduction

A suitable micro climate between the shoes and foot guarantee the comfortability and healthiness. To test the changing temperature and humidity, which are important factors in the micro climate, is a necessary procedure for the design and manufacture of footware[1~6].

The test methods used to be carried out by measuring the human's walking or motionless foot with their shoes[1]. The changing trends and interrelations of the factors, such as temperature, humidity, airflow rate, thermal resistance, wet resistance. etc, were researched[1~3]. Recently, New test methods applying the foot model or manikins to simulate the measurement have been developed[2~9]. The new methods have better feasibility and normative, however, they can not measure the moving foot and shoes' temperature and humidity.

A moving-manikins test system developed by FuJian University of Technology are introduced, which aims to measure the micro climate in footware. The comparative experiments based on the new test system and on the old method have been carried out, the results of which show that the moving-manikins test system can substitute for old test methods.

2. A Moving-manikins Test System

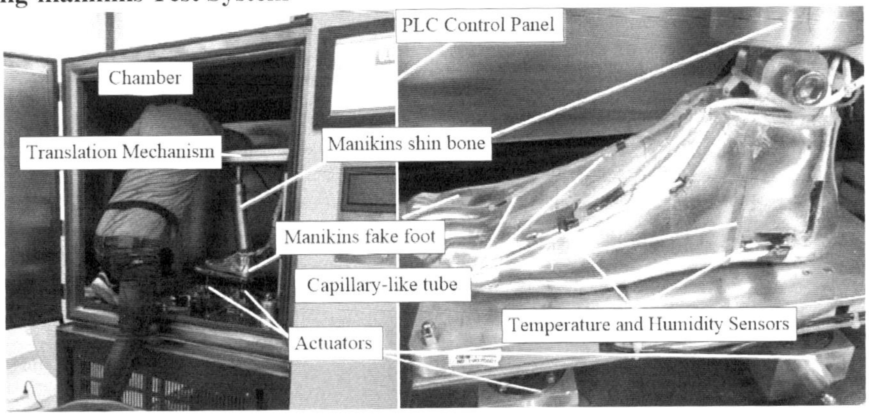

(a) (b)

Figure 1. Test chamber and moving-manikins.

As shown in Figure 1(a), the moving manikins test system are included in a chamber, in which the temperature, humidity and wind speed can be regulated by a PLC control sub-system to simulate nature environment. The temperature can be regulated from -5℃ to 50℃, with accuracy about ±0.1℃; humidity regulated from 10% to 100%,with accuracy about ±2%;wind speed regulated from 0 m/s to 4 m/s,with accuracy about ±0.2 m/s. Under the chamber, there are a set of actuators driven by step motors and mechanics sub-system, which provide two freedom motion to the bottom of manikins. In the chamber there is a translation mechanism above the manikins to provide one freedom motion. The three freedom manikins are designed to simulate gait of human being[8,9].

As shown in Figure 1(b), in the surface of manikins' fake-foot, there are six temperature and humidity sensors with accuracy of temperature about ±0.1/℃, humidity accuracy about ±2%. On the surface of the fake-foot, there are eight capillary-like tubes driven by peristaltic pump with memmert to simulate the foot's perspiration. In the fake-foot, there are several heating resistors controlled by PLC subsystem to simulate the foot's temperature. At the bottom of fake-foot, there are six pressure sensors. In the shin bone of manikins, there are a set of spring-damper-like subsystem, which can be adjusted to simulate the pressure in manikins.

3. Measuring the Temperature and Humidity

3.1. Test Methods

The human being test method (Referred to as old method) and the manikins test method (Referred to as new method) are independently carried out to compare the similarities and differences. Five sports shoes of same brand are chosen to be tested. Both the five people and the manikins are tested without socks.

3.1.1. The human being test method. The instruments of old test system includes an infrared thermal imager, a treadmill, a humidity sensing data recording instrument.

The test is carried out in an air conditioned room, where temperature is adjusted to 25±0.5℃, humidity 70% ± 2℃. Five male students with almost same weight and same foot size are chosen as subjects. The experiments are carried out in two stage, the motionless test and walking test. In the motionless stage, the foot and shoes' temperature and humidity are measured every three minutes. In the walking stage, measurements are carried out every minutes.

In each test stage, the test results of five subjects are averaged as one result, which is used to compare with that of the new test methods.

3.1.2. The human being test method. The new test method is carried out in manikins test system, as shown in Fig.1. Humidity and temperature in chamber are also regulated as 70±2% and 25±0.5℃. *The* temperature of manikins' fake foot is regulated from 36.1℃ to 37.5℃ according to test status, which simulate the temperature of human foot. The flow rate of peristaltic pump is adjusted from 0.9 ml/min to 5.5 ml/min, according to the test status. The tempo of actuators is also regulated according to the old method.

3.2. Results and Evaluation

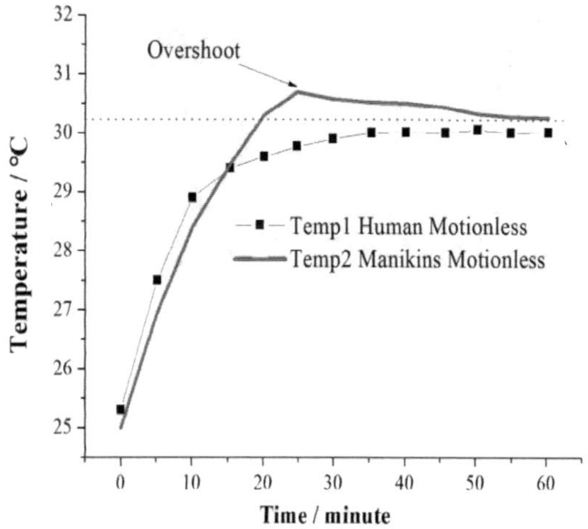

Figure 2. Temperature measured by two methods in motionless status.

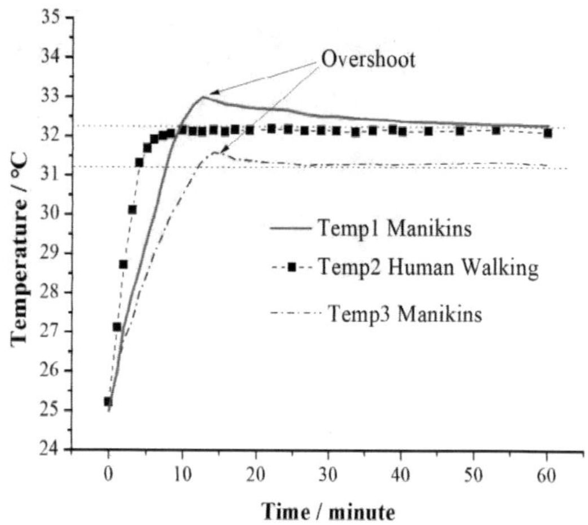

Figure 3. Temperature measured by two methods in walking status.

As shown in Figure 2, the curve marked as Temp1 Human Motionless is the records of average value of five test subjects. They sat in an air conditioned room, were tested about every five minutes. Their shoes' temperature is 29.5℃ at 30 minute and reach the stable state firstly. From 30 minute to 60 minute, the temperature of sports shoes keeps 29.5±0.2℃. The Temp2 curve is the record of Manikins Motionless. It reaches 30.2℃ at 20 minute, and reaches the peak value of 30.7℃ at 5 minute later. Then it reduces to 30.3±0.1℃ from then on. The curve of Temp2 shows an overshoot phenomenon, the reason of that may be the inertia of manikins test system, which is a mechatronic system. Though the temperature of manikins is just set as 36.1℃, its shoe's stable temperature is higher than that of human foot shoes' temperature. The reason may be that the temperature of motionless human foot in a 25±0.5℃ room is actually lower than 36.1℃, which is incorrectly supposed to be a normal value of human body.

As shown in Figure 3, the curve marked as Temp2 Human Walking is the average value measured from five spots shoes walking at a speed of 4 kilometer/hour. These values are measured about every one minute. The Temp2 Human Walking shows the shoes' temperature takes 10 minutes to reach 32.2℃, and it keeps stable temperature from then on. The Temp1 Manikins curve shows that the temperature of the shoe on the "walking" manikins' fake foot also takes 10 minutes to reach 32.2℃. After an overshoot, the temperature curve decades to 32.4℃, which is a stable value. The Temp1 Manikins curve is measured when the manikins' temperature is regulated as 37.2℃, while the Temp3 Manikins curve is the result when it is set as 36.5℃. The Temp1 Manikins curve shows the same trend, but reach the stable temperature valve at 5 minutes later and lower value at 31.3℃.

From the Figure 2 and Figure 3, there are two important conclusions. The manikins test system always show a overshoot phenomenon, the reason of which may be the inertia of mechatronic system. Another phenomenon is that, not like manikins, the human foot seems can "control" their temperature more optimal: it changes slowly when in a quiescent condition, while it changes faster when in a spots condition.

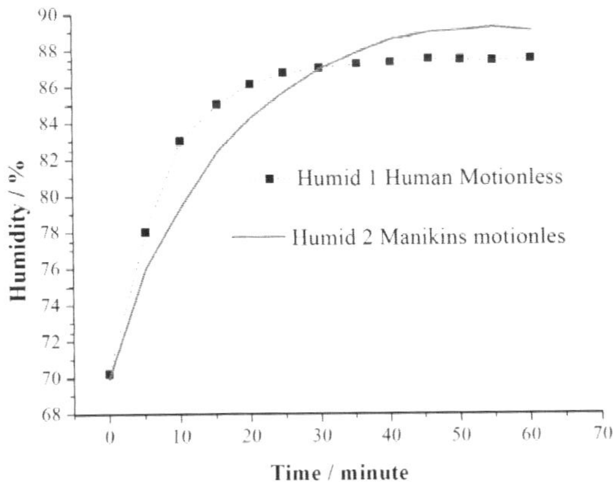

Figure 4. Humidity measured by two methods in motionless status.

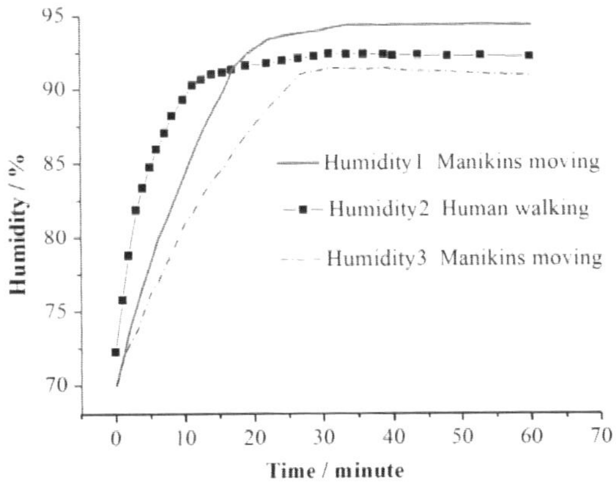

Figure 5. Humidity measured by two methods in walking status.

As shown in Figure 4, in motionless status the humidity values of human shoes are measured every 5 minutes. its curve takes 20 minutes to reach 86.5±2%, as is a stable value. While the corresponding value of manikins take almost 40 minutes to reach stable value 88.1±2%.

As shown in Figure 5, in walking status the humidity value of human shoes just takes 15 minutes to reaches the stable value 92.2±2%, while corresponding values of manikins take more than 10 minutes. The curve marked as humidity1 Manikins moving are measured as flow rate of the peristaltic pump is adjusted to 2.8 ml/min, while the curve of humidity1 Manikins moving adjusted to 2.2 ml/min. Not like in Figure 2 and Figure 3, the test results shown in Figure 2 and Figure 3 have no overshoot phenomenons. There are perhaps two reasons: one is the test accuracy, the humidity can only measure with an accuracy ±2% ; the other is that the subsystem of capillary-like tubes and the peristaltic pump is a complex hydrodynamics process, which can not be considered as a linear system.

From the above figures, it seems that the measured curves of manikins test system is not just the same as that measured by the old method. Nevertheless, the manikins' test curves are much more normative and reproducible than that of the old method.

As shown in Table 1, when the measured values above are applied to calculate the temperature resistance and wet resistance[21], it shows that the temperature resistance and wet resistance of shoes on the five walking foot have too much difference. For example, the test subject3's temperature resistance is 0.314m^{2}K/W, much higher than test subject1's 0.241m^{2}K/W. While the temperature resistance value calculated from walking manikins' shoe is 0.288±0.05 m^{2}K/W, a rather steady value. The wet resistance in table 1 also shows a remarkable individual difference, while the wet resistance of walking manikins' shoes is 175±5 m^{2}Pa/W, which shows a steady value again. It can be concluded that the old method is not suitable for testing temperature and humidity.

Table 1. The calculated temperature resistance and wet resistance .

Test subject	Thermal resistance (m^{2}K/W)	Wet resistance (m^{2}Pa/W)
1	0.241	157
2	0.294	210
3	0.314	165
4	0.276	200
5	0.262	180

4. Conclusions

Recently, many researchers have developed new test methods for measuring thermal-humidity comfort. A moving-manikins test system is firstly introduced to test thermal-humidity comfort. Experiments by the traditional method and by manikins method are carried out with purpose of comparing their repetitiveness and normalization. Based on measured values, such as the temperature and humidity, it

IOP Conf. Series: Materials Science and Engineering **964** (2020) 012007 doi:10.1088/1757-899X/964/1/012007

shows that the moving-manikins test system can perform the measurement of thermal-humidity comfort with better normative and repetitive. Though this new test system can effectively test thermal-humidity comfort of sports shoes, it still has a lot of work to do. Understanding biomechanics of a moving foot with footware and controlling the manikins system with intelligent algorithm, for examples, are expected to improve the moving-manikins test system.

References

[1] Wang X and Yuan J J et al 2015 Testing on Thermal-humidity Comfort of Sports Shoes' Micro-climate Environment in Wearing Condition China Leather vol 44(10) p 110-115

[2] Chen D Z 2016 Research on Test Method of Thermal-wet Comfort for Sport Shoes Leather Science and Engineering vol 25(03) p55-58

[3] Man X D and Yan X N 2014 Investigation on the hot and damp comfortability of human foots in socks and shoes Journal of Xi'an Polytechnic University vol 28(04) p 402-405+412

[4] Xu Y Q and Chen L et al 2018 Exploration of Measuring Water Vapor Permeability and Water Vapor Absorption of Shoes Leather Science and Engineering vol 28(02) p68-71

[5] Fan Y J 2018 Study on the measurement device for thermal comfort of socks Zhejiang Sci-Tech University

[6] Zhang Y W 2017 Research on the foot pressure, temperature and humidity system based on customization and sensing technology. DONGHUA University

[7] Hong C Y, Zhang Y F, Zhang M X, et al. 2016 Application of FBG sensors for geotechnical health monitoring, a review of sensor design, implementation methods and packaging techniques. Sensors and Actuators A: Physical, , 244: 184-197.

[8] Hu J and Wang D M et al 2016 A Multi-segment Foot Model for Gait Simulation Based on Automatic Dynamic Analysis of Mechanical Systems Journal of Biomedical Engineering vol 33(06) p 1060-1066

[9] Zhang Y and Jiang K N et al 2019 Development of Dynamic Ankle Prosthesis (review) Chinese Journal of Rehabilitation Theory and Practice vol 25(12) p 1389-1394

Climate Change in Ganjiang River Basin and Its Impact on Runoff

Yipeng Huang, Weilin Liu*, Bin Wu, Yifan Wan and Shengnan Zhu

Jiangxi Key Laboratory of Hydrology-Water Resources and Water Environment, Nanchang Institute of Technology, Nanchang, China

*Corresponding author email: weilin_liu@nit.edu.cn

Abstract. Predicting the change of runoff in the future is significant to the management of regional water resources. In order to estimate future climate change of Ganjiang river basin and explore its impact on runoff, the changes of precipitation and temperature under the three representative concentration pathways (RCP2.6,RCP4.5 and RCP8.5) were analyzed and the Soil and Water Assessment Tool (SWAT) hydrological model was established to simulate the corresponding runoff in the Basin. The statistical relationship between the factors and the predictor in the Ganjiang river basin was built based on Statistical Downscaling Model (SDSM). For three scenarios, the precipitation showed slight increase trends under RCP2.6 and RCP4.5, but a significant increase under RCP8.5; and generally decreased from July to October while increased from January to April. The temperature exhibited a rising trend under three scenarios, of which the rank was RCP8.5>RCP4.5>RCP2.6. The runoff had an increasing trend and the rank for the three scenarios was RCP8.5>RCP4.5>RCP2.6. Generally, runoff had a strong positive correlation with precipitation while a weak correlation with temperature. The flood risk in Ganjiang river may increase significantly in the future.
Keywords: SWAT model; Climate changes; Runoff; Ganjiang river.

1. Introduction

As the most effective tool for studying global climate change, the global circulation models (GCMs) has been used in the worldwide. The model has shown ability of simulation in some regions, but there are still uncertainties due to different dynamic systems, grid size, and parameterization of physics processes. Precision and resolution of the model have improved in the past few decades. In recent years, the high-resolution atmospheric model have received considerable attention[1].

The research area in this paper is Ganjiang river basin. The Ganjiang river (113.58°to 116.63°east longitude and 24.52°-28.75°north latitude) is located in the south of Yangtze River Basin and it is the seventh largest branch of the Yangtze River. It belongs to the water system of the Poyang Lake Basin with the area of 83,500 km^2. The basin covers Nanchang, Yichun, Xinyu, Pingxiang, Ji'an and Gangzhou from north to south, accounting for 51.5% of the total land area of Poyang Lake Basin. As the largest river basin in Jiangxi Province, the high-precision downscaling future climate prediction and runoff simulation of Ganjiang river basin has far-reaching significance for the national economy, ecosystem, sustainable development strategy, agriculture and food security, flood and drought prevention. To study future climate change and the impact of climate change on the runoff in Ganjiang river are conducive to a deeper understanding of the response mechanism of lake basin system to climate change, and to provide a scientific basis for adjusting water consumption structure, flood and drought prevention, especially for making the countermeasures for mitigating the impact of change on climate and runoff.

The CMIP5 model in this study is CanEsm2 from Canada, and SDSM statistical downscaling model is selected to simulate temperature and precipitation[2]. It uses MLR technology to establish statistical function relationship between factors and regional predictors. According to the assumption that the statistical relationships is still applicable in future climate change scenarios, the statistical relation was applied to the GCM future data of factors by using SWG technology. SDSM has a better simulation of temperature and evaporation, but a worse simulation of precipitation. In many areas of China, it has good application. SWAT model is a kind of distributed hydro-model. Compared with others, the distributed hydro-model takes climate and underlying surface as input factors, which can describe runoff process more reasonably and accurately[3,4]. SWAT model is not suitable for China in some modules. For example, soil classification in China is different from that in the United States, so it cannot be directly applied. For application, Chinese scholars have improved the model[5, 6]. In 2002, Wang Zhonggen, Liu Changming, etc. verified the runoff simulation function of SWAT model according to the statistics of Heihe River from 1959 to 1993, based on the meteorological data and runoff data, which is more suitable for the northwest cold region of China, but not for the simulation of single event flood process[7]. Li Hongliang proved the applicability of SWAT Model in Daqing River Basin[8]; Jin Cong verified the applicability of the model in Xinxing river.

2. Methodology

2.1. Data Sources

CMIP5 model adopted CanESM2 data that is from the Canadian Environment and Climate Change Network (https://www.canada.ca/en.html). The World Soil Database (HWSD) is derived from the FAO and the Vienna International Institute for Applied Systems. It is currently the most widely used soil data in the world. China's soil data is the 1:1000000 raster data provided by the second national land survey in Nanjing. The landuse data is the raw data downloaded from the "Western China Environmental and Ecological Science Data Center". The 1Km landuse raster data is converted from China's 2000s, 1:100,000landuse data. The meteorological data comes from China Meteorological Science Data Sharing Network from 1961 to 2005. The stations such as Ji'an, Yushu, Yichun, Guangchuan, and Nanchang were selected, the data including 50-year data of 19 elements such as pressure, relative humidity, precipitation, evaporation, wind speed, sunshine. The observed runoff data of Waizhou Station from 1961 to 2005 is from the Waizhou Hydrological Station.

2.2. Models

The SWAT model is a distributed hydrological model with strong physical mechanism, which can be used for hydrological response research in a basin simulation[6]. It has been applied in many regions such as Asia, Europe, Australia, America, etc. It is suitable for different climatic conditions and soil or landuse types. In this study the distributed hydrological model was established by SWAT based on the daily values of surface climate data from 1961 to 2005 of six climate stations and the data of soil, landuse, topography, the daily runoff from 1961 to 2005 at Waizhou Hydrological Station and NCEP. According to the data of soil, landuse and slope, 164 sub-basins were divided and 2441 Hydrological Response Units (HRU) were obtained.

In the SWAT model, the equation of water balance was given as follows[9]:

$$SW_t = SW_0 + \sum_{i=1}^{t}(R_a - Q_{surf} - E_a - \omega_{seep} - Q_{gw}) \tag{1}$$

SW_t and SW_0 are the initial and final runoff of the day i, mm; t is the number of days; Q_{surf} is the surface runoff of the day i, mm; R_a is the precipitation of the day i, mm; E_a is the evapotranspiration of the day i, mm; Q_{gw} is the return flow of groundwater at the day i, mm. W_{seep} is the amount of water infiltrated from the soil surface into the unsaturated zone, mm.

SDSM statistical downscaling model was selected, which is more categorized as a downscaling method with a coupling function coupled with a random weather generator. The model is widely used in many research fields currently. The fundamental principle was as follows[10]:

$$w_i = \alpha_0 + \sum_{j=1}^{n} \alpha_j P_{ij} ; \quad R_i^{0.25} = \beta_0 + \sum_{j=1}^{n} \beta_j P_{ij} + e_i \tag{2}$$

$$T_i = \gamma_0 + \sum_{j=1}^{n} \gamma_j P_{ij} + e_i \tag{3}$$

w_i is the probability of precipitation at day i; a, β, γ is the mode parameter; R_i is the precipitation; T_i is the temperature; e_i is the error.

3. Results and Discussion

3.1. Calibration and Verification of SWAT Model

SUFI-2 algorithm was used to calibrate the parameters, and the results showed there were four most sensitive parameters that were GW_DELAY, ALPHA_BF, CN2 and GWQMN(Table1).

Table 1. Parameter sensitivity analysis based on SUFI-2 algorithm.

Parameter	Types	Best Fitted Value	*t*-Stat	*p*-Value
GW_DELY	V_	43.13	-21.32	<0.01
ALPHA_BF	V_	0.61	4.22	<0.01
CN2	R_	0.07	1.98	0.05
GWQMN	V_	0.36	1.12	0.27

The observation sequence was divided into two parts. The first part was the calibration period (1961~1990) used to establish the statistical relationship, and the second part was the verification period (1990-2005) to test the reliability of the model. The determination coefficient(R^2) and Nash-Sutcliffe Efficiency(E_{ns}) values of the model were 0.88 and 0.84 in calibrate period, while 0.89 and 0.87 in verification period, which proved the high applicability of SWAT model in the Ganjiang river basin. As shown in Figure 1, the simulation of monthly runoff in the period of calibration and verification was accurate. The maximum simulated runoff was slightly higher than the observed, the minimum simulated runoff was slightly lower than the observed.

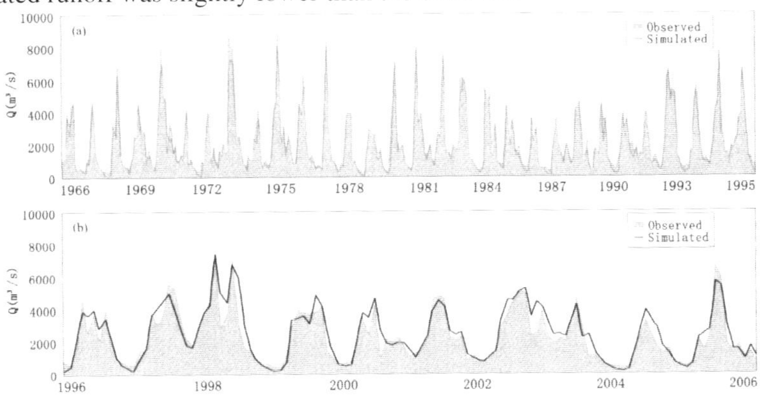

Figure 1. Monthly simulation values and observed values of Waizhou station during the calibration and validation periods.

3.2. Future Temperature Change

As shown in Figure 2, in the calibration period, the simulations of the monthly maximum temperature, minimum temperature and average temperature of the basin were close to the observed values, and the simulation of maximum temperature was closest to the observed values.

The forecast data for 2006-2100 were divided into three stages: 2020S (2006-2035), 2050S (2036-2065) and 2080S (2066-2100). As shown in Table 2, in the three scenarios, the temperature under RCP 8.5 showed a maximum increase, followed by RCP 4.5 and RCP 2.6.

The maximum temperature under RCP 2.6 increased by 1.6 ℃ in 2030S, 2.5 ℃ in 2050S and 2.51 ℃

in 2080S; under RCP 4.5, the maximum temperature increased by 1.3 ℃ in 2030S, 2.8 ℃ in 2050S and 3.1 ℃ in 2080S; under RCP 8.5, the maximum temperature increased by 1.7 ℃ in 2030S , 2.7 ℃ in 2050S and 4.1 ℃ in 2080S. The temperature had an obvious rising trend in September and October, and a slight rising trend in January and February.

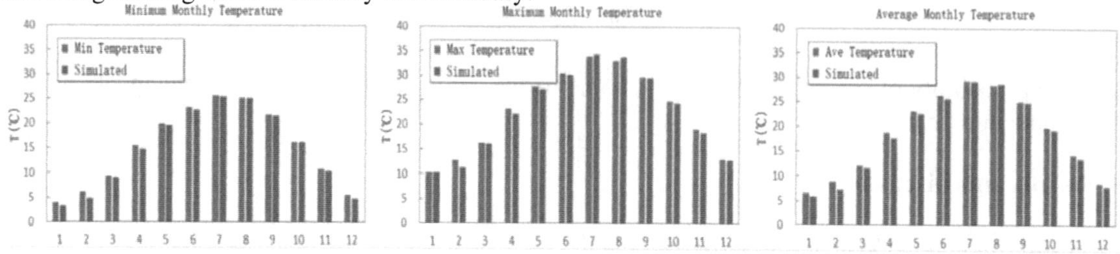

Figure 2. The comparison between observed temperature and downscaled temperature in Ganjiang river basin.

The minimum temperature under RCP 2.6 increased by 0.9 ℃ in 2030S, 1.7 ℃ in 2050S and 1.7 ℃ in2080S; under RCP 4.5 the minimum temperature increased by 0.7 ℃ in 2030S, 2.1 ℃ in 2050S and 2.3 ℃ in 2080S; under RCP 8.5, the minimum temperature increased by 1.2 ℃ in 2030S, 2.3 ℃ in 2050S and 3.6 ℃ in 2080S.

The average temperature under RCP 2.6 had an increasing trend with 1.2 ℃ in 2030S, 1.5 ℃ in 2050S and 1.6 ℃ in 2080S; under RCP 4.5 scenario, the average temperature had an increasing trend with 0.7 ℃ in 2030S, 1.7 ℃ in 2050S and 2 ℃ in 2080S; under RCP 8.5, the average temperature had an increasing trend with 1 ℃ in 2030S,2 ℃ in 2050S and 3.1 ℃ in 2080S, and it increased obviously from April to October under RCP 8.5.

Table 2. Future temperature change in the Ganjiang river basin.

Period	Minimum temperature increase(℃)			Maximum temperature increase(℃)			Average temperature increase(℃)		
	RCP2.6	RCP4.5	RCP8.5	RCP2.6	RCP4.5	RCP8.5	RCP2.6	RCP4.5	RCP8.5
2020S	0.9	0.67	1.2	1.63	2.5	2.51	1.2	1.55	1.57
2050S	1.68	2.06	2.29	1.4	2.82	3.16	0.66	1.67	1.97
2080S	1.68	2.33	3.58	1.7	2.73	4.16	1.03	1.98	3.06

3.3. Future Precipitation Change

The future precipitation in the Ganjiang river basin was shown in Figure 3. It showed that under the three climate scenarios, the simulated precipitation in the Ganjiang river basin would generally decrease from July to October, and decrease by about 20 mm per month; the precipitation would increase from January to April. Precipitation increased from RCP2.6, RCP4.5 to RCP8.5. Compared with the average annual precipitation, RCP2.6 had a fluctuating trend in 2020S, 2050S and 2080S, and the average annual precipitation was 1653mm, which had a small gap with the baseline period. The change of RCP4.5 was more complicated, and it fluctuated in 2020S and 2050S. The annual precipitation increased by 150mm in 2080S; under RCP8.5 the precipitation fluctuated in 2020S, and increased in both 2050S and 2080S, the monthly average increase was up to 10%, about 200mm .

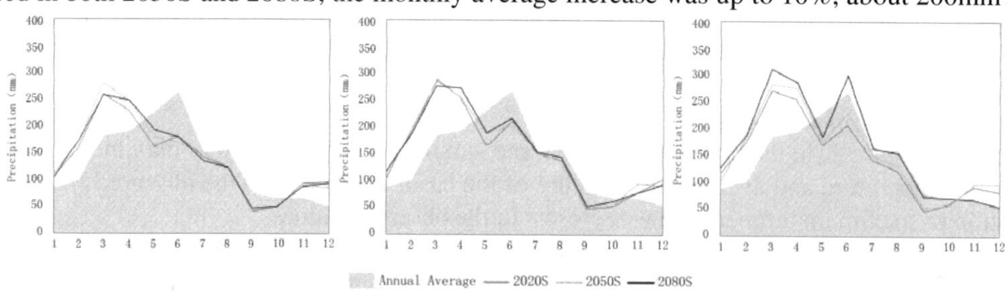

Figure 3. Comparison of monthly mean precipitation between observation and RCP data in Ganjiang river basin.

3.4. Future Runoff Change

Figure4 shows the monthly value of runoff change for the three scenarios in three periods. The maximum of runoff appeared in April under the RCP8.5 compared with the average annual runoff, and the minimum value appeared in October under the RCP8.5. It indicated that the extreme climate was more likely to appear under the RCP8.5. In 2020S, the changes in the three scenarios were not obvious, the change of runoff was very unstable, but it was in a small increase or decrease overall; in 2050S, the runoff gradually decreased in January to March, the annual average runoff increased by 1140m^3/s compared with the average annual runoff, and increased more obviously under RCP8.5. The annual average runoff increased by 1291m^3/s; in 2080S, under RCP2 .6, the runoff still had no clear increase or decrease trend. Under the RCP4.5, the runoff of simulation increased by 1137m^3/s compared with the annual average runoff. Under RCP8.5, the runoff of simulation had an increasing trend.

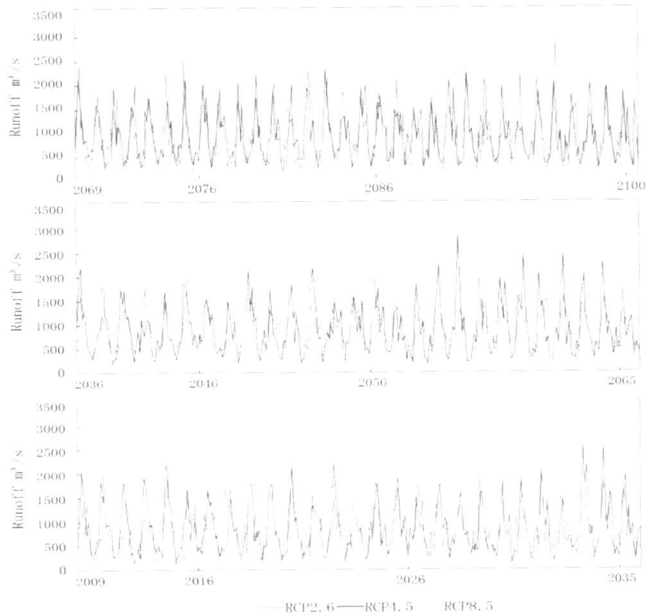

Figure 4. The simulated runoff of Waizhou station from 2009 to 2100.

Table 3. Annual runoff variation of Ganjiang river basin.

Month	Runoff (m^3/s)				Change in Runoff (%)		
	Baseline Period	RCP2.6	RCP4.5	RCP8.5	RCP2.6	RCP4.5	RCP8.5
1	273.08	664.78	638.11	704.67	143	134	158
2	462.41	987.12	1177.88	1213.17	113	155	162
3	835.41	1101.09	1148.9	1197.73	32	38	43
4	1258.27	1462.51	1598.65	1652.02	16	27	31
5	1389	1030.38	1057.52	1053.49	-26	-24	-24
6	1684.77	1055.29	1070.52	1219.93	-37	-36	-28
7	1251.82	834.25	809.54	848.01	-33	-35	-32
8	740.18	721.08	699.74	750.22	-3	-5	1
9	604.78	411.22	312.38	357.27	-32	-48	-41
10	431.34	361.42	276.69	318.89	-16	-36	-26
11	346.86	520.91	528.1	514.27	50	52	48
12	256.24	537.65	564.8	584.76	110	120	128
Max	1684.77	1462.51	1598.65	1652.02			
Min	256.24	361.42	276.69	318.89			
Mean	794.51	807.31	823.57	867.87			
Cv	0.29	0.18	0.18	0.19			

As shown in Table 3, the runoff increased from January to April and November to December, among which the runoff increased the most in January, February and December, with the maximum increase rate of 162%; the runoff decreased in May to October, and the runoff in June and July decreased the most, reaching 37%. In 2009-2100, the maximum runoff appeared in April, the minimum in October under RCP4.5, with the value of 276.69 m^3/s.

The runoff increased the most under RCP 8.5 by 10%. Under RCP 2.6, the runoff fluctuated, and the change of runoff was not obvious. The change of runoff under the RCP 4.5 was similar to that under the RCP 2.6 with a larger increase of 350 m^3/s; the increase under RCP 8.5 was much larger than the first two scenarios, especially in April and June, with an increase of 880 m^3/s. Maximum and minimum values were more likely to appear in RCP 4.5 and RCP 8.5. It possibly indicated that extreme climate was more likely to appear in these two scenarios, and low emissions were undoubtedly safer, which also reflected the importance of reducing emissions.

3.5. Runoff Response to Climate Change

Figure 5 was the diagrams of runoff and precipitation change at Waizhou Hydrological Station in the periods of 2020S, 2050S and 2080S. By comparing the precipitation and runoff of the same period, the relationship between runoff and precipitation was very close. The change of runoff was consisting with the change of precipitation. The annual runoff had no obvious change in 2020S. The annual runoff in 2050S and 2080S increased with the increase of precipitation, and increased the most in 2080S. The increase of runoff was 2080S > 2050S > 2020S under three scenarios, and the increase of runoff was RCP8.5 > RCP4.5 > RCP2.6 in three periods.

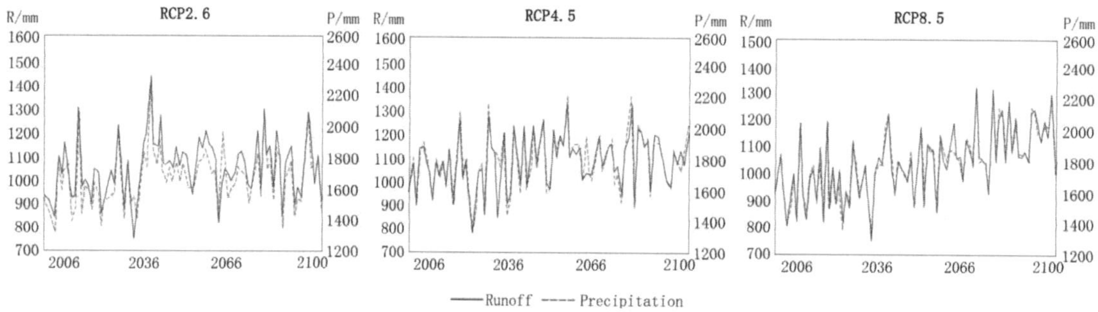

Figure 5. Runoff and rainfall of Waizhou Station under three scenarios.

Table 4 shows the design values of runoff at different frequencies (10%, 20%, 30%, 50%, 75%, 90%, 95%) in Ganjiang River Basin. Sample mean value of observed data was 9490.55, and variation coefficient (*Cv*) was 0.29. For three emission scenarios, the mean value under RCP 2.6 scenario was 10083.20 with Cv=0.12; The mean value under RCP 4.5 scenario was 10434.49 with *Cv*=0.12; Under RCP 8.5 scenario the mean value was 10964.51, and *Cv* was 0.13. It can be seen that the mean values of runoff increased according to the order of RCP2.6, RCP4.5 and RCP8.5. The prediction of future runoff had significance for preventing flood or drought in the future.

Table 4. Design values of runoff in Ganjiang river basin.

Frequency(%)	Runoff Value(m^3/s)			
	Baseline Period	RCP2.6	RCP4.5	RCP8.5
10	13132.53	13684.48	14953.47	15801.99
20	11714.47	12060.71	12961.75	13258.81
30	10750.21	10636.64	10959.49	11699.23
50	9260.7	8981.75	9138.26	9247.17
75	8530.5	8221.35	8116.94	8103.88
90	6744.59	6412.58	6539.46	6740.31
95	6389.63	5981.1	6025.01	6149.91

4. Summary and Conclusions

(1) Using SDSM model to simulate the climate in Ganjiang river basin, the temperature had a good simulation result with the explained variance of more than 70%, but the simulation of precipitation in April-June was worse than that in other months with the explained variance of 20%-40%. The future temperature of the Ganjiang river basin was overall on the rise, while the precipitation was in a fluctuating rising trend.
(2) The E_{ns} and R^2 of the SWAT model in the calibration and validation period were all over 0.8, which proved the model was applicable to Ganjiang river basin.
(3) The response of runoff to precipitation was much greater than the response to temperature. Runoff had a positive correlation with precipitation, and had a weak correlation with temperature. This is mainly because the Ganjiang river basin is an inland basin, and precipitation is the main source of recharge. The decease of precipitation leads to a decrease of runoff directly.
(4) Compared with the annual average runoff, April was the month in which runoff and precipitation got their maximum value, October was the month in which the minimum value appeared, and all occurred under RCP 8.5 scenario, which proves that the extreme climate more likely occurred under RCP8.5. The flood risk in Ganjiang river may increase significantly in the future.

Acknowledgements
This work was financially supported by the Science and Technology Research Program of Jiangxi Education Department(GJJ170980), National Natural Science Foundation of China(52069015), the Superiority Science and Technology Innovation Team Construction Plan of Jiangxi Province(20171BCB24012), and Jiangxi Science and Technology Department Science and Technology Plan Project (20192BAB206047).

References

[1] H, L.Y., Y. YQ, and Z.W. P, *Review of Development and Application of High Resolution Global Climate System Model.* Advances in Earth Science, 2016. **31**(03): p. 258-268.
[2] Seung - Hwon Hyun, Sang - Wook Yeh, Se - Yong Song, et al. *Understanding Intermodel Diversity When Simulating the Time of Emergence in CMIP5 Climate Models.* Geophysical Research Letters, 2020, **47**(17):n/a-n/a.
[3] Jingrui Wang, Litang Hu, Didi Li, et al. *Potential Impacts of Projected Climate Change under CMIP5 RCP Scenarios on Streamflow in the Wabash River Basin.* Advances in Meteorology, 2020, 2020
[4] Xiaoying Yang, Ruimin He, Jinyin Ye, et al. *Integrating an Hourly Weather Generator With an Hourly Rainfall SWAT Model For Climate Change Impact Assessment in the Ru River Basin,* China. Atmospheric Research, 2020, 244
[5] C, S., A.J. G, and W.J. R, *Validation of the Swat Model on a Large Rwer Basin With Point and Nonpoint Sources.* Jawra Journal of the American Water Resources Association, 2010. **5**(37): p. 1169-1188.
[6] C, W. and C.X. W, *Simulation of Hydrological Effects on Vegetation Restoration of Degraded Mountain Ecosystem with SWAT Model.* Mountain Research, 2008(01): p. 71-75.
[7] G, W.Z., L.C. M, and H.Y. B, *The Theory of SWAT Model and its Application in Heihe Basin.* Progress in Geography, 2003. **22**(1): p. 79-86.
[8] L, L.H., *Study on The Impacts of Land Use/Land Cover Change to the Hydrological Base on the SWAT Model.* 2007, Hebei Normal University.
[9] W, J.Y., *Principles and Practice of Distributed Hydrological Models for River Basins*, in *China Water & Power Press.* 2005, China Water & Power Press: Beijing.
[10] Q, C., X.Z. X, and J.X. H, *Comparison of Two Statistical Downscaling Methods in the Taihu Basin.* Resources Science, 2012. **34**(12): p. 2323-2336.

Design and Research on Full Bolt Connected Box Beam-Column Joints of Cantilever Short Beams with Attachments Connecting Plate

Xiaobo Wang and Yajie Yan*

Civil Engineering College, Taiyuan University of Technology, China

*Corresponding author email: yanyajie.tyut.edu.cn

Abstract. Aiming at the assembly connection of box beams and columns, this paper proposes a connection that can realize rapid on-site assembly. The joint is mainly composed of two parts, a box-shaped column welded with a short cantilever beam and structural reinforcement measures and a box-shaped beam attached with connecting plates. The connecting plates are used as accessories of the box beam and shipped with the box beam. When installing, the outer connecting plate is rotated by 180° to the designated position, the inner connecting plate is slid into the designated position, and then these members including the box beam and the short cantilever beam are connected by bolts. The joint is simple and easy to install. It can not only realize rapid assembly on site, but also transmit force directly and reliably. It is an excellent joint for assembled steel structure.
Keywords: Full-bolt connected; Double-shear; Rotate.

1. Introduction

In the field of assembly fabricated connections, many scholars have carried out numerical, theoretical and experimental research and have achieved tremendous results. Wang Yan[1] et al. studied the mechanical properties of the inner sleeve connection combined by bolts (high-strength bolts + opposite tension bolts). The joint stiffness increases with the increase of the sleeve thickness, but the sleeve gap will reduce the stiffness. It is believed that the manufacturing accuracy should be improved as much as possible to reduce the gap. Li Liming[2] et al. designed a fully bolted outer sleeve H-shaped steel beam-column joint. They found that the joint has good seismic performance and energy dissipation capacity by researching on the seismic performance of the joint. To a certain extent, increasing the thickness of the outer sleeve can effectively improve the seismic performance and stiffness of the joint, but because this joint is restricted by the closed section of the square steel tube column, it is necessary to open installation holes in the column and sleeve to complete the installation so that the degree of assembly is not high. Liu Xuechun[3] et al. studied the seismic performance of a modular joint connected by a square steel tube column and H-section steel beam with full bolts. The joint was tested and studied by changing the thickness of the flange plate, the bolt diameter, the number of cover bolts and the gap between the bolt rod and the bolt hole. The results of finite element analysis show that the number of cover bolts is the main factor that affects the energy consumption of the joint. The more the number of cover bolts, the stronger the energy consumption of the joint. Based on the theory of damage control, Zhang Ailin[4] and others designed a beam-column connection joint (semi-rigid node) that can realize full bolt connection with a cantilever short beam. They adjusted the thickness of the flange connection cover, the middle row bolt spacing and other parameters to control the stiffness of the flange cover plate, so as to use the flange cover plate to dissipate energy and ensure that the main components of the beam and column maintain elasticity without damage, which can realize rapid repair after the earthquake.

Prefabricated buildings have the advantages of convenient and fast installation, high degree of industrialization, green and energy-saving. The state attaches great importance to the development of prefabricated buildings and vigorously promotes prefabricated buildings. The box-shaped column has the same moment of inertia in the two main axis directions and has good bending resistance. It is widely used in the frame structure system. Under the same geometrical dimensions, the box-shaped beam has better bearing capacity than the H-shaped beam. The smooth surface of the box-shaped beam have great advantages in fire protection and decoration. At present, the box-shaped column-box beam connection is mainly be welded, which cannot meet the requirements of prefabricated buildings. The box beam-column connection proposed in this paper has the following advantages: the full-bolt connection can achieve full assembly; As an accessory, the connecting plate is shipped with the box beam components, which reduces the types of components and is conducive to rapid assembly; because the box beam section is closed, the bolt is generally single-shear, but the joint realizes double-shear, which can greatly reduce the number of bolts and component materials.

2. The Structures of the Full-bolted Connected Box Beam-column Joint

2.1. The Details of the Beam End

The beam end mainly consists of two parts. The first part is a box-shaped beam with installation holes and bolt holes at the corresponding positions. Besides, the channel steel structure prepared for the inner connecting plates is welded on the inside of the beam. The second part is the connecting plate, including the inner connecting plate and the outer connecting plate. The inner connecting plate is placed on the welded channel steel structure and temporarily fixed with bolts. the outer connecting plate is fixed on the surface of box-shaped column with temporary fixing bolts. The fixing bolts are arranged on the center of the bolt group, so that the outer connecting plate can be connected to the column end by rotating 180°. The inner and outer connecting plates leaves the factory as a whole part with the box beam, which is conducive to the integration of components and resource allocation, and speeds up the construction. Figure 1 is a detailed display of beam end parts. Figure 2 is the rendering picture of the beam end.

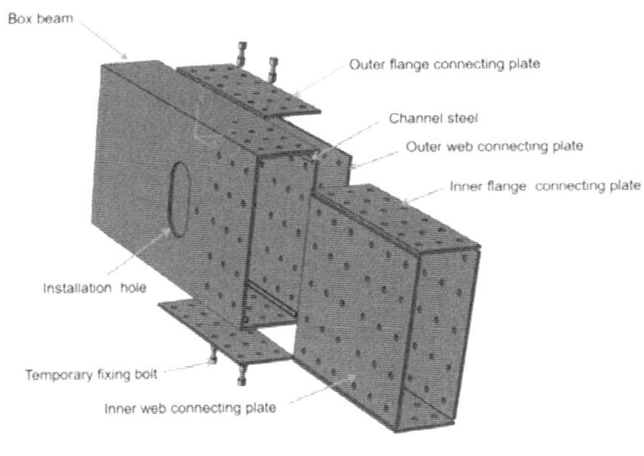

Figure 1. Schematic diagram of beam end structure.

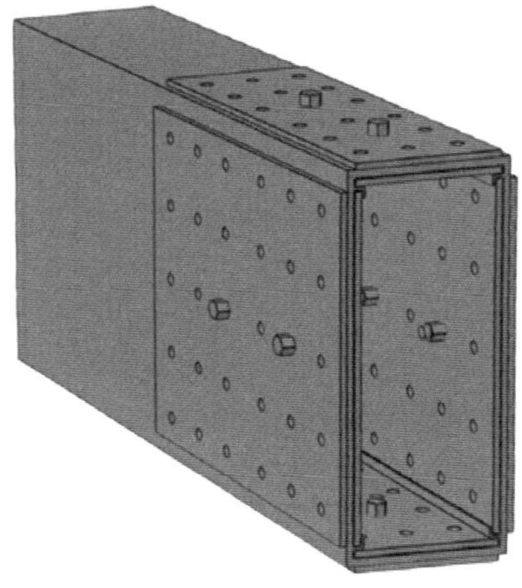

Figure 2. Beam end renderings.

2.2. The Details of the Column End

The column end is composed of a box cantilever short beam, a box-shaped column and strengthening structural measures which consists of flange reinforcement and web reinforcement. First, the box-shaped short beam with bolt holes and installation hole is welded to the box-shaped column. And then the stiffening plate(stiffener) is welded on the inner side of the column at the flange position of the box-shaped short beam. At the same time, reinforcements including flange reinforcement and web reinforcement are arranged at the junction of the cantilever short beam and the box column to ensure that structural damage does not occur here. The details are displayed in Figure 3.

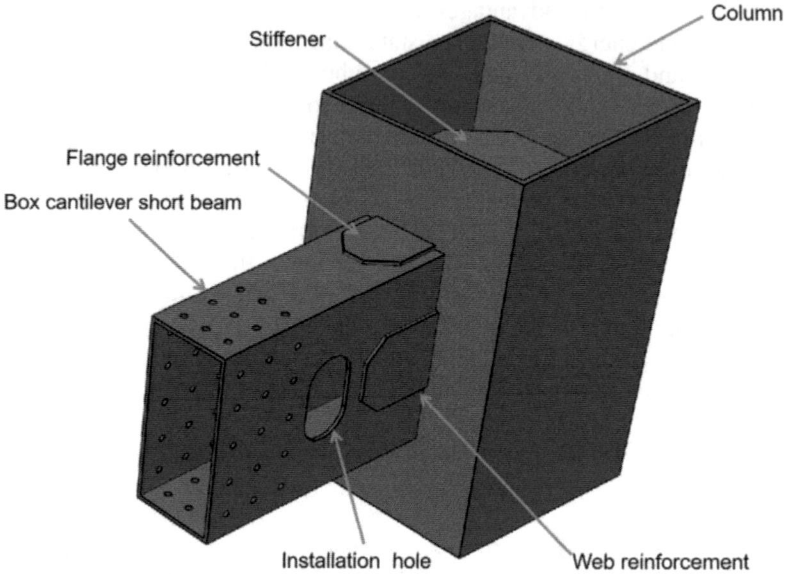

Figure 3. Schematic diagram of column end structure.

2.3. Splicing and Installation

The first step is to splice the beam end and the column end and place them in the designated position which is showed in Figure 4. The second step is to remove the non-center temporary fixing bolt, rotate the outer web connecting plate with the center bolts center by 180°to the designated position, remove the center bolt, and slide the inner web connecting plate to the designated position through the installation hole, so that the bolt holes of the outer web connecting plate, box beam, cantilever short beam and inner web connecting plate are aligned. This step can be seen in Figure 5. The third step is to install high-strength bolts through the installation hole. The box beam can be safely fixed on the column end after the inner and outer web plate is installed completely and the corresponding auxiliary equipment is removed which is showed in Figure 6. The installation method of the inner and outer flange connecting plates is the same as that of the web connecting plates. The installation completed by finishing the flange connecting plates. The final step is to weld the cover plate at the installation hole to prevent the inner side of the box beam and short box beam from rusting. Figure 7 is the finished status of installation.

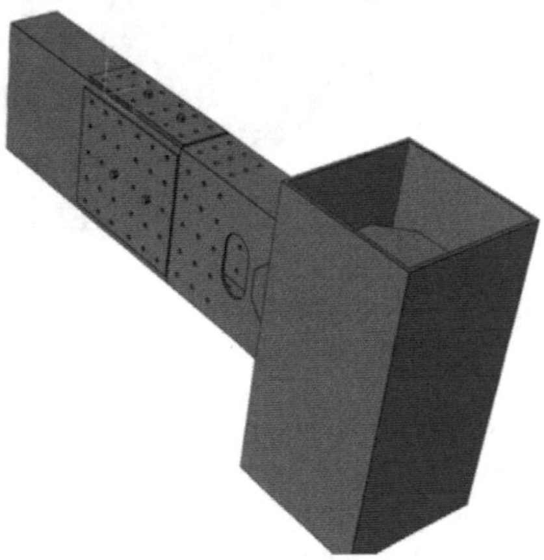

Figure 4. Schematic diagram of the first step.

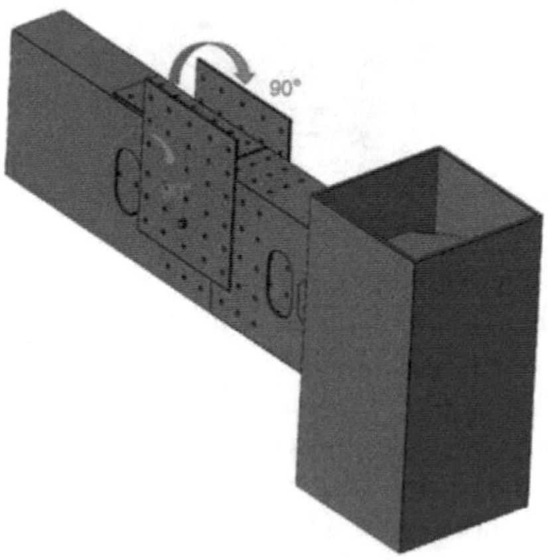

Figure 5. Schematic diagram of second step.

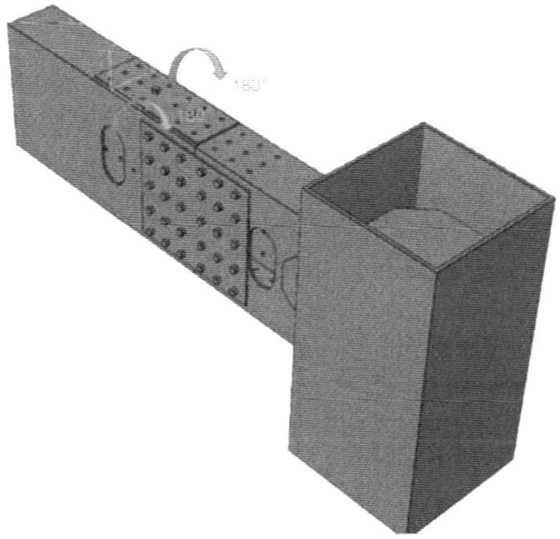

Figure 6. Schematic diagram of the third step.

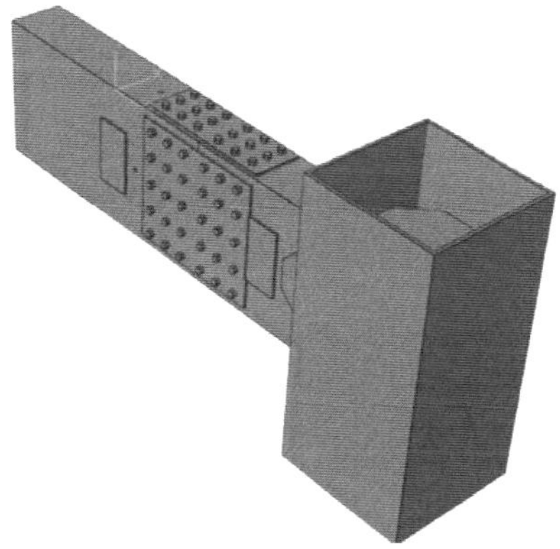

Figure 7. Installation completion.

3. Conclusion

Nowadays, prefabricated buildings are developing vigorously. Steel structure has huge advantages as an excellent structural form of prefabricated buildings. The connection of steel structure is big problem of fabricated steel structure which is worthy of in-depth study and exploration.The connection between box column and box beam needs to be developed and researched. Based on previous research, the connection in this paper proposes a fabricated connection suitable for box column and box beam. The joint mainly has the following advantages:

(1) The structure is fully connected by bolts, which can realize rapid assembly, shorten the construction period and speed up the construction progress.

(2) The connecting plate is shipped with the main part, which is conducive to component protection, processing and transportation. At the same time, it reduces the types of components and facilitates.

(3) The connecting plate can be positioned by rotating 180° or sliding, which is convenient.

(4) The key parts are welded in the factory, which can guarantee the welding quality.

(5) The force transmission is direct and reliable.

(6) Although the section is closed, double-shear is realized, which can greatly save materials.

But there are also the following problems:

(1) The junction of joint are relatively weak, which is harmful to structural stability and the torsional stiffness.

(2) Factory production requires high precision.

(3) The installation hole weakens the cross section and reduces the integrity of the component.

Acknowledgments

I was fortunate enough to come to Taiyuan University of Technology two years ago and I met a group of very lovely persons here. First of all, I would like to thank Mr. Yan for his careful guidance and patient cultivation. Secondly, I would like to thank my classmates Lin Jianpeng and Du Bin for their help and support and finally thank my parents for their support and understanding.

References

[1] Wang Yan, Ma Qiangqiang, Yang Songsen 2016 Mechanical Properties of Beam-Column Joints Using Inner Sleeve Composite Bolts in Fabricated Steel Structure *Journal of Tianjin University(Science and Technology)* vol 49 p 73-79

[2] Li Liming, CHEN Yi-yi, Li Ning, CAI Yu-chun 2010 Seismic performance of outer-shell connection of cold-formed square tubular column and II-shaped beam *Journal of Jilin University(Engineering and Technology edition)* vol 49 p 67-71

[3] X.C. Liu, Z.W. Yang, H.X. Wang, A.L. Zhang, S.H. Pu, S.T. Cai, L. Wu 2017 Seismic performance of H-connection beam to HSS column connection in fabricated structures. *Journal of constructional steel research* vol 138 p 1-16
[4] Zhang Ailin, Li Ran, Jiang Ziqin, Yang Xiaofeng 2018 Finite element analysis of static behavior of PBCSC with double flange cover plates. *Industrial construction* vol 48 p 30-36

Performance Analysis and Optimization of Prefabricated Frame Connection Nodes

Jinping Zhuang* **and Shuqing Dong**

Department of Civil Engineering, Fujian University of Technology, Fujian, China

*Corresponding author email: 19781208@fjut.edu.cn

Abstract. Structural connection and seismic capacity were key technologies in assembly reinforced concrete structure. In this paper, the connection form, transmission force and the seismic performance of fabricated connection nodes were analyzed. Based on the analysis, a kind of assembled steel pipe restrained energy dissipation frame joint with good seismic performance was put forward to realize the seismic requirement of "strong node, weak component", " strong shear and weak bending ". Besides, The influence on the core area is reduced by plastic hinge migration. At last, the construction process of the connection node is introduced.
Keywords: Assembly; Frame structure; Joint; Seismic performance; Optimization.

1. Introduction

During the 13th Five-Year Plan period, China will take the development of green building and intelligent building as the direction. Prefabricated reinforced concrete structure has the advantages of high production efficiency and green environmental protection. In addition, the standardized and refined construction of prefabricated components can also improve the reliable guarantee of project quality, which is a construction method with promotion value. Among all kinds of structural systems, frame structure is suitable for prefabricated construction because of its flexible plane layout, and beam, column and other components are easy to be standardized and typed. America, Europe, Japan and so on began to use prefabricated frame structure as early as the 1950s, has developed to today's super-high prefabricated frame structure.

2. Connection Form and Transmission Force Analysis of Fabricated Frame Structure

2.1. Fabricated Frame Structure Connection Form

Frame connection node is the key factor for the safety and reliability of prefabricated frame structure. In the great earthquake of Tangshan, China in 1976, the prefabricated structure was generally severely damaged and the damage was mainly concentrated in the joints. In order to ensure the rationality of the joint, the researchers studied the methods and structural measures of the joint from the aspects of mechanical performance and construction convenience [1-5]. The assembled frame structure can be divided into four types according to the position of beam-column split, the connection method of members and the position of reinforcement joints, etc. : the joint inside form, the beam mid-span form, the integrated form of beam-column joint, and the fully assembled form.

The internal connection form of beam-column joint refers to the connection of precast beam in the core area of the joint, in which the core area and the composite beam and plate are all made of cast-in-place concrete, and the precast upper column is connected with the precast lower column through the grouting sleeve. The internal connection form of beam-column nodes is also one of the recommended forms in the current national standards. The purpose of "equivalent in-situ" is achieved by casting frame nodes in situ. Its advantage is that the component is easy to make, the site template installation amount is less, the disadvantage is that the joint reinforcement is complex, the construction quality is difficult to guarantee.

The difference between the connection form of beam mid-span and the connection form of beam and column joints is mainly that the connection position of precast beam is moved out from the core area of the node and the connection section is set near the middle of the span. The connection section adopts the traditional method of cast-in-place concrete. Its advantage is to reduce the difficulty of reinforcing steel bar construction at the joints, but its disadvantage is to increase the workload of beam bottom formwork.

The integrated form of beam-column joints is that "node core area – beam-end" adopts the form of factory prefabrication, with perforation holes reserved in the core area. After the steel reinforcement of the prefabricated lower column passes through the perforation hole in the reserved core area and is connected with the grouting, it is then connected with the prefabricated upper column through the grouting sleeve. Cast-in-place concrete is used to connect the precast beam with the core section of the precast joint. Its advantage is that node core precast, construction quality can be guaranteed; The disadvantage is to increase the beam bottom formwork.

The fully assembled type means that the connection between the precast column and the precast beam is connected through the grouting sleeve or the steel element embedded in the precast member. Its advantages are high precast rate, no concrete pouring on site, and its disadvantages are high precision and small allowable deviation of steel components. For example, although the bolt connection installation is fast and agile, but in the production accuracy requirements are high, small allowable deviation, transportation and installation process is prone to bending, damage and pollution. Because the steel components are embedded in precast concrete components, it is very difficult to replace the embedded parts if the processing is not accurate or the steel components are damaged during transportation.

2.2. Force Transfer Performance of Fabricated Frame Connection Nodes

The transmission path and mechanism of frame nodes are directly related to the form of frame nodes, and the transmission mechanism of frame nodes in different connection forms is not the same. At present, the force transmission mechanism of beam-column joints is divided into rigid connection, semi-rigid connection and hinge joint according to the structure form and rotation capacity of the joints. The bending resistance and shear resistance of frame joints are two important indexes to reflect the reliability of joint connections. The bending resistance reflects the main indexes to resist deformation or failure [6-9]. The shear resistance is a key factor to ensure that no sliding occurs between each other.

At present, there is no obvious difference between cast-in-place structure and fabricated structure in the research methods and indexes of the bending performance of beam-column joints, and researchers mainly study the influence of the form and construction of joints on the bending performance of joints. Wang Tao (2013) [6] studied the influence of bolt diameter, grade, slip coefficient of anti-slip surface and adding some T-shaped parts at the beam-column joint on the joint bending performance by combining finite element method with static bending test. Chen Jian-jia (2014) [8] made a scale specimen of steel sub-beam penetration joint in the proportion of 1:2 and carried out static loading test on it to study the stress process and failure mechanism of the joint. Zhang Pei (2007) [9] analyzed the mechanical properties of three kinds of beam-column joints, which were slotted with tooth, slotted with external steel plate and planted with bent steel bar after straight slotted without tooth.

The shear properties of nodes usually include the shear properties of the core area of nodes and the shear properties of the connections of prefabricated components. The shear strength of core area is basically the same as that of cast-in-place concrete structure. But some scholars use steel fiber concrete instead of ordinary concrete in the core area, so as to improve the shear resistance of the core area. Vasconez 's test results show that the incorporation of steel fiber into concrete in the core area can not only reduce the amount of stirrup in the core area [7], but also improve the shear resistance and energy dissipation capacity of the core area. 3% steel fiber volume content can reduce the amount of stirrup in node core area by 50%. And the beam midspan connection form and beam column joint integration form, the connection of the shear is mainly artificial gouge, keyway, planting bars to ensure the way.

2.3. Seismic Performance of Fabricated Joint

The experimental results of most researchers at present show that the seismic performance, including ductility and energy consumption, can be equivalent to that of cast-in-place[10-16].

The Precast Seismic Structural System Research Program (PRESS), a cooperative project between the United States and Japan, proposed three ductility nodes of prestressing technology. The test results show that, using the framework of these ductility nodes, the lateral deformation can reach 1/25 of the height of the structure, and the structure can even maintain the elastic state under the horizontal load exceeding the ultimate wind pressure. Ertas(2006)[10] made 5 fabricated beam-and-column joints and carried out low-cycle loading tests, including 1 cast-in-place node, 1 welded node with bracket column, 1 bolted frame node and 2 cast-in-place integral nodes. The test results showed that: these prefabricated joints showed good strength, ductility and energy consumption, etc. In addition, some researchers use the good ductility and energy consumption of steel to improve the energy consumption or rotation capacity of the joint by setting up connecting steel elements in the joint area. Huang Xianghai (2005)[11] proposed a new type of fully precast prefabricated concrete frame joint. This new type of frame joint is to use the legs or dark legs with high vertical bearing capacity as the vertical load-bearing member and adopt the steel plate with good ductility as the horizontal force transmission member, so as to effectively transfer the shear force and bending moment. Li Xiangmin (2013)[14] used the connection node embedded with low yield and high ductility connecting rod in the node core area and effectively anchored through the middle flange. The test results showed that: the connecting rod first yielded and had sufficient plastic deformation, and the beam and column basically remained elastic, so as to achieve the purpose of ductility connection design. Cheng Wanpeng (2015)[15] proposed to place steel bones locally in the connection area of members and the core area of beams and columns, and conducted a hysteresis test study on the hysteretic performance of partially fabricated steel bone concrete frame structures. The research results showed that the bearing capacity and energy dissipation capacity of the joints were significantly better than that of cast-in-place concrete structures.

However, individual researchers also point out that the ductility and energy consumption of prefabricated joints may not reach the level of cast-in-place. Kaya(2009)[16] showed that, although the strength of non-bonded post-tensioned prestressing joint is basically the same as that of the cast-in-place joint, its energy consumption is not as high as that of the cast-in-place joint. The effect of form and construction on ductility and energy consumption of various fabricated joints is the focus of attention.

3. The Utility Model Relates to an Assembled Steel Tube Constrained Energy Dissipation Joint

In order to further improve the application range of precast prefabricated concrete frame structure in earthquake zone, better connection mode is an urgent problem to be solved, and is also a hot research topic at present. Put forward a kind of prefabricated steel pipe constraints of the core concrete connection nodes, prefabricated in the factory and the connection node in the core area reserved through hole, prefabricated under column of reinforced through the core area reserved through hole, down from a precast connection node in the middle of the grouting hole, grouting slurry filling of precast column and precast connections under the connection between the overflow from the through hole up after layer, ensure the reliability of reinforced and the shear bond; Then the reinforcement is connected with the precast upper column through the grouting sleeve. The joint between the precast joint and the precast beam is cast-in-place steel fiber self-compacting concrete. Precast joint structure schematic is shown in figure 1, and this kind of prefabricated steel tube - beam end constraint core concrete cast-in-situ steel fiber reinforced concrete beam-column connection mode referred to as "prefabricated steel pipe constraint link node energy consumption", based on the node of precast frame structures is called "prefabricated steel pipe constraint energy dissipation frame structure".

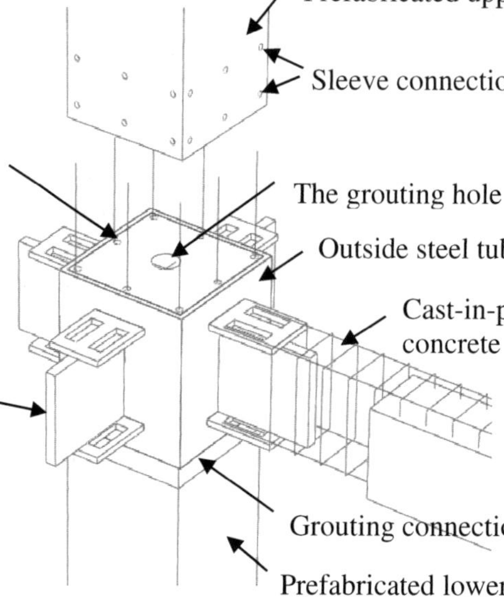

Figure 1. Assembled steel tube constrained energy dissipation joint.

The steel pipe provides lateral constraint for the core concrete and improves the strength and ductility of the concrete, thus realizing the seismic design concept of "strong joint and weak member". Steel-fiber self-compacting concrete combines the advantages of traditional steel-fiber concrete and self-compacting concrete. Adding steel fiber into self-compacting concrete not only enhances the energy dissipation capacity and ductility of concrete, but also restrts the shrinkage of self-compacting concrete. Cast-in-place steel fiber self-densifying concrete at the connection part of the beam end can better exert the energy dissipation capacity of the plastic hinge at the beam end and delay the degradation rate of the plastic hinge stiffness, so as to achieve the purpose of performance-based seismic resistance. In addition, the prefabricated node core area is embedded with outreaching steel elements to improve the shear capacity of the beam end, so as to achieve "strong shear and weak bending", and at the same time, the plastic hinge is moved outward to reduce the adverse impact of plastic hinge on the core area of the node. On the construction level, the longitudinal reinforcement of the surrounding beam can be connected by the overhanging steel elements, which overcomes the problem of dense reinforcement in the core area of the traditional beam and column joints, and at the same time, the good fluidity of steel fiber self-compening concrete reduces the difficulty of concrete placement at the beam end.

4. The Construction and Installation Process of the Assembled Steel Tube Constrained Energy Dissipation Frame Structure

The frame structure based on the fabricated steel tube constrained energy dissipation joint is called the fabricated steel tube constrained energy dissipation frame structure. The assembling process are mainly: core connection node prefabrication, precast beams and columns, three components such as factory to transport to the construction, installation of prefabricated column connection and installation of prefabricated core nodes under installed to install prefabricated composite beams, composite plate - casting precast beams connected with core sections - casting composite beams and composite plate - on the installation of prefabricated column - connection installed on a layer of prefabricated core node - >... . Figure 2 is the connection diagram of the connection node of the prefabricated lower column and the prefabricated core area, Figure 3 is the connection diagram of the prefabricated beam and the core area, and Figure 4 is the connection diagram of the prefabricated upper column and the core area.

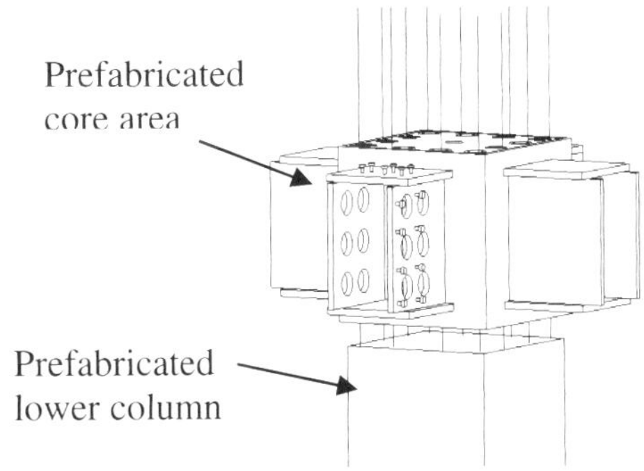

Figure 2. Schematic diagram of connection node between prefabricated lower column and prefabricated core area.

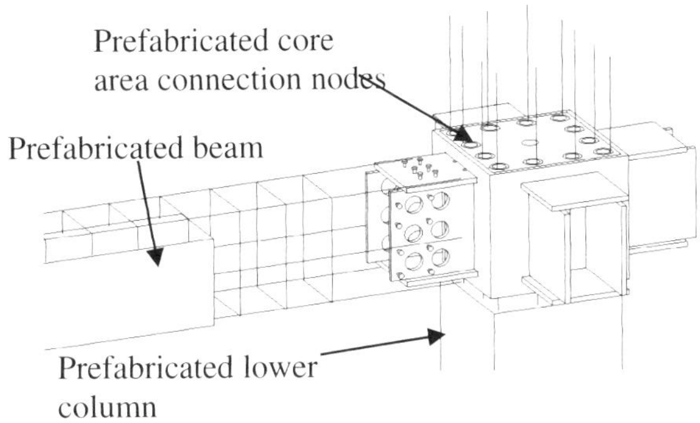

Figure 3. Schematic diagram of connection between precast beam and core area.

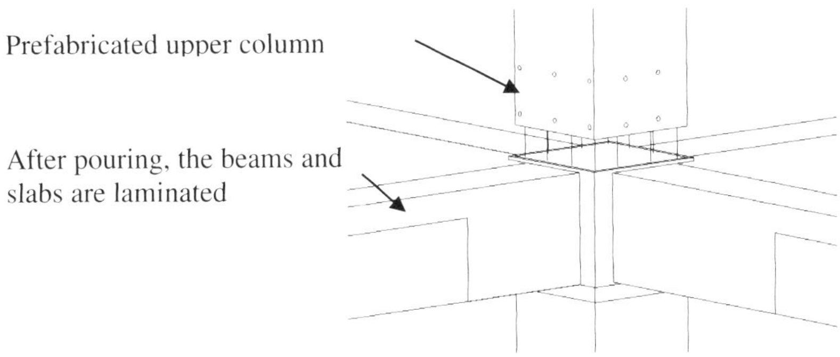

Figure 4. Schematic diagram of connection between prefabricated upper column and core area.

5. Conclusion

At present, there are many forms of fabricated connection nodes, and the main goal is to achieve the effect of "equivalent to cast-in-place". Based on the assembly characteristics and the advantages of high performance concrete and steel structure, it is the trend of future development to study the assembly-type joint with better seismic performance and more efficient construction.

In the assembled steel pipe constrained energy dissipation joint, the steel pipe not only provides lateral constraint for the core concrete, but also improves the strength and ductility of the concrete, thus realizing the seismic design concept of "strong joint and weak member". In the core area of prefabricated joints, the embedded protruding steel element can improve the shear capacity of beam end, realize "strong shear and weak bending", and at the same time move the plastic hinge outward to reduce the adverse impact of plastic hinge on the core area of joints.

Acknowledgements
This work is supported by the National Natural Science Foudation of (51678152) and the Research Development Foudation of Fujian University of Technology(GY-Z13119)

References

[1] Industrial Standard of the People's Republic of China. Jgj1-2014, Technical Specification for prefabricated Concrete Structure [S]. Beijing: China Building Industry Press, 2014.
[2] Gao xiangling, xu longbiao, li jie, et al. Test and mechanical performance analysis of prefabricated concrete beam-column joints [J]. Journal of huazhong university of science and technology (natural science edition),2016,44(10):47-52.
[3] Thinh D T, Kusunoki K, Tasai A. Experimental Study on a New Precast Unbonded Post-tensioned Beam-column Jonit System with Shear Bracket [J]. Journal of Structural and Construction Engineering, 2009,76(641): 1249-1257.
[4] Ousalem Hassane, Tshikawa Yuji , Kimura Hideki, et al. Seismic Performance and Flexural Stiffness Variation ofAssembled Precast High-Strength Concrete Beam Jointed atMid-Span Using Transverse Bolts[J]. Journal of Advanced Concrete Technology, 2009.7(2): 205-216.
[5] Zhang Yiping, Ping Wei, Zhang Chunlei, et al. A study of a new prefabricated concrete framework post-pouring integrated bear-column joint [J]. Sichuan Architectural Science Research, 2017, 43(03): 110-115.
[6] Lu Hao. Research on semi-Rigid Joints of Prestressed prefabricated concrete frame Structures: [Master's Thesis]. Hangzhou: Zhejiang University, 2016.
[7] Vasconez, Rosa M., Antoine E. Naaman and James K. Wight. Behavior of HPFRC connections for precast concrete frames under reversed cyclic loading [J]. PCI Journal, 1998, Vol. 43(6): 58-71.
[8] Chen Jian-Jia. Research on the mechanical properties of joints between steel beam and concrete beam (column) [D]. Master of Engineering Thesis of South China University of Technology, 2014.
[9] Zhang Pei. Experimental Study on the Shear Capacity of existing Reinforced concrete columns connected by rear beams [D]. Master degree of Harbin Institute of Technology, 2007.
[10] Ertas, Onur, Sevket Ozden,etal. Ductile Connections in Precast Concrete Moment Resisting Frames[J].PCI Journal,2006(3):66-76.
[11] Ma Qiangqiang. Research on the Mechanical properties of composite bolted joints in fabricated beam-column: [Master's Thesis]. Qingdao: Qingdao University of Technology, 2016.
[12] Zhou Z B, PAN Z H. Seismic performance analysis of fully prefabricated hybrid structure with frictional spliting energy dissipation nodes [J]. Building Structures, 2014, 44(13): 72-76. (in Chinese)
[13] Zhou Yuling. Experimental Research on Hybrid prefabricated Prestressed concrete frame Joints: [Master's Thesis]. Nanjing: Southeast University, 2016.
[14] Li xiangmin, gao rundong, xu qingfeng. Experimental study of precast prefabricated concrete frame ductile joints [J]. Journal of central south university (natural science edition),2013,44(8): 3453-3463.
[15] Cheng wanpeng, song yupu, wang jun. Experimental study on the seismic performance of the middle joints of prefabricated steel-reinforced concrete frame beams and columns [J]. Journal of dalian university of technology, 2015,55 (2): 171-178.
[16] Kaya M , Arslan A S. The Effect of the Diameter of Prestressed Strands Providing the Post-tensioned Beam-to-column Connections[J]. Materials & Amp Design,2009,30(7):2604-2167.

FEM Analysis of Hole-opened Metallic Plate Connections in Prefabricated Energy-dissipation Joint

Mingbei Tuo[1], Di Wang[2], Guiyun Yan[2,*], Xiaofei Xiao[2] and Liwei Xu[2]

[1]CSCEC Strait Construction and Development Co., Ltd, Fuzhou 350115, China
[2]College of Civil Engineering, Fujian University of Technology, Fuzhou 350118, China

*Corresponding author email: 2190603020@smail.fjut.edu.cn

Abstract. The prefabricated structure is vigorously promoted due to its obvious superiorities of high-efficiency construction rate and environment-friendly construction scheme. However, this fast-growing new structural system is still facing the threat of earthquakes, especially its beam-to-column connecting joints which are the weaker link in the entire structural system all the time. This study investigates the failure behaviour of hole-opened metallic plate connections in prefabricated energy-dissipation joints and its finite element models were provided. Parametric study was performed to obtain the influence of metallic-plate size on the bearing capacity, deformation performance, and ductility. Based on the results of numerical simulation, the failure modes of specimens were focused on weakened sections. Moreover, proposed design sizes were given, comprising that the weakened length a/L of holes is from 0.25 to 0.55 and the weakened width b/B is from 0.3 to 0.5. The suggested value of width to thickness ratio is 0.2 and the value of the gap is within 0.5 to 2 mm.
Keywords: Prefabricated hole-opened metallic plate connection; Failure behaviour; Finite element model; Metallic-plate size; Parametric study.

1. Introduction

Prefabricated beam-to-column connecting joints are vulnerable to earthquake threats causing the failure of structures [1-2]. Researchers are eager to focus on the material and construction of joints to improve the seismic performance of beam-to-column connecting joints. With the proposal of new joints used new materials, such as the fibre-reinforced polymer system, glulam-concrete composite beam-to-column joints, and shape memory alloy [3-4], joints adopted new materials exhibited better performance on seismic behavior. On the other hand, considering the construction of beam-to-column joints, the connection method of unbonded post-tensioned prestressed concrete beams and columns using friction devices were provided and a new ductile flexural prefabricated beam-column joint connection for high seismic zones was developed [5-6]. A kind of prefabricated concrete beam-to-column connection joint with grout sleeves was developed and it indicated that the new precast joint could achieve a similar seismic performance of cast-in-place joints [7]. Simultaneously, installing dampers regarded as energy-dissipation components at beam-to-column joints bring the possibility of fast recoverability-function after the slight earthquake damage. Wang et al. [8] use MR dampers to test the feasibility of nonlinear seismic control of RC frame structures, results show that MR dampers are effective for the damage control of structures under earthquake action, and the performance of semi-active control and passive control is better. Based on the thinking of passive control, a steel joint with a damper [9] was used to replace the cast-in-place concrete beam-to-column joint. Experiments show that the joint has better hysteretic performance and higher energy dissipation,

and its ductility is higher than that of the integral joint. A precast concrete connection with all-steel bamboo-shaped energy dissipaters [10] appears good self-centering ability. Some corresponding finite element models were also performed to analyze the mechanical properties of joints [11-12].
This work presents a finite element model of the hole-opened metallic plate connection as the key role in prefabricated energy-dissipation joints which was shown in figure 1. Proposal of hole-opened metallic plate connections is to transfer damage to the outside of the beam-to-column joints and undertakes the tensile and compressive axial force generated by the decomposition of bending moment. Based on the numerical simulation, the failure modes were observed. Besides, several parametric analyses were performed to study the influence of hole-opened metallic plate connections on bearing capacity, deformation performance, and ductility. The suggested design values of hole-opened metallic plate connections were obtained.

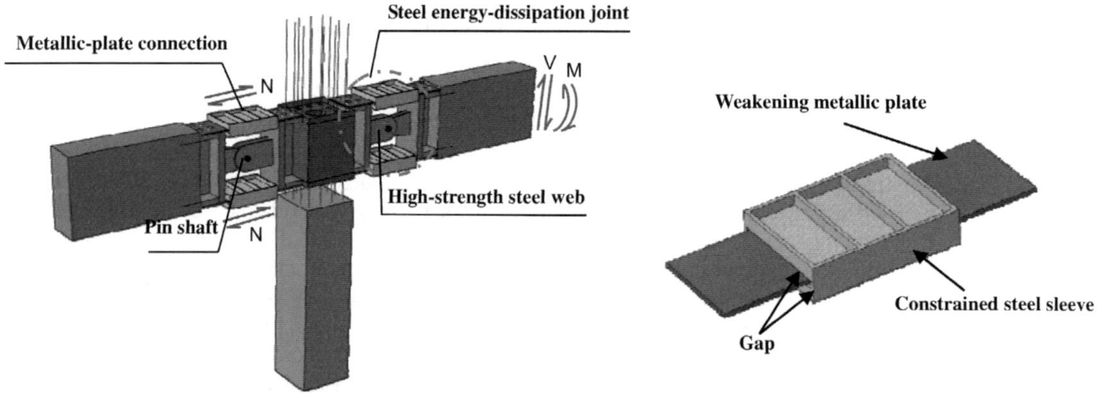

Figure 1. Innovative prefabricated energy-dissipation joint (unit: mm).

2. Numerical Modeling of SPDs

A numerical simulation was established to obtain an innovative numerical model of specimens by the software ABAQUS. ABAQUS was used to establish the numerical model of hole-opened metallic plate connections to simulate its strength performance under axial reciprocating load. The established numerical model considers the influence of factors such as material, geometrical nonlinearity, material damage degradation, and connection contact.

2.1. Constitutive Relationships

The ideal constitutive relationship can be described as the bilinear hardening model. Where E_s represents the initial stiffness of the steel at the elastic stage; f_y is the yielding strength of the steel; After the steel yielded, the stiffness at the strengthening stage was taken as 1/100 of the initial stiffness ($0.01E_s$) for strengthening. f_u is the ultimate strength of the steel, the stress no longer increases when the steel reaches the ultimate strength f_u.
Considering damage degradation of constitutive relationship, ductile damage and damage evolution were added to properties of the metallic plate. When the cumulative plastic damage reaches 0.95, models stop working.

2.2. Elements and Interaction

The finite element model of the hole-opened metallic plate connection is composed of two parts: the constrained steel sleeve and the metallic plate. Unit of the metallic plate and the constrained steel sleeve are both C3D8R elements. The contact relationship is set to set and face-to-face with hard contact. The tangential friction coefficient is taken as 0.1. The loading end releases axial displacement and restricts other displacements, and the fixed end is fixedly connected which is limited displacement and rotation in all directions. It's necessary to apply the initial imperfection according to the first five buckling modes. The amplitude is 1/1000, 3/5000, 1/5000 of the effective length of the metallic plate. FEM was shown in figure 2.

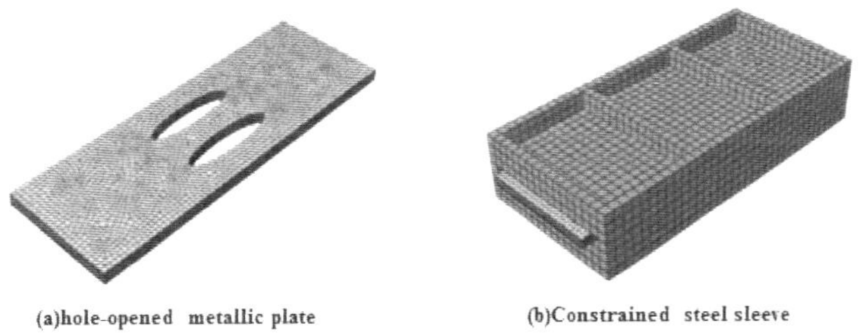

Figure 2. Finite element models.

2.3. Failure Modes

The failure modes of hole-opened metallic plate connections were shown in figure 3. Specimen E-10-200 was took as an example. "E'' represents the elliptical open hole. "10" refers to the thickness of steel plates and "200" refers to the weakening length of the open hole. The unit of numerical value is both millimeters. In the final failure mode, the middle hole-opened weakening part of metallic plates all exhibited cracks and fractures. After the stress released, the maximum stress of three specimens was all concentrated on two sides of the weakening open hole. The distribution of the maximum stress presents a certain relationship with the weakening length of the open hole.Only specimen E-10-200 appeared the out-plane buckling phenomenon. The maximum stress was concentrated at two sides of hole-opened weakened metallic plates. The resistance to deformation of specimen E-10-200 was weaker than specimen E-6-100 and E-10-100. It indicated that the increase of the hole-opened weakening length could improve the ductility performance of hole-opened metallic plate connections, but it was also easy to cause the phenomenon of out-plane lateral buckling under compression.

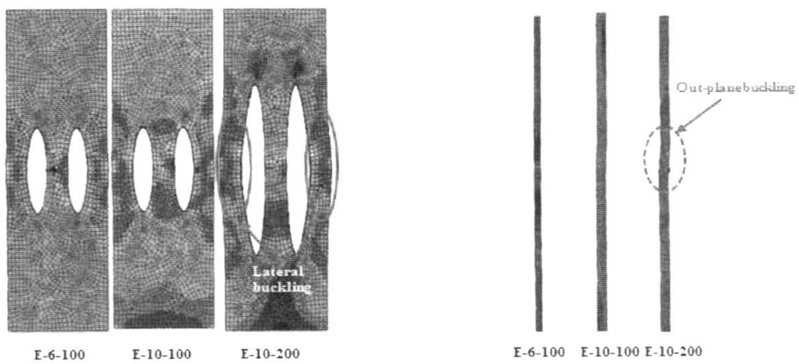

Figure 3. Failure modes of numerical modelling.

3. Parametric Analysis of Hole-opened Metallic Plate Connections

3.1. Hole-opened Weakening Size

Hole-opened weakening dimension is defined as the ratio of hole-opened length to steel-plate length in specimens' longitudinal direction and the ratio of hole-opened width to steel-plate width in specimens' transverse direction. Considering the influence of different hole-opened weakening dimensions on the hysteretic performance of dampers, parametric analysis from twenty-five different weakening-dimension metallic plate was carried out. "*B*" is the width of metallic plates and "*L*" is the length of metallic plates. "*a*" represents the weakening length of holes and "*b*" represents the weakening width of holes. The effective length of metallic plates is 370mm. Observing the number, the first letter "X" represents the weakening size; the second letter "E" represents the elliptical open hole; the first number represents the serial number that changes with b/B; the second number represents the serial number that changes with a/L. Specific dimensions and examples are listed in table 1 and it listed all examples. Figure 4 exhibited the related sizes of hole-opened metallic plate connections.

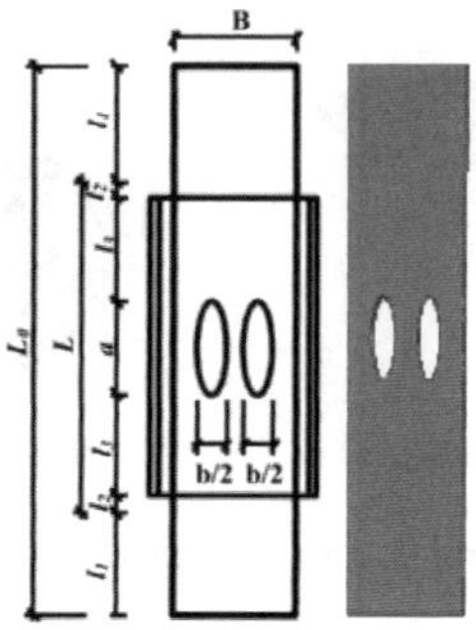

Figure 4. Dimension diagram of the metallic plate connection.

Table 1. Parametric analysis examples based on the change of hole-opened weakening size.

B=125mm *L*=370mm		*a/L*				
t=10mm	*g*=1mm	0.135	0.270	0.405	0.541	0.676
	0.24	X-E-1-1	X-E-1-2	X-E-1-3	X-E-1-4	X-E-1-5
	0.32	X-E-2-1	X-E-2-2	X-E-2-3	X-E-2-4	X-E-2-5
b/B	0.4	X-E-3-1	X-E-3-2	X-E-3-3	X-E-3-4	X-E-3-5
	0.48	X-E-4-1	X-E-4-2	X-E-4-3	X-E-4-4	X-E-4-5
	0.56	X-E-5-1	X-E-5-2	X-E-5-3	X-E-5-4	X-E-5-5

Table 2 exhibits the influence of of hole-opened weakening size on ductility. As shown in figure 5(a), it can be seen that when *b/B* is constant, the value of *a/L* is between 0.270 and 0.541, compared to the value of *a/L* which is 0.135 and 0.676, the ductility of the example is relatively good, and the bearing capacity is not much different from that of the calculation example XE-3-1. In figure 5 (b), it can be seen from the figure that when *a/L* is constant, the value of *b/B* is 0.24~0.56 the hole-opened weakening dimensions of specimens exhibits a certain effect on bearing capacity, deformation capacity, stiffness, energy dissipation, and ductility. When the hole-opened length and the hole-opened width are larger, bearing capacity of SPDs is lower but its ductility and energy-dissipation presents better. Based on the analysis results of different size parameters, recommended hole-opened sizes are given, including that the weakening length *a/L* of the hole is from 0.25 to 0.55, the weakened width *b/B* is from 0.3 to 0.5.

Table 2. Influence of hole-opened weakening size on ductility.

Mean value of ductility		*a/L*				
		0.135	0.270	0.405	0.541	0.676
	0.24	6.896	6.928	7.531	11.091	10.312
	0.32	6.378	7.784	8.651	9.104	9.556
b/B	0.4	6.244	8.462	8.577	10.158	10.238
	0.48	7.044	7.720	8.961	10.124	9.846
	0.56	7.435	9.262	9.152	9.335	11.283

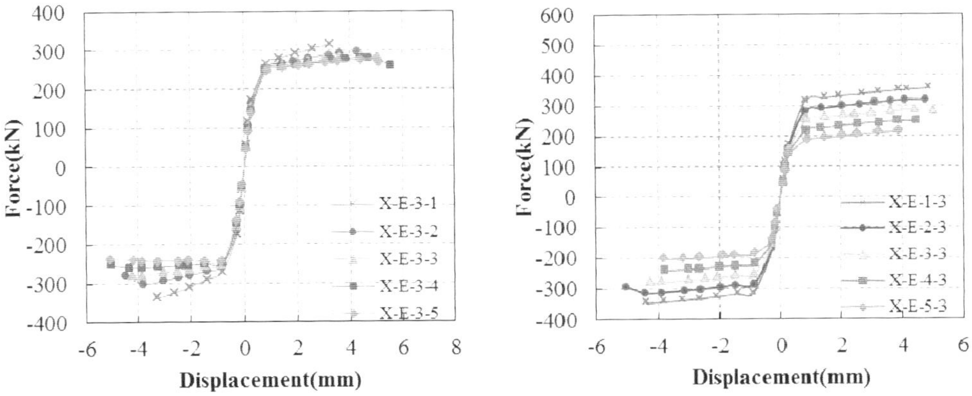

(a)Comparison of skelton curves of K-E-3-n (b)Comparison of skelton curves of K-E-n-3

Figure 5. Effect of hole-opened weakening size on mechanical properties.

3.2. Width-to-thickness Ratio

Observing the number, the first letter "K" represents the width-to-thickness ratio; "E" represents the elliptical open hole; the first number indicates the serial number; the second number indicates the specimens' number. The first series maintains the section thickness of t=10mm and changes the section width B; the second series keeps the section width B=125mm unchanged and changes the section thickness of metallic plates.

Table 3. Parametric analysis examples of width-to-thickness ratio.

First Series	B/t	Section Thickness /t (mm)	Section width/B (mm×mm)	a×b (mm×mm)	Second Series	B/t	Section Thicknes s/t (mm)	Section width/B (mm)	a×b (mm×mm)
K-E-1-1	7.5	10	75		K-E-2-1	20.83	6	125	
K-E-1-2	10	10	100		K-E-2-2	15.63	8	125	
K-E-1-3	12.5	10	125	100×25	K-E-2-3	12.50	10	125	100×25
K-E-1-4	15	10	150		K-E-2-4	10.42	12	125	
K-E-1-5	17.5	10	175		K-E-2-5	8.93	14	125	

Table 4. Effect of width thickness ratio on mechanical properties.

First Series	Initial Stiffness /K_0 (kN·mm^{-1})	Bearing Capacity /F_u (kN)	Ductility Factor/μ	Damping coefficientζ	Second Series	Initial Stiffness/ K_0 (kN·mm^{-1})	Bearing Capacity /F_u (kN)	Ductility Factor/μ	Damping coefficientζ
K-E-1-1	242.82	93.97	10.829	0.456	K-E-2-1	258.76	178.88	8.69	0.498
K-E-1-2	326.54	195.38	8.148	0.484	K-E-2-2	342.59	232.03	8.89	0.486
K-E-1-3	426.22	298.71	9.211	0.466	K-E-2-3	426.22	298.71	9.21	0.466
K-E-1-4	540.34	386.18	8.414	0.491	K-E-2-4	488.35	353.42	8.40	0.493
K-E-1-5	624.32	476.47	8.332	0.494	K-E-2-5	593.24	415.26	8.68	0.494

IOP Conf. Series: Materials Science and Engineering **964** (2020) 012011 doi:10.1088/1757-899X/964/1/012011

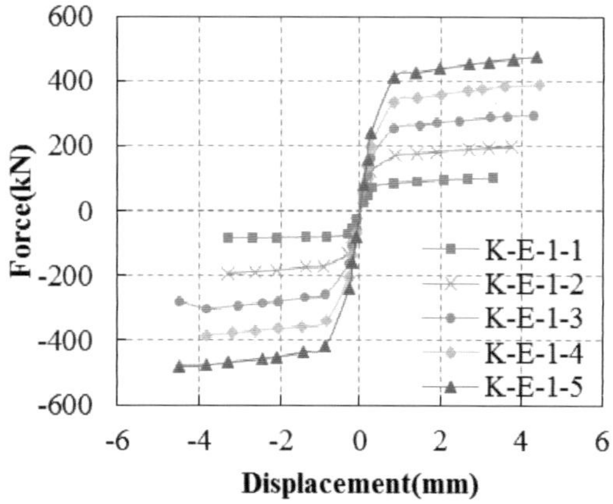

(a)First series of width-to-thickness ratio

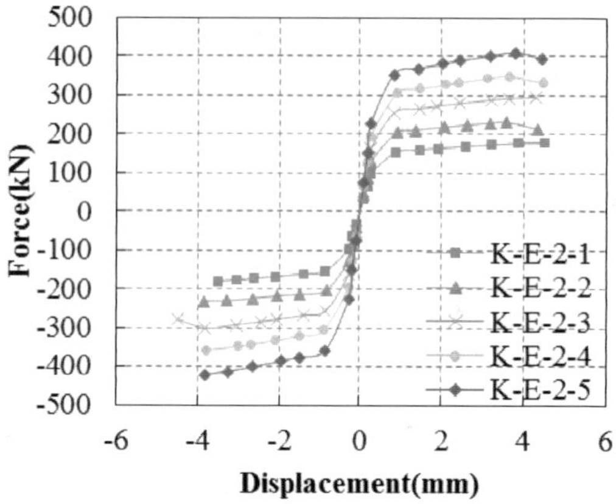

(b)Second series of width-to-thickness ratio

Figure 6. Effect of width to thickness ratio on skeleton curves.

As shown in figure 6 and table 4, when the thickness of metallic plates is10 mm, bearing capacity becomes higher with the increase of width to thickness ratio of sections. When the width of metallic plates is constant, the initial stiffness, bearing capacity, and the equivalent viscous damping coefficient all decrease with the increase of the width-to-thickness ratio, the ductility coefficient μ is greater than 8.40 and the varietal range is not large. The ductility of specimens is good. The recommended value of the width-to-thickness ratio is from10 to 15.

3.3. Gap between Constrained Sleeve and Metallic Plate

The gap between the metallic plate and the constraining sleeve in the plate-thickness direction cause relative sliding. Once the gap becomes too large, it may affect the hysteretic performance of hole-opened metallic plate connections. Examples of the gap are listed in Table 5.

As shown in figure 7, skeleton curves are coincident and the growing trend of bearing capacity is the same. With the increase of the gap, the restraining effect of the sleeve is weakened, and the pressure bearing capacity gradually decreases in the process of compression. When the gap is 0.5mm, the tensile and compressive bearing capacity of hole-opened metallic plate connections is symmetrical, and the working performance is the best. In summary, the gap between the metallic plate and the constraining sleeve should be controlled within 0.5 to 2 mm.

Table 5. Parametric analysis examples of the gap.

Example	Section dimension/ B×t (mm×mm)	Hole-opened dimension/ a×b (mm×mm)	Gap/g (mm)
V1-G-1			0.5
V1-G-2			1
V1-G-3	125×10	100×50	2
V1-G-4			3
V1-G-5			4
V1-G-6			5

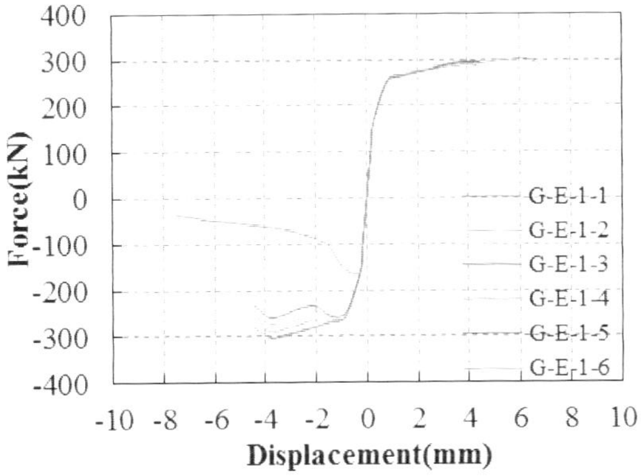

Figure 7. Parametric analysis examples of the gap.

4. Summary

Some conclusions have been obtained.

- Based on the simulation of finite element models of hole-opened metallic plate connections, they exhibited reasonable failure modes. The failure of specimens were focused on weakening sections which exhibited fractures and cracks. It successfully achieves the plastic controllability.
- Weakening hole-opened size, width-to-thickness ratio, and the gap between constrained sleeve and metallic plate are important parameters that affect the bearing capacity, ductility and energy-dissipation performance of hole-opened metallic plate connections. The weakening length *a/L* of the hole is from 0.25 to 0.55. The weakened width *b/B* is from 0.3 to 0.5. The width-to-thickness ratio is form10 to 15 and the gap is within 0.5 to 2 mm.

Acknowledgments

The research presented herein was supported by the National Natural Science Foundation of China (Grant no.51878174), the National Natural Science Foundation of Fujian Province of China (Grant no.2019J01778), the Fuzhou Science and Technology Plan Project (Grant no.2018-G-60) and the The Scientific Research Project of Fujian Institute of Technology (GY-Z0601). The writers wish to express their sincere gratitude to the sponsors.

References

[1] Savoia M, Buratti N, Vincenzi L 2017 *Eng. Struct.* **137** 162-180.
[2] Korkmaz HH and Tankut T 2005 *Eng. Struct.***27(9)** 1392-1407.
[3] Nassereddine A, Youcef SY, Sofiane A. 2019 *Structures* **20** 353-364
[4] Fadi O, Raafat E.. 2017 *Eng. Struct.* **151** 745-760
[5] Koshikawa T. 2017 *Eng. Struct.* **138** 170-180.
[6] Parastesh H, Hajirasouliha I, Ramezani R. 2014 *Eng. Struct.* **70** 144-157.
[7] Yan Q S, Chen T Y, Xie Z Y. 2018 *Eng. Struct.* **155** 330-344.
[8] Li Z X , Lv Y , Xu L H , et al. 2013 *Eng. Struct.* **49** 248–263.
[9] Li Z H, Qi Y H, Teng J. 2020 *Eng. Struct.* **209** https://doi.org/10.1016/j.engstruct.2020.110217.
[10] Wang C L, Liu Y, Zheng X L, et al. 2019 *Eng. Struct.* **178** 298-308
[11] Magliulo G, Ercolino M, Cimmino M et al. 2014 *Construction and Building Materials* **69** 271–284
[12] Bahrami S, Madhkhan M, Shirmohammadi F, et al. 2017 *Eng. Struct.* **132** 808-821.

Lightweight Design of Palletizing Manipulator Based on Structural Performance Influence Analysis

Haiyan Hua* and Xiaolei Yan

Laboratory of Internet of Things, College of Mechanical and Automotive Engineering, FuJian University of Technology, Fuzhou, FuJian, China

*Corresponding author email: yukihua810608@fjut.edu.cn

Abstract. In order to better coordinate the structural performance with lightweight goal, multi-stage lightweight strategy based on structural performance influence analysis is proposed. The framework of knowledge-guided lightweight design with multiple behavior routes was constructed firstly. Some behavior routes, such as initial structural design, lightweight optimization combined with structural performance influence analysis and knowledge-guided lightweight optimization, were designed in this section. Secondly, the decomposition of decision task and the multi-stage lightweight decision based on structural performance influence knowledge were discussed. Finally, the palletizing manipulator with parallelogram leveling mechanism was taken as an example to apply proposed strategy. Comparing with the initial optimization model in the first stage, the total weight of the components after multi-stage lightweight has been reduced by 23.37%, but the maximum stress of total manipulator only increased by 10.3Mpa, which shows the significant effect of lightweight.
Keywords: Palletizing manipulator; Multi-stage lightweight design; Multiple behavior routes; Multiple working conditions; Influence analysis.

1. Introduction

Palletizing manipulator is an important equipment in automatic production, which is mainly used for grasping and stacking of materials. In palletizing operation, it is expected to have sufficient strength and stiffness under the loading condition, but also have the light and flexible structure to reduce energy consumption. To meet these expectations for palletizing manipulator, many researches have been made on structural performance analysis and optimization. In some studies, the size optimization of the arm and the topology optimization of the linkage have been carried out to reduce the weight without affecting the strength and stiffness of the structure[1][2]. However, the more design features, variables, working conditions, and performance constraints considered, the more complex the lightweight becomes[3]. Additive manufacturing technology can reduce the difficulty of manufacturing for the complex structure[4][5], but the complexity of the design process caused by the increase of structural features also needs to be solved urgently. Due to the intricate relationships among parameters, structural performance and lightweight goals, it is necessary to seek a breakthrough in design process to improve the lightweight efficiency and effectiveness. Therefore, a multi-stage lightweight strategy based on structural performance influence analysis is discussed in this paper.

2. The Framework of Knowledge-guided Lightweight Design with Multiple Behavior Routes

In order to realize the lightweight design with the guidance of structural performance influence analysis, a framework of knowledge-guided lightweight design is constructed and shown in figure 1, which includes nine behavior modules, thirteen behavior channels and a knowledge base. This framework is especially suitable for the lightweight design of part family with different sizes and similar structures, because some knowledge can be reused. On the basis of this framework, multiple behavior routes are easy to be combined, some of which are shown in table 1.

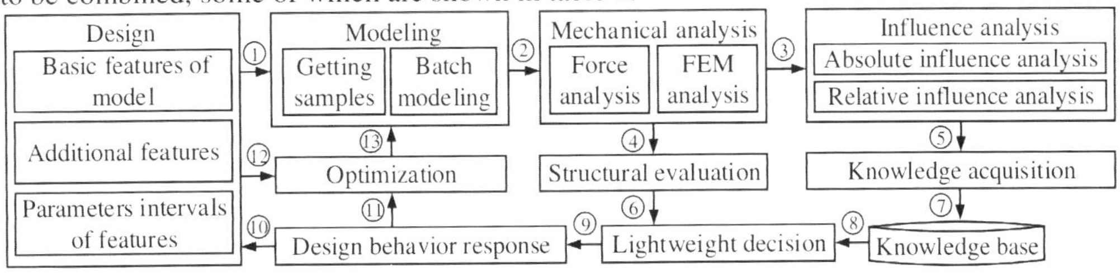

Figure 1. The framework of lightweight design behaviors.

Table1. The behavior routes of lightweight design.

No.	Task	Behavior route
Route1	Initial structural design	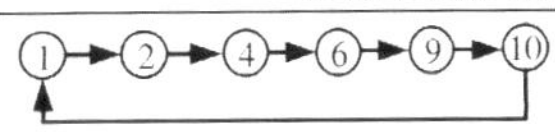
Route2	Lightweight optimization combined with structural performance influence analysis	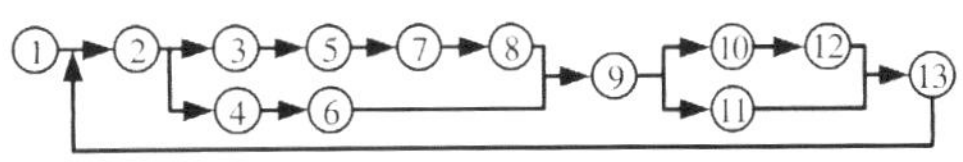
Route3	Knowledge-guided lightweight optimization	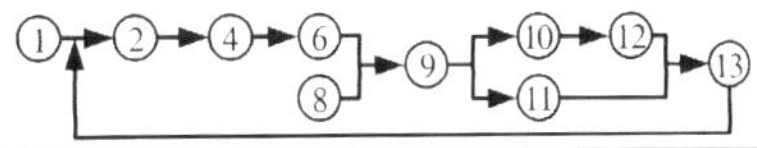

The main task of route 1 is to design and optimize the initial palletizing manipulator according with the requirements of structural performance under working conditions. Firstly, the basic features of model as well as the intervals of structure parameters are determined in the design module. Secondly, the 3D parametric model established by Creo to generate the solid samples in the modeling module. Thirdly, by using Creo and ANSYS, every 3D sample is realized force analysis and finite element analysis respectively. After evaluating the structural performance of samples, the elites are chose from samples. If meet the requirements, the initial design is completed. Otherwise, the behaviors whether to reselect the basic features, adjust parameter intervals or change the class of model are determined by the lightweight decision module. After that, according the response of behaviors, the route returns to the design module to update the design behaviors.

Route 2 is used for combining knowledge acquisition with optimization by executing the influence analysis module, knowledge acquisition module and optimization module in the cycle. The influences of structural parameters on the structural performance and the weight of model are analysed by Spearman correlation analysis, and then the relative influence is calculated. After that, the knowledge acquisition module is used for evaluate the adjustment priority for every structural parameter. The detail method of acquiring the structural performance influence knowledge had been illustrated in our past work[6]. When getting the response of design behaviors by lightweight decision, one branch is to update the optimization variables as well as their parameter ranges, or adding additional features by channel 10. The other branch is to execute some searching behaviors, such as competition and elites saving, generating new individuals and so on, to acquire the new samples under the guidance of the structural performance influence knowledge. Transforming these new samples to 3D models, the optimization enter into the next cycle.

Route 3 is to realize knowledge-guided lightweight optimization. Since the structural performance influence knowledge is relatively stable within a certain range, the influence analysis does not need to be performed in every cycle. In this route, the influence analysis module and knowledge acquisition module are not executed, but the knowledge is directly obtained from the knowledge base. Furthermore, the optimal route after the lightweight decision module is the same as route 2.

3. Multi-stage Lightweight Decision Based on Structural Performance Influence Knowledge

Due to the large number of components, features, structural parameters and working conditions, the lightweight design for overall palletizing manipulator is a hard work. Considering the differences of influence on lightweight, it is worthwhile to break down the task. The complex decision task is described by a decomposition tree, which shown in figure 2. According to this decision tree, it is easy to divide the process of lightweight into several stages with their models recombined adaptively. That is, you only need to change the decision behaviors to make the lightweight model in every stage changed flexibly.

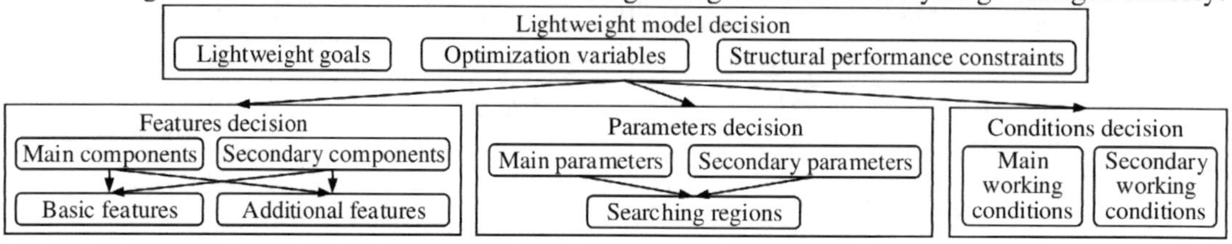

Figure 2. The decomposition tree of decision task.

In this paper, according to the classification of features, the dangerous conditions for every component and the structural performance influence knowledge, there are five stages in the lightweight task. In the first stage, the basic features of main components for palletizing manipulator are taken as the objects, and the structural performance characteristics under dangerous conditions are considered. The hierarchical optimization of structural parameters with regions adjustment is carried out under the guidance of the adjustment priority. In the second stage, the additional features of the main components are designed and optimized. In the third and fourth stages, the lightweight of the basic and additional features for the secondary components are carried out respectively. In the fifth stage, the local features are optimized and the result of lightweight with best comprehensive structural performance is identified.

4. Application

4.1. The Task of Lightweight Optimization

The palletizing manipulator with parallelogram leveling mechanism is taken as an example in this paper. The basic model of this manipulator, which composed of the rear arm, driving link, drag link, first parallel link, forearm, triangle joint, second parallel link, and end gripper, is shown in figure 3.

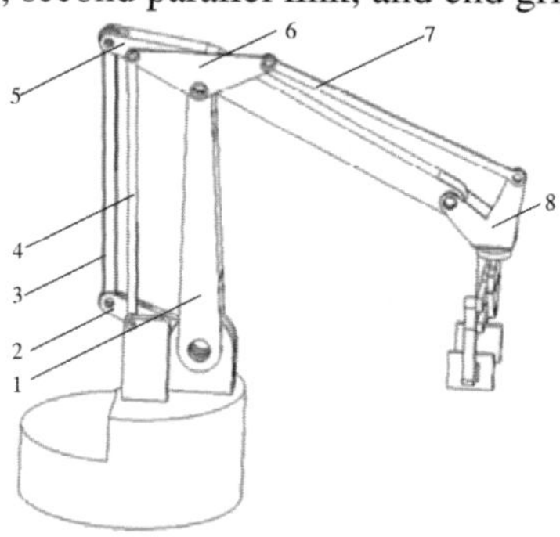

1-rear arm, 2-driving link, 3-drag link, 4-first parallel link, 5-forearm, 6-triangle joint, 7-second parallel link, 8-end gripper

Figure 3. The basic model of palletizing manipulator.

In this example, there are seven components of the palletizing manipulator taken as the lightening objects. According to the working process, four extreme positions of palletizing manipulator are considered in the structural performance analysis, which are the highest, the lowest, the nearest and the farthest. On this basis, the lightweight optimization is carried out by taking the minimum weight of palletizing manipulator as objective function and the maximum stress less than the allowable value under four working conditions as structural performance constraints. The optimization model is as follows:

$$min W = \sum_{i=1}^{N} W_i \tag{1}$$

$$\text{s.t.}\begin{cases} x_{ij\min} \le x_{ij} \le x_{ij\max} & i=1,2,\cdots N;\ j=1,2,\cdots J_i \\ \sigma_{i\max} \le [\sigma_i] & \sigma_{i\max} = max\{\sigma_{i1\max},\sigma_{i2\max},\sigma_{i3\max},\sigma_{i4\max}\} \end{cases} \tag{2}$$

Where N is the number of components; W is the total weight of palletizing manipulator; W_i is the weight of the ith components, which depend not only on the size of features, but also on the basic features and additional features of the model decided in different stages of optimization. All of variables are the positive integer in this paper. x_{ij} is the jth variable of the ith component; J_i is the number of variables for the ith component considered in optimization; $x_{ij\max}$ and $x_{ij\min}$ are the maximum and minimum values for x_{ij} respectively. Variables as well as their interval constraints are determined under the guidance of influence knowledge. $\sigma_{i\max}$ is the maximum stress under four working conditions;$[\sigma_i]$ is allowable value of stress for the ith component.

4.2. Multi-stage Lightweight Design Based on Structural Performance Influence Analysis

4.2.1. The first stage lightweight design for main components. As the main load-bearing components for palletizing manipulator, the rear arm and the forearm are designed in the first stage. By analysing under four typical working conditions, it is found that the highest working condition is the most dangerous condition for both components. Considering the efficiency and effectiveness for analysing many samples, the adaptive meshing method is adopted in the cycle and tens of thousands of finite elements are formed in each sample. Through the loop execution of route 1, the features of components and the intervals of parameters are determined. The rear arm contains 13 structural parameters, and the forearm contains 15 structural parameters. Because the influence knowledge may vary with parameter intervals, the route 2 and route 3 execute alternately in this stage.

By influence analysis, the sensitive factors for rear arm mainly include thickness of plates, the distance of side plates, distance from the arc plate to the hole and the external circle of the holes. The stress nephograms of excellent rear arm acquired by route 1 is shown in figure 4, and that of the best rear arm in first stage is shown in figure 5. Comparing the weight of the model, the rear arm in figure 4 is 67.72kg, while that in figure 5 is 55.85kg. It is illustrated that the weight has been reduced by 17.53% after the first stage of lightening, but the maximum stress only increases by 8.71Mpa and less than the allowable value of 145Mpa. It is also found that the dangerous positions mainly appear in the bigger reamed hole and the edge of side plates near the bigger reamed hole. when the weight difficult to be lightened by optimizing the structural parameters of basic features, it is necessary to enter the next stage.

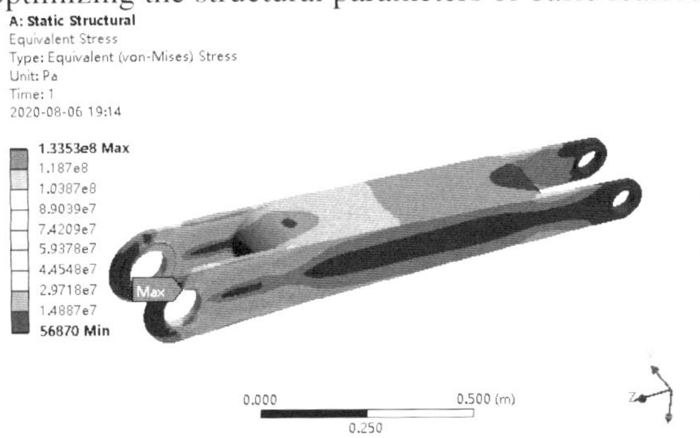

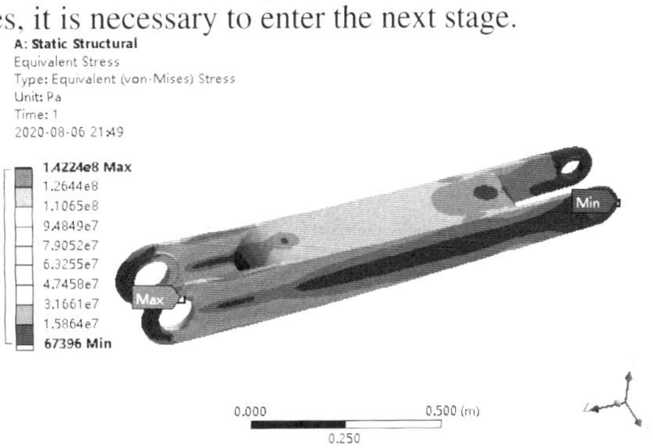

Figure 4. Excellent rear arm acquired by route 1. **Figure 5.** Best rear arm in the first stage.

Similarly, the first-stage lightweight of forearm is carried out, and the stress nephograms of forearms are shown in figure 6 and figure 7. By influence analysis, it is found that the maximum stress of the forearm always occurs near the hole connected with the rear arm, and the geometrical parameters of the side plates have significant influence on the maximum stress and the weight. Figure 6 also shows that the maximum stress is 81.16Mpa, while that in figure 7 is 112.65Mpa. Comparing to figure 6, although the stress increased, the weight of model has been reduced from 59.77kg to 44.82kg.

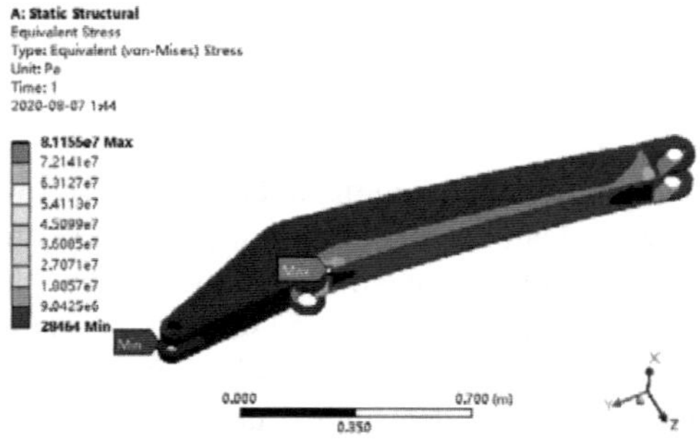

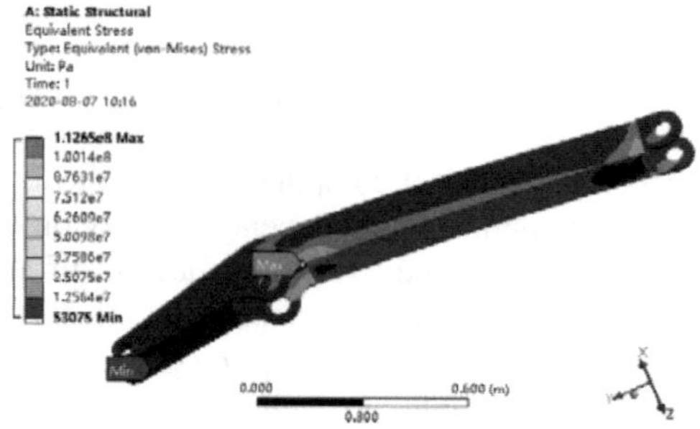

Figure 6. Excellent forearm acquired by route 1.

Figure 7. Best forearm in the first stage.

4.2.2. The second stage lightweight design for main components. The second stage mainly focus on the design and optimization of additional features for the main components based on the lightweight results of the first stage. According to the stress distribution, the geometric shape, size and position distribution of additional features are determined to remove the material from the model. Stress nephogram of the best components in this stage are shown in figure 8 and figure 9 respectively. Figure 8 illustrates that the maximum stress of the best rear arm is 144.09Mpa. Comparing to figure 5, the weight of the rear arm has been reduced from 55.85kg to 49.14kg by removing local material from the side plates. Similarly, the weight of the forearm has been reduced from 44.82kg to 40.29kg, but the maximum stress only increases by 1.57Mpa.

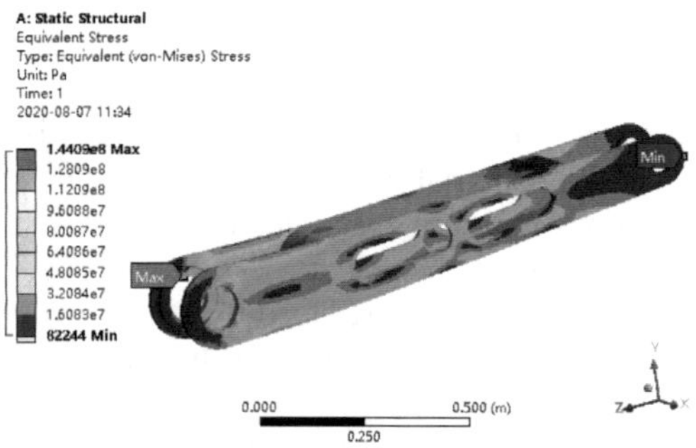

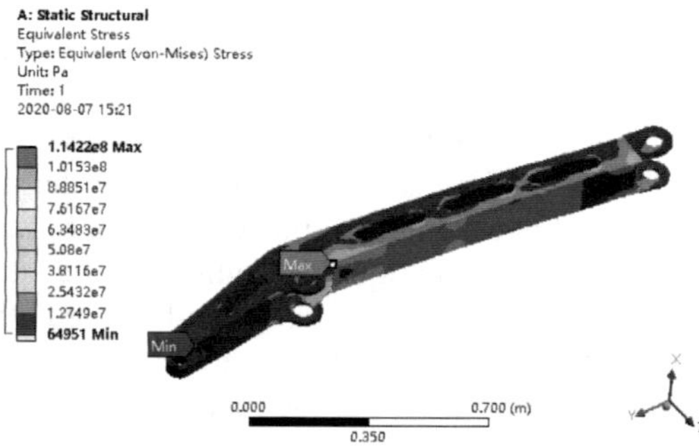

Figure 8. Stress nephogram of best rear arm.

Figure 9. Stress nephogram of best forearm.

4.2.3. The third stage and the fourth stage lightweight design for the secondary components. In the third and the fourth stages, the lightweight of the secondary components are carried out. The dangerous condition of secondary components may be different from that of main components, so it is also necessary to execute route 1 and optimize the basic characteristics firstly. Among these components, the driving link and triangle joint are considered to remove material in the fourth stage. Stress nephogram of the excellent drive link and triangle joint are shown in figure 10 and figure 11 respectively.

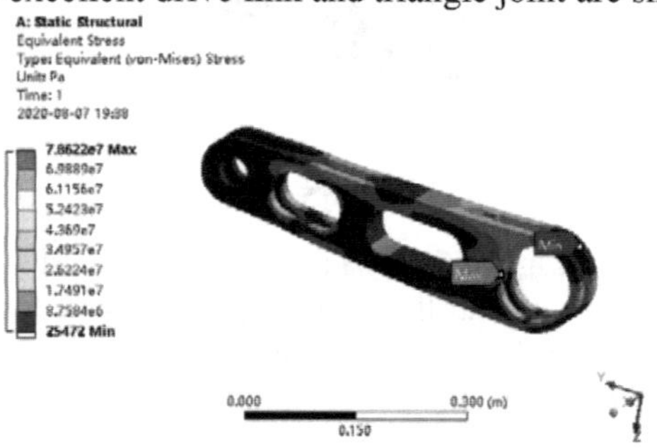

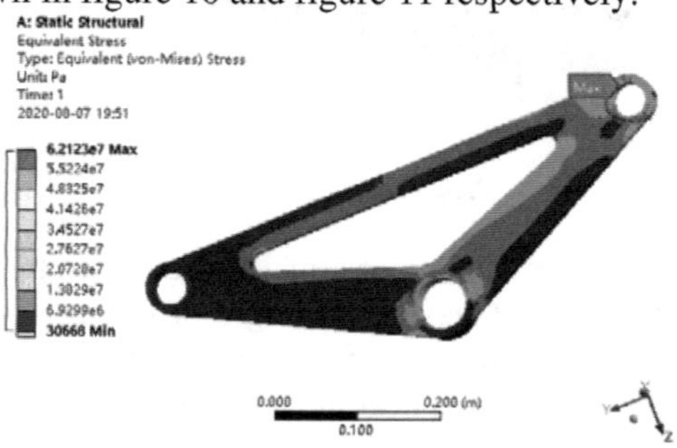

Figure 10. Stress nephogram of driving link.

Figure 11. Stress nephogram of triangle joint.

4.2.4. The final result of lightweight for palletizing manipulator. In the fifth stage, the local features are adjusted. It is more important that when the change of values for some features make the increase ratio of maximum stress in component much more than the reduction ratio of weight, the behavior is no longer wise. In addition, for the slender parts, the allowable stress is often set less than that of the main component. After multi-stage optimization, the final results of palletizing manipulator under four typical working conditions are shown in table 2. Comparing with the initial optimization model in the first stage, the total weight of the final components has been reduced from 188.14kg to 144.17kg. It is indicated that the total weigh reduced by 23.37%, but the maximum stress of total mechanism only increased by 10.3Mpa.

Table 2. The final results of lightweight for palletizing manipulator.

Name	Stress under four typical condition(MPa)				Weight
	the highest	the farthest	the nearest	the lowest	(kg)
rear arm	143.83	81.28	28.91	9.64	49.14
driving link	45.30	78.62	60.31	69.08	12.00
drag link	23.56	33.12	31.54	15.15	18.00
first parallel link	44.20	47.10	114.95	47.10	5.89
forearm	114.22	60.63	69.82	47.53	40.29
triangle joint	62.12	10.11	20.72	10.53	9.89
second parallel link	119.70	21.86	25.56	26.93	8.96

5. Conclusion

For complex mechanism, optimizing all of components at the same time is a hard work. By taking the multi-stage lightweight design for palletizing manipulator as example, the conclusions are as follows:

- The structural performance evaluation and influence analysis are very important in the process of lightweight design. It is found that some influences closely relate to basic features of model as well as their parametric modeling schemes. The reproduction and change of geometric features in the basic models are more convenient by behavior decomposition.
- The multi-stage lightweight strategy is benefit to simplify the design task. Actually, behavior decomposition improves the flexibility of combining more behavior schemes, which also facilitates the increase and decrease of behaviors.
- In contrast, when the value of parameter changes within a certain range, the influence of parameter on structural performance and lightweight are relatively stable. Therefore, the methods of knowledge expression and utilization in lightweight design will be further studied in the future to improve the reusability of knowledge in similar structures.

Acknowledgements

This project supported by the National Natural Science Foundation of China (No.51905100), the Natural Science Foundation of Fujian Province (2019J05111), and the Scientific Research Foundation Project of FuJian University of Technology (No. GY-Z14075). The authors would like to acknowledge the support of Digital Fujian Industrial Manufacturing IOT Lab and Fujian Key Laboratory of Automotive Electronics and Electric Drive (Fujian University of Technology).

References

[1] Wei F. 2016. Structure Analysis and Optimization of small Palletizing Robot (*Tianjin University of Science &Technology*). p44-55.

[2] Zhang D J. 2018. Design and Optimization of Arm Structure of Bag Palletizing Robot (*Harbin Institute of Technology*). p33-40.

[3] Guo J. 2017. Finite Element Analysis and Structure improvement of Flour Bag Palletizing Robot (*Shandong Jianzhu University*).p40-50.

[4] Chen J W, Chen Q P, Hu X L, et al. 2020. Lightweight Design of Small Arm for Palletizing Manipulator Facing 3D Printing.(*Chinese Journal of Construction Machinery*).vol 1 p45-50,55.

[5] Mohamed A, Erich M, Kemajl S. 2015. New Concept for Design and Control of 4 Axis Robot Using the Additive Manufacturing Technology. (*Procedia Engineering*), vol 100 p1364-1369.

[6] HUA H Y, LIN H. 2019. Automatic Acquisition of Stress Distribution Characteristic and Influence Knowledge for Large-scale Component (*Journal of Mechanical Strength*) . vol 41(1) p117-124.

IOP Conf. Series: Materials Science and Engineering **964** (2020) 012013 doi:10.1088/1757-899X/964/1/012013

Overall Equipment Efficiency Measurement System Based on Raspberry Pi

Timo Rautio* and Antti Järvenpää

Kerttu Saalasti Institute, University of Oulu, Pajatie 5, Nivala 85500, Finland

*Corresponding author email: timo.rautio@oulu.fi

Abstract. A complete system with the associated hardware, software and graphical user interface for Overall Equipment Efficiency measurement was developed in this work and then thoroughly tested on a factory environment for a period of over three years. The system was designed to be easily utilized on a typical production line, where the products can be identified with a simple sensor, which could also be deployed without interfering the line itself and the related electronics. This work shows that an extremely cost-effective solution based on popular Raspberry Pi architecture can be effectively utilized for all the required tasks needed for a functioning independent system. Data collection electronics with the related database for storing the data and a web server for the graphical user interface and reporting, were all integrated on a single enclosure. In addition, an analysis software was developed to get the benefit from the collected data, and the software was further developed along with the graphical user interface based on the feedback from the collaborating factory. During the test period, over three million products were correctly detected, and all the stoppages were recorded. The developed system revealed important bottle necks on the production line.

1. Introduction

To survive in the harsh competitive environment, every manufacturer must optimize their production to get the maximum efficiency and to minimize cost. Every reasonably sized company has a way of measuring their efficiency in real time, but SMEs (Small and Medium-sized Enterprise) do not necessarily have effective means to get accurate data about their day-to-day effectiveness. Today, one of the most widely used methods for measuring the production line efficiency is the Overall Equipment Efficiency (OEE) that was first introduced as a metric for Total Production Maintenance (TPM) by Nakajima [1]. In its simplest form, the OEE represents the percentage the production line was running in relation to the available time it should have been running. To give a better and more detailed view of the production process, Nakajima proposed six big losses categorized in three groups. These are collected in Table 1, where the relation of the losses to the efficiency components (availability, performance and quality) is referenced and defined in terms of equations based on the production metrics. While the OEE can be used to develop the production, some pitfalls need to be considered first. While it may be intriguing to compare the OEE between different machines and platforms, it is only a measure of internal efficiency and should not be directly compared to other production lines [2]. It is important to keep the focus on the production development and not to get carried away by nice graphs easily acquired with an OEE system [3]. There is also some variation in the definition of OEE, which makes comparing the OEE of one plant to another even more dangerous and pointless [2]. The big picture of the whole process should be carefully considered when planning actions to improve one OEE component [4]. Improving the quality by one percent can lead to bigger loss in the performance, for example.

Table 1. The definition of OEE components by production line metrics and 6 big losses defined in [1].

Efficiency component	6 Big losses
$Availability = \frac{Actual\ production\ time}{Planned\ production\ time}$	Breakdowns Setups/Adjustments
$Performance = \frac{Total\ number\ produced}{Planned\ number\ produced}$	Reduced Speed Idling / Minor Stoppages
$Quality = \frac{Accepted\ quality\ produced}{Total\ number\ produced}$	Defects / Rework Yield

Measuring the availability, performance and quality of a given production line can be utilized to define a more representative figure for the OEE. This is obtained by multiplying the components using the equation:

$$\boldsymbol{OEE = Availability * Performance * Quality} \quad (1)$$

In this equation, OEE can never be more than 100% like any of its products. Availability is the actual running time in relation to the planned production time. This is reduced by breakdowns and unplanned setups and adjustments. Planned maintenance breaks or the workers' break times agreed on the employment contract do not affect the availability but everything in addition to them does. Performance is defined similarly as the actual number of products produced in relation to the planned number of products.

If the production line is running its maximum speed with no idle run, the performance will be 100%, but every minor stoppage or reduction in speed will decrease it. Quality factor considers all the faulty products that have had to be reproduced. This is defined as the ratio of accepted quality products to the total number of produced parts.

Many new automated production lines are readily equipped with necessary software capable of giving the most important metrics of the production in real time. Depending on the customer requirements, these might also include the OEE measurement. Many times, especially in older systems, the amount of information can be extremely limited. In these cases, a specially designed system for the OEE measurement is required. Such a system usually involves building some sort of sensoring and related data acquisition, capability to store and analyze the data from the sensors and implementing a user interface to utilize the whole system. Collecting the data accurately and identifying the production losses will define how accurately the OEE can be calculated [5].

Raspberry Pi computer has become widely known, and its sales have reached 25 million units already by March 2019 [6]. The small size accompanied with a reasonable performance and ability for wireless networking make it an attractive solution for many different applications. Its performance has increased gradually after the first launch of the Raspberry Pi 1, and at the time of writing, the newest version is the Raspberry Pi 4. This has created even more possibilities and innovations. Raspberry Pi 3 was used as a basis of an autonomous robot that could be controlled by an android application on phone [7]. Salih & Omer used a Raspberry Pi 3 model B as a video server that was able to store video with audio [8]. The first version (Raspberry Pi 1) already had the capability to work as a station to count people based on face recognition [9]. Small size and price have made Raspberry Pi an attractive solution for home automation requirements as well [10].

2. Measurement System for the OEE

After an evaluation of the requirements for the system, a Raspberry Pi 2 computer was decided as the basis for the system. Raspberry Pi 2 was the latest version of the Raspberry Pi family at the time this system was designed. It had all the necessary I/O-connections and the capability for ethernet and wireless networking and some additional hardware already at the market like an inexpensive 7" display. It was estimated to be powerful enough to handle the data collection in addition to the required web server and data analysis software. A simplified block diagram of the system consisting of a printed circuit board (PCB), enclosure with the display and the Raspberry Pi itself are presented in figure 1. The associated hardware, software, and user interface development for the OEE measurement system will be presented on the following chapters in more detail.

a) b)

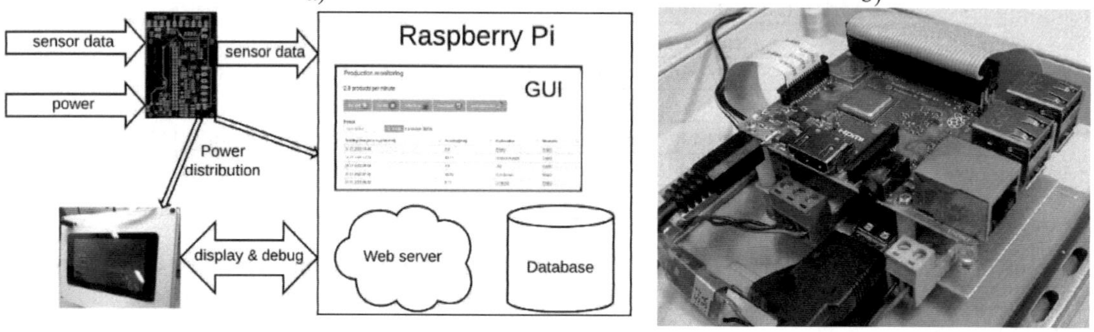

Figure 1. a) A simplified block diagram of the measurement system b) Raspberry Pi with the circuit board for sensor connections connected at the back of the 7" display.

2.1. Data Collection Unit Hardware

Basic settings for the OEE measurement can be made with just one sensor, but some additional inputs were added at the PCB for possible future expansion needs. These wouldn't have to be directly related to OEE measurement but could be temperature or other condition monitoring etc. All four inputs were optically isolated and the power supply for the system was settled at 12 volts to support a wide range of sensors. It was intended from the beginning that the system could be installed to any manufacturing line without interfering the electrics or electronics of the factory in any way so that only a power connection would be needed.

The complete system was assembled on an enclosure with only a power supply cable and a sensor wire as the outside wiring. For the evaluation and development purposes, this enclosure was also equipped with a 7" touch screen and a USB connector to connect a mouse or a keyboard. The screen has been specifically designed for the Raspberry Pi and has mounting points ready for securing the computer on top of pillar supports. The PCB for this work was designed so that it could be also connected to these same mounting points and the Raspberry Pi on additional pillar supports on top of it, resulting on a sandwich type installation shown in the figure 1b. This way the measurement unit size could be minimized, and it can be fitted in a very small enclosure when the screen is not needed.

2.2. Analysis and Reporting Software

MySQL database was installed on the Raspberry Pi for the data storage purposes, and a simple command line program was developed to catch and store the sensor data to the database. The program was automatically started when the computer was switched on. To get the benefit from the collected data, an analysis software is needed to calculate the figures associated with the OEE measurement and all other relevant metrics that can be obtained with the system. Most importantly, it is used daily to monitor the production and is an essential tool for the decision making on how to improve the production. A Windows desktop application was developed for this purpose and the main window is presented in figure 2. The two calendars are used to select the dates of interest and the program calculates the OEE with its components and various other metrics for the range automatically. Graphics show the daily or hourly (if only one day chosen) number of parts produced, the length of stoppages and the reproduction numbers at one glimpse.

More detailed view of the stoppages for the chosen period can be seen with a click of a button. The report presents them in a descending order with the total time and number of stoppages for each reason. These can be utilized to get a detailed view of where the time was consumed when the production was not running. Weekly reports will show if the actions followed have had the desired outcome or if there is a need for further development. Real time view of the production can be seen with the help of takt time and current day's production numbers.

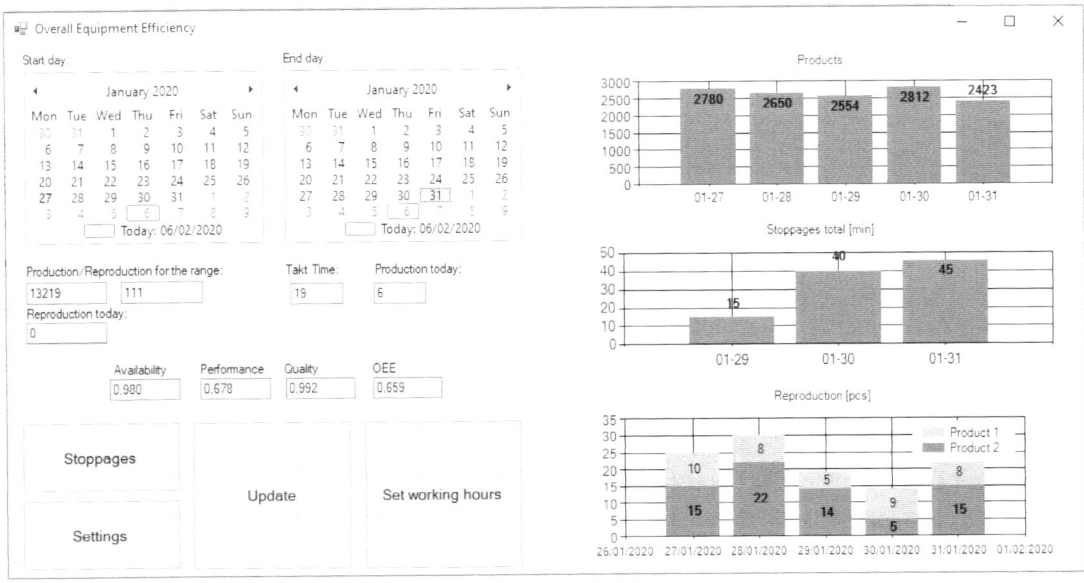

Figure 2. The main window of the OEE calculation and analysis software.

2.3. Graphical User Interface

Lighttpd web server was installed with PHP packages as a basis for the graphical user interface (GUI) development. One of the key features of a high usability OEE measuring system is a clear and intuitive GUI, which can be effectively used to collect the data from the users and to relate it to the sensor data. Such an interface needs to be designed to minimize the interferences to the daily work and the effort to feed the necessary input to the system. After some optimization, the GUI took the form shown in figure 3. Web page interface was chosen since it is easily accessed by handheld devices like tablets or mobile phones, also reducing the time needed to interact with the system.

The most important feature is the list of the latest stoppages. If no production is observed after a predefined time, the system creates a new event on the list indicating the user needs to interact. To minimize the time needed to feed this information, a ready-made list opens when a new stoppage is clicked. This list (editable by the management) includes the most common reasons for the stoppages and only in special cases they need to be manually typed. Only the last five stoppages are listed (uncertainty of the input increases if a lot of time has passed since the stoppage), and the reasons can be fed at any point while the stoppage is showed at the list. Forgotten or incorrect inputs can be corrected by the management from the analysis software presented earlier.

Quality component calculation is handled based on the number of reproduced parts stored by the users. The factories that took part in this study, made all the reproduction only once a day, usually at the end of last shift. The GUI was designed so that the reproduction numbers could be given at any time of the day, again interfering as little as much the production itself. The reproduction could be also counted automatically, but this would require user input anyway as the sensor cannot separate reproduction.

The interface shown in figure 3 also shows some of the optional elements designed. These include the buttons that can be used to follow the shift time and the length of coffee and meal breaks. Takt time can be also presented if it is desirable that the workers see the speed of the production at any time. Finally, the system can be also used to store the production with product numbers, a feature that is not necessarily readily available on the production line, especially in an aged one.

IOP Conf. Series: Materials Science and Engineering **964** (2020) 012013 doi:10.1088/1757-899X/964/1/012013

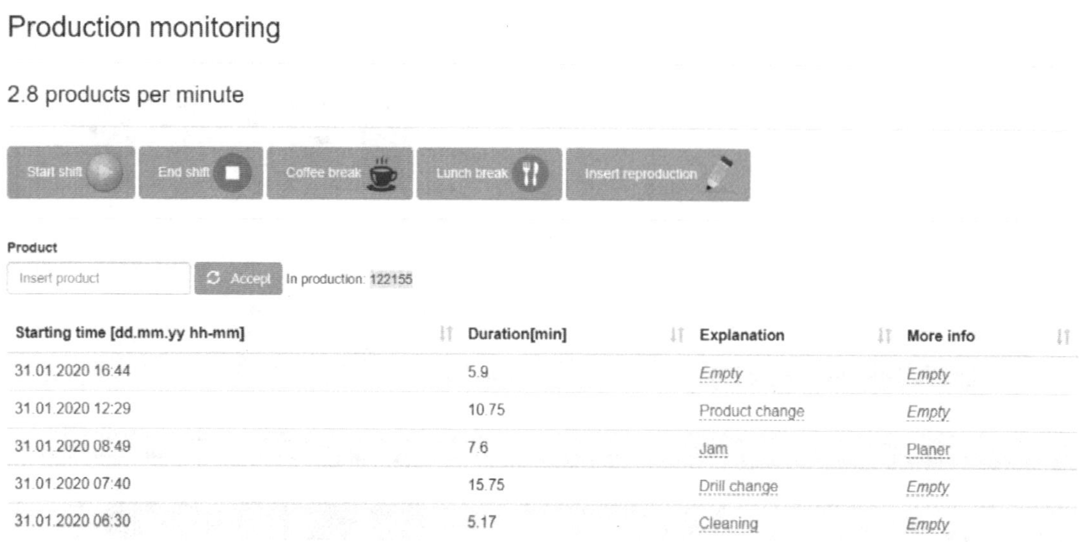

Figure 3. User interface to collect data from the stoppages, breaks and reproduction.

3. Experimental Runs at Factory Environments

Both factories we had the opportunity to run the system in were wood processing lines, where the products move on conveyor belts. This made the installation easy for the measurement system as only the availability of power at the installation location was needed to be confirmed and only one sensor was needed to follow the products in the production line. A bracket was fabricated for the sensor and it was installed at the end of the production line on both cases to record every product passing by. Stoppage at any point of the line stops the whole line making one sensor feasible to monitor and count simultaneously. The enclosure was secured to a wall so that it was not in the way of daily operations on the line.

One of the systems was left on a factory line and has now been running for over three years recording over three million products during the years. Uptime of the system has been well over 99.99% as only one instance was recorded when the system had stopped for a reason that was left uncertain. The issue was fixed by a simple power off – power on maneuver and has not repeated since. Another issue emerged after the system had registered over two million samples. All products were stored in one database table, one line each, and at that stage the select statement from the table took several seconds and was impairing the user experience. This was solved by converting the data to hourly when the day ended. This way no relevant data was lost, and the table size reduced by 99% making the select time about 0.3 seconds.

Both the GUI and the analysis software were developed during the first couple of months of use with the operators and the management to reach the versions presented earlier in this paper. This is the way most software development should be made, in close contact with the end user listening and reacting to the feedback. One of the biggest changes to the original GUI was how the reproduction numbers were collected. At first, there was a button which was used when the reproduction was started and stopped, but it was noted that it was forgotten on some instances and it would be easier to just type the daily numbers as they were already precisely known by the workers. An option to the software settings was also added, enabling the management to change the number of working hours planned for each week to get an accurate presentation of the availability component from the system.

As a result of putting the measurement system to use, the factory could find some of the underlying reasons for the stoppages and found ways to enhance the production. Unfortunately, these cannot be discussed here in detail due to a non-disclosure agreement.

4. Conclusions

A complete system with the associated hardware, software and GUI for the OEE measurement was developed and then thoroughly tested for a period of over three years in real factory environments. This work showed that a very cost-effective solution such as Raspberry Pi can be used in a factory environment successfully. The system was found to be extremely reliable and able to handle all the processing related to data storing, analysis and updating the GUI. The main purpose of the work was to help the factory to analyze their own production in a cost-effective manner. It was shown that no special and expensive hardware or complicated software development is required to build a system that can be used to improve the efficiency of manufacturing industry.

Acknowledgements

The authors would like to acknowledge the financial support received from the Council of Oulu Region and the European Union (European Regional Development Fund) for the "KuTuA" project.

References

[1] Nakajima S 1988 *Introduction to TPM* (Cambridge MA: Productivity Press)
[2] Jonsson P and Lesshammar M 2001 *Int. J. Oper. Prod. Manag.* vol 21 no 11 pp 1404-16
[3] Ishikawa K 1982 *Guide to Quality Control.* (Tokyo: Asian Productivity Organization)
[4] Konopka J and Trybula W 1996 *Nineteenth IEEE/CPMT International Electronics Manufacturing Technology Symposium* pp 137-140
[5] Hedman R, Subramaniyan M, and Almström P 2016 *Procedia CIRP* vol 57 pp 128-133
[6] Piltch "A Raspberry Pi Celebrates 25 Millionth Sale as 7th Anniversary Arrives", *Tom's Hardware*, https://www.tomshardware.com/news/raspberry-pi-25-million-sold,38724.html, accessed 1 February 2020.
[7] Pujari S S, Patil M S, Ingleshwar S S 2017 Remotely controlled autonomous robot using Android application. *International Conference on I-SMAC (IoT in Social, Mobile, Analytics and Cloud) (I-SMAC)* IEEE pp 588-593
[8] Salih F and Omer M S A *2018 International Conference on Computer, Control, Electrical, and Electronics Engineering (ICCCEEE)* pp 1-4
[9] Parthornratt T, Burapanonte N and Gunjarueg W *2016 International Conference on Electronics, Information, and Communications (ICEIC)* pp 1-5
[10] Pavithra D and Balakrishnan R 2015 *Global Conference on Communication Technologies (GCCT)* pp 169-173

Numerical Calculation of Two-qubit Rabi Model

Peihua Qu and Zhanyuan Yan*

Department of Mathematics and Physics, North China Electrical Power University, Huadian Road, Baoding, 071003, China

*Corresponding author email: yanzhanyuan@ncepu.edu.cn

Abstract. The purpose of this paper means to study the numerical solutions of two-photon Rabi model of two-qubit system. We described the Hamiltonian in a collective space of Fock space and spin-1 space. To calculate the dynamics, the initial state is set to a normal squeezed state and we calculate the exact solution of time evolution of the initial state. And a time evolution of average angular momentum is set for examples at the last part of this paper.
Keywords: Rabi model; Two-qubit system; Numerical calculation; Quantum computation.

1. Introduction

The quantum Rabi model (QRM) is extremely useful in many physical realms, including quantum computation, quantum information, quantum optics and so on. The prototypical method was proposed in 1963, which called Jaynes-Cummings model (JC model), describing the interaction between a two-level system(qubit) and a single mode field with rotating-wave approximation (RWA) terms [1]. Because of its solvability, the JC model was massively applied to describe experiments in different technologies. However, the JC model tends to be unsuitable when the coupling strength of the interaction becomes comparable with the transition frequency of the system. The problem can be solved accurately by using the QRM, on the contrast, which introduces the counter-rotating-wave approximation (CRWA) terms. The quantum Rabi model describes several interesting phenomena, for examples, generation of photons [2], the cold-atom based implementation [3] and the entanglement generated by "vacuum fluctuations" [4]. The open-form analytical solution for the QRM, a non-integrable model, has been derived by Braak in 2011 [5].

A major current focus in the QRM is to calculate various interactions or multi-qubit systems. A well-known superradiant phase transition was discussed in the Dicke model, which is a famous type of Rabi model with N-qubit system [6]. The two-qubit system leads to varieties of implementations, such as prototype of superconducting circuits [7] and quantum gates [8].

The purpose of this paper is to provide a numerical solution of two-qubit quantum Rabi model with mathematical tools. By calculating the energy spectrum and the eigenstates of the model, we present a dynamical evolution of angular momentum.

2. Methodology

2.1. Hamiltonian

The Hamiltonian of two-qubit QRM can be set as

$$\hat{H} = \omega\hat{a}^{\dagger}\hat{a} + \Omega\hat{J}_x + g\hat{J}_z(\hat{a}^{\dagger} + \hat{a}) \tag{1}$$

where ω is the frequency of the harmonic oscillator, $\hat{a}^{\dagger}$ and $\hat{a}$ are the creation and annihilation operators. Ω is the atomic transition frequency in collective spin-1 states, in which angular momentum operators can be explained as $\hat{J}_x = (\hat{\sigma}_{1x} + \hat{\sigma}_{2x})/2$ and $\hat{J}_z = (\hat{\sigma}_{1z} + \hat{\sigma}_{2z})/2$, where $\hat{\sigma}_{1x}$ means the Pauli operator in direction x of the first qubit, etc.. Symbol g represents the qubit-oscillator coupling strength.

The eigenstates of angular momentum $|j_x\rangle$ can be chosen as

$$\begin{cases} |1_x\rangle = |\uparrow\uparrow\rangle \\ |0_x\rangle = \frac{1}{\sqrt{2}}\left(|\uparrow\downarrow\rangle + |\downarrow\uparrow\rangle\right) \\ |-1_x\rangle = |\downarrow\downarrow\rangle \end{cases} \tag{2}$$

the state $|\uparrow\rangle$ and $|\downarrow\rangle$ are eigenstates of the Pauli operator $\hat{\sigma}_x$, and the 1-spin state $|j_x\rangle$ satisfies

$$\hat{J}_x |j_x\rangle = j_x |j_x\rangle \tag{3}$$

and

$$\begin{aligned} \hat{J}_z |1_x\rangle = \hat{J}_z |-1_x\rangle = \frac{1}{\sqrt{2}} |0_x\rangle \\ \hat{J}_z |0_x\rangle = \frac{1}{\sqrt{2}}\left(|1_x\rangle + |-1_x\rangle\right) \end{aligned} \tag{4}$$

In the space of photon number designated states $|n\rangle$ associated with angular momentum states $|j_x\rangle$, i.e. $|n\rangle \otimes |j_x\rangle$, the Hamiltonian can be expanded as

$$\hat{H} = \begin{pmatrix} H_0 & H_{01} & 0 & \cdots \\ H_{01} & H_1 & H_{12} & \cdots \\ 0 & H_{12} & H_2 & \cdots \\ \vdots & \vdots & \vdots & \ddots \end{pmatrix} \tag{5}$$

where H_n and H_{mn} are the diagonal matrix and off-diagonal matrix in the Hamiltonian, which can take the form

$$H_n = \begin{pmatrix} n\omega - \Omega & 0 & 0 \\ 0 & n\omega & 0 \\ 0 & 0 & n\omega + \Omega \end{pmatrix} \tag{6}$$

and

$$H_{mn} = \begin{pmatrix} 0 & \sqrt{n/2}g & 0 \\ \sqrt{n/2}g & 0 & \sqrt{n/2}g \\ 0 & \sqrt{n/2}g & 0 \end{pmatrix} \tag{7}$$

2.2. Energy Spectrum and Eigenstates

The energy spectrum and eigenstates can be expressed by the Schrödinger equation

$$\hat{H}\left|\psi_k\right\rangle = E_k\left|\psi_k\right\rangle \tag{8}$$

The eigenstate in equation (8) can be expressed with a useful form as

$$\left|\psi_k\right\rangle = \sum_{n=0}^{\infty}\left(C_{n,-1_x}^k\left|n,-1_x\right\rangle + C_{n,0_x}^k\left|n,0_x\right\rangle + C_{n,1_x}^k\left|n,1_x\right\rangle\right) \tag{9}$$

where C_{n,j_x}^k are associated coefficients calculated by

$$C_{n,j_x}^k = \left\langle n, j_x \middle| \psi_k\right\rangle \tag{10}$$

2.3. Average Photon Numbers

With the expression of an eigenstate $\left|\psi_k\right\rangle$, the mean photon number can be obtained as

$$\begin{aligned}\left\langle\hat{n}\right\rangle_k &= \left\langle\psi_k\right|\hat{n}\left|\psi_k\right\rangle \\ &= \sum_{n=0}^{\infty} n\left(\left|C_{n,-1_x}^k\right|^2 + \left|C_{n,0_x}^k\right|^2 + \left|C_{n,1_x}^k\right|^2\right)\end{aligned} \tag{11}$$

2.4. Dynamics

The dynamical evolution can also be calculated by the eigenstate presented above. The initial state can be set to an arbitrary coherent state described by a complex number α, such as

$$\begin{aligned}\left|\varphi(0)\right\rangle &= \left|\alpha,-1_z\right\rangle \\ &= \mathrm{e}^{\left(\alpha\hat{a}^{\dagger}-\alpha^{*}\hat{a}\right)}\left|0\right\rangle\otimes\left|-1_z\right\rangle \\ &= \mathrm{e}^{-|\alpha|^2/2}\sum_{n=0}^{\infty}\frac{\alpha^n}{\sqrt{n!}}\left|n\right\rangle\otimes\left|-1_z\right\rangle\end{aligned} \tag{12}$$

Where α^* means the conjugate complex number of α. The angular momentum state $\left|-1_z\right\rangle$ satisfies

$$\left|-1_z\right\rangle = \frac{1}{\sqrt{2}}\left|-1_x\right\rangle + \frac{1}{2}\left|0_x\right\rangle + \frac{1}{\sqrt{2}}\left|1_x\right\rangle \tag{13}$$

The initial state can be substituted by $\left|\psi_k\right\rangle$

$$\left|\varphi(0)\right\rangle = \mathrm{e}^{-|\alpha|^2/2}\sum_{n=0}^{\infty}\frac{\alpha^n}{\sqrt{n!}}\sum_k\left(\frac{1}{2}C_{n,-1_x}^k - \frac{1}{\sqrt{2}}C_{n,0_x}^k + \frac{1}{2}C_{n,1_x}^k\right)\left|\psi_k\right\rangle \tag{14}$$

By using the equation (8), the time evolution of the state can be described as

$$\begin{aligned}\left|\varphi(t)\right\rangle &= \mathrm{e}^{-i\hat{H}t}\left|\varphi(0)\right\rangle \\ &= \mathrm{e}^{-|\alpha|^2/2}\sum_{n=0}^{\infty}\frac{\alpha^n}{\sqrt{n!}}\sum_k\left(\frac{1}{2}C_{n,-1_x}^k - \frac{1}{\sqrt{2}}C_{n,0_x}^k + \frac{1}{2}C_{n,1_x}^k\right)\mathrm{e}^{-iE_k t}\left|\psi_k\right\rangle \\ &= \sum_{n=0}^{\infty}\sum_k R_n^k(t)\left|\psi_k\right\rangle\end{aligned} \tag{15}$$

where

$$R_n^k(t) = \mathrm{e}^{-|\alpha|^2/2} \frac{\alpha^n}{\sqrt{n!}} \left(\frac{1}{2} C_{n,-1_x}^k - \frac{1}{\sqrt{2}} C_{n,0_x}^k + \frac{1}{2} C_{n,1_x}^k \right) \mathrm{e}^{-iE_k t} \tag{16}$$

Take the form back to the $|n\rangle \otimes |j_x\rangle$ space, function (15) can be rewritten as

$$\begin{aligned} |\varphi(t)\rangle &= \sum_{n=0}^{\infty} \sum_k R_n^k(t) \sum_{m=0}^{\infty} \left(C_{m,-1_x}^k |m,-1_x\rangle + C_{m,0_x}^k |m,0_x\rangle + C_{m,1_x}^k |m,1_x\rangle \right) \\ &= \sum_{n=0}^{\infty} \sum_k \sum_{m=0}^{\infty} \left(\beta_{n,m,-1_x}^k |m,-1_x\rangle + \beta_{n,m,0_x}^k |m,0_x\rangle + \beta_{n,m,1_x}^k |m,1_x\rangle \right) \end{aligned} \tag{17}$$

where

$$\beta_{n,m,j_x}^k = R_n^k C_{m,j_x}^k \tag{18}$$

The dynamical characteristics can be calculated by function (17). For an example, we calculated the time evolution of $\langle \hat{J}_z \rangle$, which is

$$\begin{aligned} W(t) &= \langle \varphi(t) | \hat{J}_z | \varphi(t) \rangle \\ &= \frac{1}{\sqrt{2}} \sum_{n=0}^{\infty} \sum_k \sum_{m=0}^{\infty} \left(\beta_{n,m,-1_x}^k {}^* \beta_{n,m,0_x}^k + \beta_{n,m,0_x}^k {}^* \beta_{n,m,-1_x}^k + \beta_{n,m,0_x}^k {}^* \beta_{n,m,1_x}^k + \beta_{n,m,1_x}^k {}^* \beta_{n,m,0_x}^k \right) \end{aligned} \tag{19}$$

3. Results

The Hamiltonian can be diagonalized effortlessly with the help of mathematical tools, like Matlab or Mathematica, so that we can get the energy spectrum and eigenstates in a numerical way. However, an exact solution demands the arithmetic of infinite n, which is unreachable by the tools. A reasonable settlement is making the scale of the matrix, i.e. the maximum n in the progress, big enough to let the solution of low energy states meet exact solutions. In this paper, the result was calculated by self-made Matlab programs and the maximum n was chosen by 20 to make the solutions virtually stabilized. Figure 1 shows the energy spectra with two sets of conditions, which are $\Omega = \omega$ and $\Omega = 2\omega$.

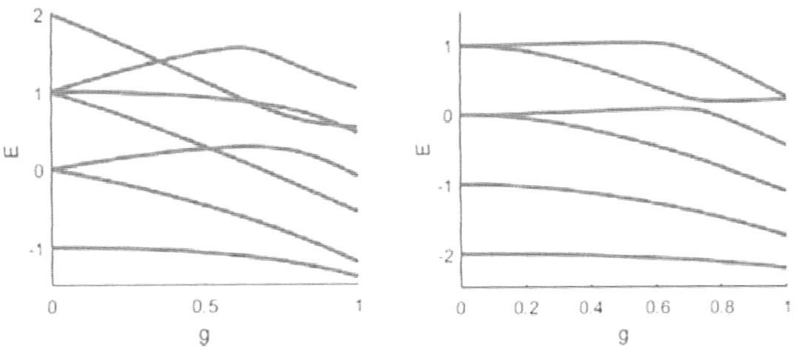

Figure 1. The numerical energy spectrum of the two-qubit QRM. The left is set as $\Omega = \omega$ and the right one is set as $\Omega = 2\omega$.

And the mean photon numbers can be displayed as Figure 2, which takes the same parameters as Figure 1.

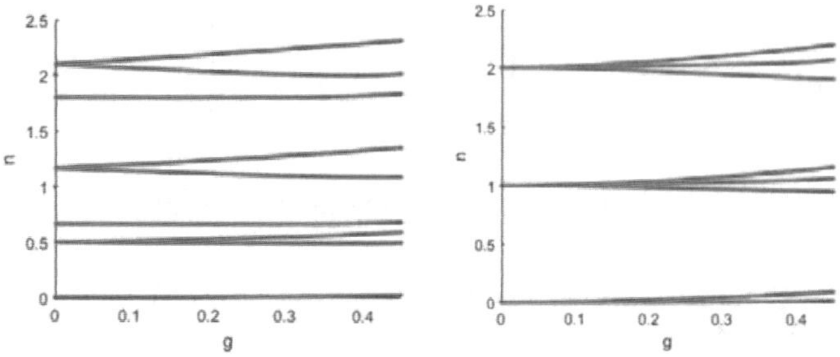

Figure 2. The mean photon numbers of the two-qubit QRM. The left is set as $\Omega = \omega$ and the right one is set as $\Omega = 2\omega$.

Finally, we calculated the time evolution of $\langle \hat{J}_z \rangle$ in the Figure 3 by $\Omega = 2\omega$, $g = 0.3\omega$ and $\alpha = 3$.

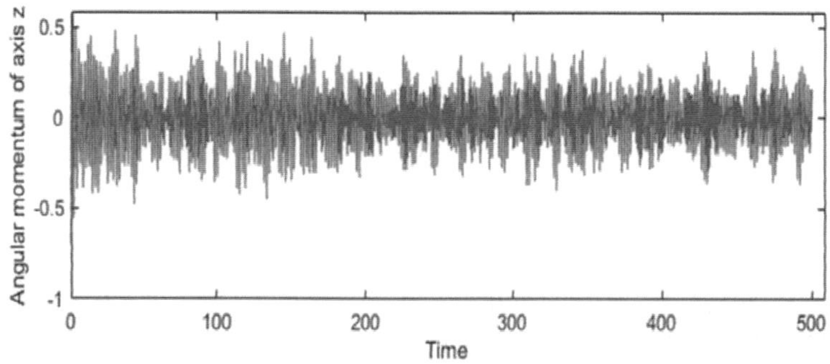

Figure 3. The time evolution of $\langle \hat{J}_z \rangle$ as a function of time for $\Omega = 2\omega$, $g = 0.3\omega$ and $\alpha = 3$. The unit of time is taken by $\Omega t / 2\pi$.

4. Conclusion

The solutions of two-qubit Rabi model can be calculated reasonably in a numerical way, which can be used as the exact solutions of the model. The energy spectrum and the mean photon number shows the static properties. The dynamics can also be expressed by the time depended states. Methodology in this paper provides the principle of how to build a program to calculate the parameters of two-qubit Rabi model. With choosing a keen cut-off size of the Hamiltonian matrix, results tend to be regarded as the exact results, and can be expected to portray some physical apparatus like quantum circuits. Moreover, the result can also be extended to more complicated situations, such as two-photon interaction in a two-qubit system.

References

[1] Jaynes E T , Cummings F W . Comparison of quantum and semiclassical radiation theories with application to the beam maser[J]. 1963, 51(1):89-109.

[2] Werlang T , Dodonov A V , Duzzioni E I , et al. Rabi model beyond the rotating wave approximation: generation of photons from vacuum through decoherence[J]. Physics, 2008, 78(5):4948-4954.

[3] Schneeweiss P , Dareau A , Sayrin C . Cold-atom-based implementation of the quantum Rabi model[J]. Physical Review A, 2018, 98(2):021801.

[4] Ng H T , Burnett K . Vacuum Fluctuations induced Entanglement between Two Mesoscopic Systems[J]. New Journal of Physics, 2007, 10(12).

[5] Braak D . Integrability of the Rabi Model[J]. Physical Review Letters, 2011, 107(10):100401.

[6] Dicke, R. H . Coherence in Spontaneous Radiation Processes[J]. Physical Review, 1954, 93(1):99-110.

[7] Felicetti S , Rossatto D Z , Rico E , et al. Two-photon quantum Rabi model with superconducting

circuits[J]. Phys.rev.a, 2018, 97(1):013851.
[8] Yang C P , Chu S I , Han S . Simplified realization of two-qubit quantum phase gate with four-level systems in cavity QED[J]. Physical Review A, 2004, 70(4):044303.

IOP Conf. Series: Materials Science and Engineering **964** (2020) 012015 doi:10.1088/1757-899X/964/1/012015

Damping Characteristics of Three-dimensional Co-continuous Network Graphite / Cast Steel Composites

Shengshan Feng [1,2,*], Chunyi Zhan [1,2], Qingqing Bian [1,2], Weifeng Wu [1,2], Dong Huang [1,2], Xiaoming Zhang[1,2], Jianli Chen [1,2], Dunbai Zhou [1,2], Wenjiang Chen [1], Zebin Wei [1], Xihui Ling [1], Xiaoming Ou [1], Jinghui Li [1], Guiran Feng [1] and Junhua Qin [1]

[1]College of Mechanical and automobile Engineering, Guangdong Polytechnic of Science and Technology, Zhuhai, China
[2]Zhuhai Key Laboratory of Advanced Equipment Manufacturing and Material Processing, Zhuhai, China

*Corresponding author email: fss8172@gmail.com

Abstract. The infiltration casting method is used for the preparation of three-dimensional co-continuous network structure of graphite/cast steel composite material in atmospheric pressure. Using modal analysis method, flake graphite cast iron specimen and composite sample are compared about damping factors. The results show that amplitude attenuation of the composite material is more quickly than that of flake graphite cast iron specimens. Supported in two rigid points state, the first step damping ratio of the composite material is 2.1 times of flake graphite cast iron. At the same time, its tensile strength is 1.54 to 1.68 times that of flake graphite cast iron, its hardness reaches more than 86%, and its bulk density is reduced to less than 94%. The composite not only has good vibration damping and anti-friction performance, but also has comprehensive mechanical properties exceeding the level of flake graphite cast iron. It is possible to replace flake graphite cast iron for manufacturing important mechanical equipment castings such as fuselage and guide rail to obtain better anti-wear / vibration damping effect.

1. Introduction

In the field of mechanical equipment, now flake graphite cast iron is commonly used as the material for manufacturing important parts such as all kinds of fuselage and guide rail. Because of the existence of graphite, the flake graphite cast iron has good wear resistance and vibration damping at the same time, its damping attenuation coefficient is 6 to 10 times that of steel. However, with the rapid development of machining technique, all kinds of machine tools are developing towards high precision, high speed and heavy large size. The dynamic load of various important and heavy large mechanical equipment are getting larger and more complex, and related parts produce more and more serious vibration, fatigue and wear, ordinary flake graphite cast iron can no longer meet the requirements of modern machine tools. In order to reduce the harm, materials with high strength, damping and wear resistance should be used[1]. In recent years, non-metal matrix composites such as resin mineral composites, artificial granite composites and epoxy resin concrete filled structure have been developed. Although these non-metal matrix composites have good damping performance, they have low strength and poor thermal expansion property, making it difficult to be extended for the application[2-4]. Metal matrix composite is the main development of low vibration and noise material[5-6].

The three-dimensional continuous network structure metal matrix composite material is a research field of new composite material that has been paid more and more attention by material researchers at home and abroad over the past decade. The composite has a completely different spatial topology from the traditional composite material, that is, the metal matrix phase and the composite phase (or modified phase) are continuous (connected) in the three-dimensional space, and exhibit an interleaving network structure. The structure makes this material have very unique mechanical, physical and chemical properties. It also has isotropic properties, and has the potential to be used as anti-friction/ damping material[7-8]. We took the lead in studying the three-dimensional network structure silicon carbide / flake graphite cast iron atmospheric pressure casting composite at home and abroad, hoping that it has excellent wear resistance and good damping property, which can be used to replace flake graphite cast iron to make important castings such as fuselage and guide rails. The relative wear resistance of the three-dimensional network structure silicon carbide/flake graphite cast iron composite we prepared is as high as 4.7 times to 7.9 times of the flake graphite cast iron, its amplitude attenuation is faster than that of flake graphite cast iron, the first-order damping ratio is 2.3 times that of flake graphite cast iron, and its damping effect is obvious. However, mechanical properties such as strength is not as good as flake graphite cast iron, and it is difficult to meet the performance requirements of more and more developed and heavy mechanical equipment[9-10].

For this reason, we first proposed the following conception of research at home and abroad: using cast steel as the metal matrix phase and flake graphite as the composite phase (modified phase), by using the infiltration casting method to prepare the three-dimensional co-continuous network graphite / cast steel composites. The composite has good technical feasibility as a new anti-wear/vibration damping composite.

2. The Test Sample Preparation Method and Basic Performance

The conventional method for preparing the three-dimensional network structure ceramic / metal composite is pressure casting infiltration method at present. The metal liquid overcomes the permeation resistance under pressure, so that the casting infiltration can be completed. The technological process of pressure infiltration method is complicated, it is inconvenient to operate and control, and the production cost is high.

First, we prepare the three-dimensional co-continuous network graphite preform by using foam plastic precursor coating method, then coat the surface of graphite preform skeleton with metal, and prepare the three-dimensional co-continuous network graphite / cast steel composite by using the infiltration casting method. The method for preparing composite material by penetrating metal melt into the porous preform without exterior load under normal pressure can simplify the preparation technology, reduce the production cost, and easy to realize industrialization production.

In this study, the macro-morphology of the composite material specimen prepared by normal pressure casting infiltration method is shown in Figure 1. The preparation conditions are as follows: the pore size of the foam ceramic preform used is 8ppi, the surface is treated with Cu-Cr-Ti metallization coating, the grade of molten steel is ZG200-400, and the sand casting process of furan resin is used.

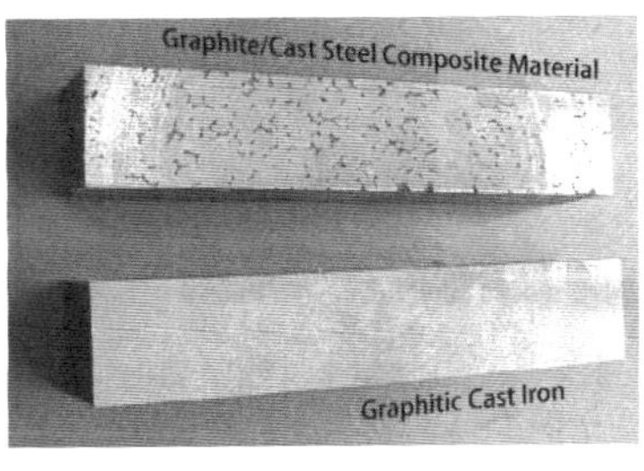

Figure 1. Macro-morphology of three-dimensional network structure of graphite/cast steel composite material and HT150 flake graphite cast iron specimen.

The basic performance of the prepared new three-dimensional network structure graphite / cast steel composite material specimen after heat treatment are shown in Table 1.Its tensile strength exceeds that of flake graphite cast iron and reaches more than 200MPa, which is 1.54-1.68 times of flake graphite cast iron; Its hardness is close to flake graphite cast iron, reaching more than 86% of it, and the bulk density is 6.7~6.8g / cm^3, which is reduced to below 94% of flake graphite cast iron.

Table 1. The base parameters of the composite material and HT150 flake graphite cast iron specimen.

Material type	Composites	Composites	Composites	flake graphite cast iron
Metallization process of foam	Untreated	Ni-Cr-Fe	Cu-Cr-Ti	--
Average tensile strength (MPa)	192	236	257	153
Hardness (HBW)	147~158	154~168	158~172	178
Average density (g/cm^3)	6.70	6.78	6.78	7.22

Note: (1) The diameter of the parallel working section of the tension test specimen is 20 mm, the nominal length is 60 mm, and the corner radius R is 25 mm.
(2) The heat treatment of the composite material specimen is normalizing (heating to 850°C for heat preservation for proper time and then air cooling), the purpose is to refine the structure of the steel and improve its strength and hardness.
(3) The heat treatment of the flake graphite cast iron specimen is annealing (heating to 500°C for heat preservation for proper time, furnace cooling to 200°C and then air cooling), the purpose is to remove thermal stress and increase strength. But the hardness is reduced.

3. Experimental Method for Vibration

We conducted a vibration comparison experiment on flake graphite cast iron specimen and three-dimensional continuous network graphite/cast steel composite material with the same geometric dimension using modal analysis to identify the modal parameters of the two material under the same support condition. In this way, the damping characteristics of the three-dimensional continuous network graphite / cast steel composites were evaluated preliminarily.

The modal analysis is a method of modal analysis of structure by microcomputer, the steps are as follows:

(1) Pulse or steady random simulation test was conducted on the structure to obtain the force signal by the force sensor, and obtain the vibration response signal at each measuring point by the vibration sensor;

(2) The force and the vibration response signal on the measuring point were parallel sampled, and converted into digital signal, which is stored in the microcomputer.

(3) The real part and imaginary part of the transfer function of each measuring point were calculated by the microcomputer, and the amplitude-frequency response curve was obtained;

(4) The amplitude-frequency response of the transfer function of each measuring point should be lumped average, and the frequency point of each peak was used as the estimated value of the modal frequency.

(5) The modal frequency and the modal damping ratio corresponding to the minimum value of the objective function E were found by the optimization method to calculate the modal shape.

The basic principle of modal analysis technique is as follows: the mechanical dynamic characteristics H(ω) and the external force Y(ω) and its corresponding response X(ω) have the following relationship: H(ω) = Y(ω) / X(ω). H(ω) is also called a mechanical transfer function or frequency response function in here. After discretization, the mechanical dynamic characteristics can be described by a n-order matrix differential equation:

$$M\ddot{x}(t) + C\dot{x}(t) + Kx(t) = f(t) \tag{1}$$

in which f(t) is the n-dimensional excitation force vector; $x, \dot{x}$ and $\ddot{x}$ are the n-dimensional displacement, velocity, and acceleration response vector respectively; M, K, and C are the mass, stiffness, and damping matrix of the mechanical structure respectively. They are usually real symmetric n-order matrices.

According to equation (1), the natural frequency and the corresponding vibration mode of the mechanical structure can be obtained. The vibration mode is arranged in column the mode vectors in matrix form (mode matrix) by column:

$$\Phi = [\phi_1, \phi_2, \ldots\ldots, \phi_n] \tag{2}$$

and can be obtained:

$$\Phi^T M \Phi = [m_r] \qquad \Phi^T K \Phi = [k_r] \qquad \Phi^T C \Phi = [c_r] \tag{3}$$

in which: m_r is generalized mass, k_r is generalized rigidity and c_r is generalized damping.

The modal parameters are used to express the frequency response function of the mechanical structure:

$$H(\omega) = \sum_{r=1}^{n} \frac{\phi_r \phi_r^T}{m_r[(\omega_r^2 - \omega^2) + j^2 \xi_r \omega_r \omega]} \tag{4}$$

Then the element of the ith row and j[th] column of the H (ω) matrix are:

$$H_{ij}(\omega) = \sum_{r=1}^{n} \frac{\phi_{ir} \phi_{rj}}{m_r[(\omega_r^2 - \omega^2) + j^2 \xi_r \omega_r \omega]} \tag{5}$$

In which: $\omega_r^{\ 2} = k_r / m_r$ is the rth modal frequency, $\xi_r = c_r / 2m_r \omega_r$ is the rth modal damping ratio and ϕ_r is the rth vibration mode.

From the above analysis, it can be seen that a frequency response function of a mechanical structure $H_{ij}(\omega)$ contains its information on the modal frequency and damping ratio. Therefore, if a typical frequency response function of a mechanical structure is measured, its modal frequency and damping ratio can be obtained from the frequency response function. The damping ratio refers to the ratio of the damping coefficient to the critical damping coefficient.

In this experiment, three supporting methods were used to test the damping characteristics of the material: the specimen state of supported over two rigid points (see Figure 2), the specimen state of supported over level soft foam (see Figure 3), and the specimen state of two points handing (see Figure 4).

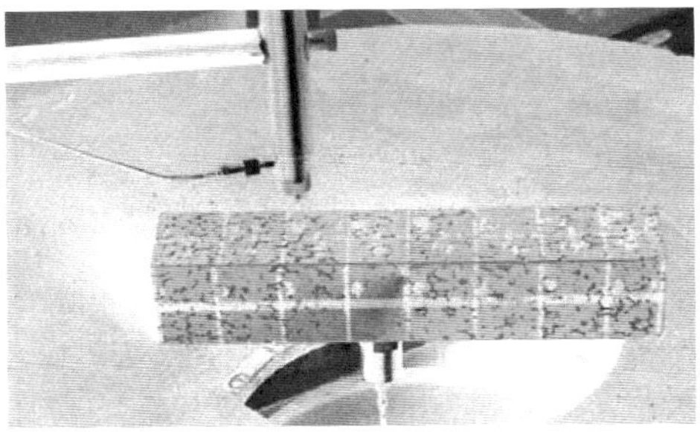

Figure 2. The specimen state of supported over two rigid points.

Figure 3. The specimen state of supported over level soft foam.

The principle block diagram of the vibration testing system is shown in Figure 5.

The frequency response function testing process is as follows: use a force hammer to excite the vibration at point A, and the force sensor on the hammer feels the force signal, which is amplified by the amplifier and input to channel 1 of the UT3316FS-ICP. At the same time, the acceleration sensor is used to measure the acceleration response at point B. The signal is amplified by the amplifier and input to channel 2 of UT3316FS-ICP, and then collected and analyzed by UT3316FS-ICP to obtain the

IOP Conf. Series: Materials Science and Engineering **964** (2020) 012015 doi:10.1088/1757-899X/964/1/012015

frequency response function from point B to point A. Finally, UT3316FS-ICP is used to curve fit the frequency response function to obtain the damping ratio.

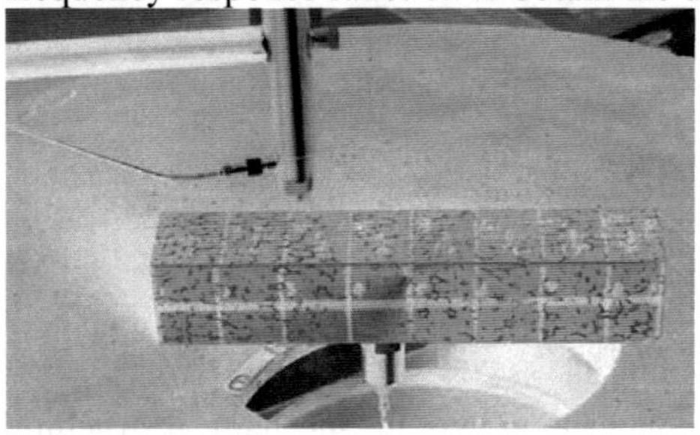

Figure 4. The specimen state of two points handing.

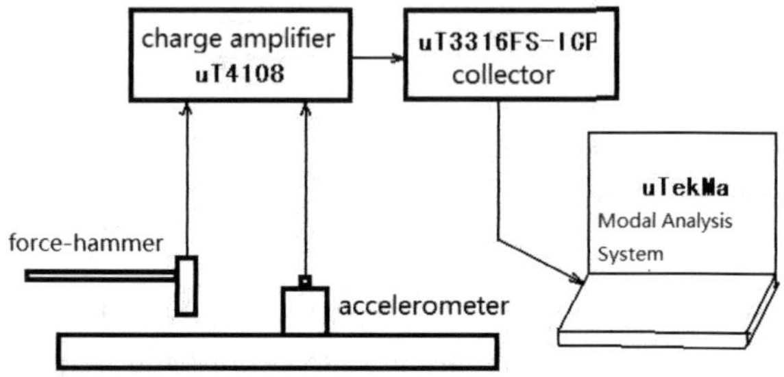

Figure 5. The principle block diagram of the vibration testing system.

4. The Experimental Results of Vibration

The measured damping ratio of the three-dimensional continuous network graphite/cast steel composites and flake graphite cast iron specimen is shown in table 2. The ratio of the damping ratio of the specimen is shown in Table 3.

Table 2. The modal frequency and damping ratio of the specimen.

Test status	**Material type**	**The first step**	**The second step**	**The third step**
supported over two points	Composites	0.2108	1.4324	0.7630
	Flake graphite cast iron	0.1004	1.4290	0.1004
supported over foam	Composites	0.1411	2.2789	0.3524
	Flake graphite cast iron	0.1176	1.8991	0.3204
two points handing	Composites	0.1807	1.0996	0.3366
	Flake graphite cast iron	0.1004	0.2972	0.1683

Table 3. The ratio of damping ratio (composite materials / flake graphite cast iron).

Test status	The first step	The second step	The third step
supported over two points	2.1	1.0	7.6
	1.2	1.2	1.1
two points handing	1.8	3.7	2.0

It can be seen from Table 2 and Table 3: under different supporting states, the damping ratio of the three-dimensional network structure graphite/cast steel composites specimen is larger than that of flake graphite cast iron.

Under free condition, the impulse response time history of each specimen is shown in Fig. 6.

Under the two points rigid supporting state, the amplitude attenuation of the composites is faster than that of the flake graphite cast iron specimen. The first step damping ratio is 2.1 times that of the flake graphite cast iron and the damping effect is more significant than the two points handing and foam supporting state. Since the modal frequency difference between the composites and the flake graphite cast iron is small, the difference in weight is not large, and the difference in stiffness is not large, so it can be concluded that the damping ratio increase of the composite material is mainly due to the material damping coefficient increase.

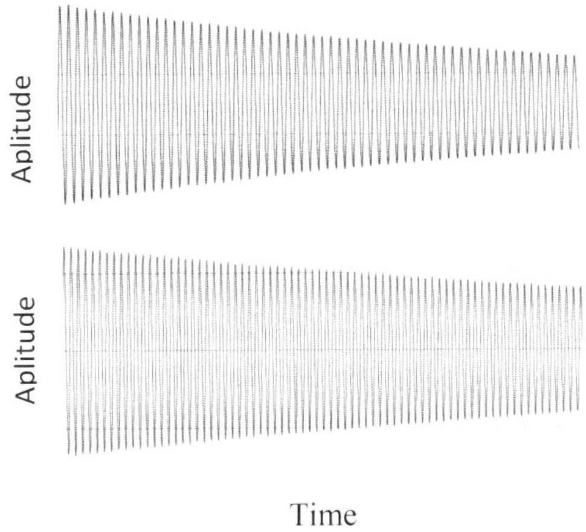

Figure 6. The impulse response time history chart of composite materials and flake graphite cast iron The picture above—composite materials the picture below—flake graphite cast iron.

5. Mechanism Analysis of Damping

Flake graphite cast iron has good damping performance because the flake graphite in the flake graphite cast iron cuts the metal matrix apart (see Figure 7), and when subjected to vibration, it will produce small plastic deformation to consume vibration energy. Moreover, the stiffness and elastic modulus of graphite are both smaller than that of metal, the interface of internal graphite and graphite/metal matrix causes sliding during vibration, and causes internal friction, which converts the vibration energy into heat energy, thereby consuming vibration energy. The factors affecting the damping size include the carbon content of cast iron and the shape of graphite, etc. The higher the carbon content, the longer and more coarse the flake graphite, the better the damping performance[11-12].

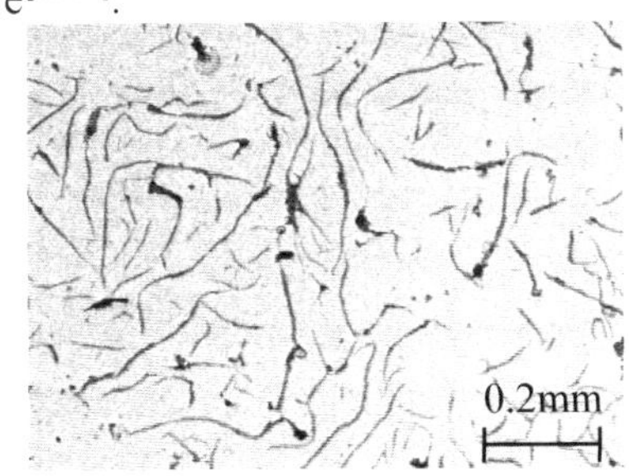

Figure 7. Graphite content and morphology in flake graphite cast iron (matrix is not corroded).

The internal structure of the three-dimensional co-continuous network graphite/cast steel composite material has the following two characteristics:

(1) The material has a three-dimensional continuous network structure formed by graphite network (see Figure 8).Its volume ratio is more than two times of flake graphite in flake graphite cast iron, and its graphite skeleton is longer and more coarse than flake graphite in flake graphite cast iron;

(2) The network graphite skeleton generates channels due to the ablation of the plastic precursor during molten steel pouring (see Figure 9).

Figure 8. The fracture surface morphology of composite.

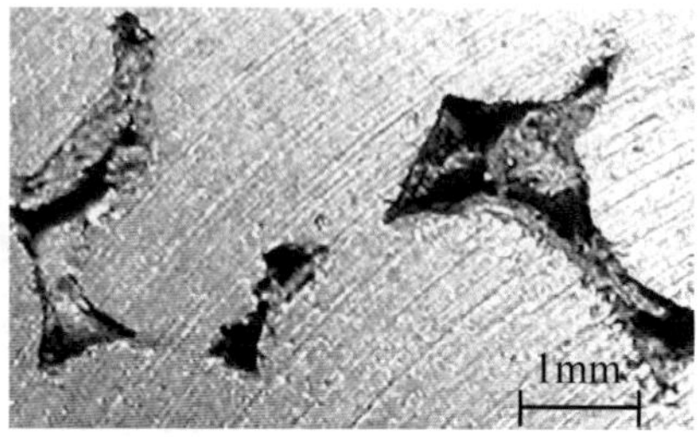

Figure 9. Morphology of the core channels of network graphite skeleton in composite.

The separation of the metal matrix caused by the graphite network, and the channels of the graphite skeleton caused by the ablation of the plastic precursor, can make the composite material easier to produce small plastic deformation when subjected to vibration, the interface of the interior of graphite and the graphite/metal matrix interface are easier to slip, which significantly hinders the propagation of vibration waves, making the three-dimensional co-continuous network structure graphite/cast steel composites absorb more vibration energy, thus showing a higher damping ratio.

6. Conclusion

(1) The damping ratio of the three-dimensional co-continuous network structure graphite/cast steel composite material is greater than that of gray cast iron. Supported in two rigid points state, amplitude attenuation of the composite material is more quickly than that of flake graphite cast iron specimens, the first step damping ratio of the composite material is 2.1 times of flake graphite cast iron, the damping effect is more significant than the two points handing and foam supporting state.
(2) The damping mechanism of the three-dimensional co-continuous network structure graphite/cast steel composite material is: the separation of the metal matrix caused by the graphite network in the composite material, and the channels of the network graphite skeleton caused by the ablation of the plastic precursor, can both effectively hinder the propagation of vibration waves.
(3) The tensile strength of the three-dimensional co-continuous network graphite/cast steel composite material is 1.54 to 1.68 times that of flake graphite cast iron, the hardness reaches 86% or more of flake graphite cast iron, and the bulk density reduces to 94% or less. It not only has good vibration and friction reduction performance, but also has comprehensive mechanical properties that exceed the level of gray cast iron. It not only has good vibration damping and anti—friction performance, but also has comprehensive mechanical properties exceeding the level of flake graphite cast iron. It is possible to replace flake graphite cast iron for manufacturing important mechanical equipment castings such as fuselage and guide rail to obtain better anti-wear / vibration damping effect.

Acknowledgments
In this paper, the research was supported by Science and Technology Planning Project of Guangdong Province of China (Grant No. 2015A010105035), the Science and Technology Innovation Strategy Project of Guangdong (Grant No. pdjh2020b0969), Guangzhou Science and Technology Plan Project (Grant No. 201707010172) and Guangdong Education Department Characteristic Innovation Project (Grant No. 2017GKTSCX016).

References

[1] Qiang Wenjiang, Wu Chengjian. *Metal material(3rd edition)* [M]. Beijing: the publishing house of metallurgical industry, 2016
[2] PAN Yuejian, LI Duanneng,WAN Zhenghai.Research on Vibration Damping Performance of High Speed Ceramic Milling Machine for Resin Concrete Filled Structure[J]. *Machine Tool & Hydraulics*,2020(7):64-67
[3] Wang Tao. Preparation and Property Study of Carbon Fiber Reinforced Resin Mineral Composite for Machine Tool[D]. *Shandong University*,2014

[4] Hu Yanhua. Study on Preparation Process and Key Properties of Artificial Granite Composites[D]. *Zhongyuan University of Technology*,2013
[5] Wang Jingfeng, Wei Wenwen, Pan Fusheng,etc. New Development and Prospect of Research on Metallic Damping Materials[J]. *Materials Review*,2009(13):15-19
[6] Li Ning, Teng Jing, Wen Yuhua, etc. Damping Alloy for Vibration and Noise Reduction and Its Engineering Applications[J]. *Functional Materials Information*,2009(4):16-19
[7] Dachn.G.S, Breslin.M.C. Co-continuous composite materials for friction and braking applications[J]. *Journal of the Minerals Metals and Materials Society*, 2006,58(11): 87-92
[8] Lavecchia.G.M. Co-continuous A1/Al2O3 composite produced by liquid displacement reaction relationship between microstructure and mechanical behavior[J]. *Journal of Materials Science*, 2003,(38):3567-3577
[9] Feng Shengshan,Wang Zejian, Liu Qingfeng,etc.Research Status of Three-dimensional Co-continuous Network Metal-ceramic Composites[J].*Development and Application of Materials*, 2009, 24(1): 60-68
[10] Wang Zejian. Preparation and characteristic of three-dimensional co-continuous network SiC ceramic / cast iron composite material[D]. *Hubei University of Technology*,2009
[11] Tang Hu, Peng Jinhua, Yang Yitao. Relationships among Internal Friction, Resonant Frequency and Microstructure of Cast Grey Iron with Pearlite Base[J]. *Shanghai Metals*,2011,33(4):23-26
[12] Peng Jinhua, Tang Hu, Chen Meng, etc. Measurement and Analysis on Low Frequency Internal Friction Characteristic of Gray Iron Specimen[C]. *China Foundry Conference 2010*, 2010

IOP Conf. Series: Materials Science and Engineering **964** (2020) 012016 doi:10.1088/1757-899X/964/1/012016

Positively Charged Gold Nanoparticle - Graphene Oxide Nanohybrid Modified Electrode for Electrochemical Detection of 5-hydroxytryptamine

Suxing Luo[1], Qingsong Mou[1], Xiaoxue Luo[2] and Yuanhui Wu[1,*]

[1]Department of Chemistry and Chemical Engineering, Zunyi Normal College, Zunyi, 563006, P. R. China
[2]Hainan Tropical Ocean University, Sanya, 572022, P.R. China

*Corresponding author email: yhwull@126.com

Abstract. In this study, positively charged gold nanoparticle - graphene oxide (AuNPs-GO) was synthesized and constructed a sensor for the detection of 5-hydroxytryptamine(5-HT). Various technologies were used to characterize the AuNPs-GO nanohybrid and the proposed sensor. The sensor showed a sufficient linear range from 0.5 to 60 μM 5-HT and the limit was 0.05 μM (S/N = 3). Furthermore, the sensor has been successfully utilized to determine 5-HT in real samples.

1. Introduction

5-Hydroxytryptamine (5-HT), plays the critical roles in controlling of endocrine secretion, mood, sleep, etc [1]. Hence, creating a rapid detection method for 5-HT would be meaningful in diagnosis of some diseases and elucidating the functions of 5-HT in some neurological disorders [2]. Until now, a lot of methods were developed to detect the concentration of 5-HT, including electrochemical sensors, liquid chromatography, chemiluminescence. Among these methods, electrochemical sensors which are fast response, low cost, high sensitivity, has attracted considerable attention for 5-HT determination.

With the advantage of specific surface area, strong mechanical strength and fast electrode reaction kinetics, graphene oxide (GO) has been widely developed to fabricate electrochemical sensors. The integration of zero-dimensional gold nanoparticle (AuNPs) with two-dimensional GO hybrid materials that combine the characteristics of each material as well as have their unique properties [3]. In addition, GO exhibits negative charge when dissolved in water. Although citrate-capped AuNPs which carry negative charges was extensively used currently, it is inappropriate for negative charged AuNPs to be integrated with GO, due to the electrostatic repulsion.

In this study, we constructed a novel AuNPs-GO/GCE sensor based on the GO- AuNPs (positively charged) for sensitively measuring 5-HT. CTAB-capped AuNPs, which always take positive charges were in-situ growth on the surface of GO. Consequently, the AuNPs-GO/GCE new sensor was constructed. At last, the fabricated sensor was successfully utilized to quantify 5-HT in real samples.

2. Materials and Methods

2.1. Experimental Materials

All chemical reagents in this work were analytical or guaranteed reagents. All the solutions used throughout were completely dissolved in HPLC water.

2.2. Synthesis of AuNPs (CTAB-capped) -GO Nanocomposite

The overall process of AuNPs -GO nanocomposite preparation was shown as follows [4]. Firstly, a 3.5 mL mixture containing $HAuCl_4 \cdot 4H_2O$ (3.5×10^{-4} mol/L) and CTAB (0.1 mol/L) was prepared. Immediately, 1.5 mL ice-cold $NaBH_4$ (0.01 mol/L) was added to the obtained solution. After being shaken violently for 20 seconds, the solution became light brown and was stored as a seed solution. Secondly, to in-situ synthesis of positively-charged AuNPs - GO nanocomposite, 80 µL CTAB (0.2 mol/L) and 80 µL $HAuCl_4{\cdot}4H_2O$ (2.4×10^{-2} mol/L) were mixed, followed by addition of 9.4 mL0.8 mg/mL graphene oxide dispersion. Then, 0.6 mL of 0.10 mol/L L-AA solution was added to this mixture. Subsequently, 52 µL Au seed solution mixed into the growth solution and was shaken for 10 s, and the solution turned brownish red. The solution was centrifuged and washed. The residue was heated at 60 ℃ for a period of 8 h.

2.3. Electrochemical Experiment Procedure

Prior to modification, bare GCE was treated according to the reference [5]. Then, 6 µL AuNPs -GO dispersion (0.8 mg/mL) was carefully deposited onto the surface of the treated GCE and dried naturally. The electrochemical tests were conducted using a 2273 electrochemical workstation. All experiments were carried at 25℃.

3. Results and Discussion

3.1. SEM Characterization of AuNPs-GO Nanocomposite

The surface morphology of the nanocomposite was investigated using SEM. As shown in Figure 1, after in-situ growth of AuNPs on GO, the AuNPs were homogeneously and densely attached to the surface of the layered GO sheets and the average size of AuNPs was about 12 nm. Also, EDX spectrum of the nanocomposite confirmed presence of AuNPs (not shown).

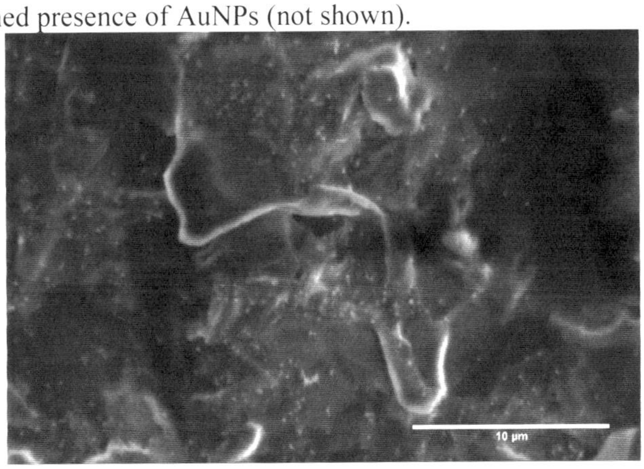

Figure 1. SEM images of AuNPs /GO.

3.2. EIS and Zeta Potential Characterization of AuNPs -GO/GCE

The zeta potentials were measured to study the surface charge of GO and AuNPs-GO (Figure 2). Meanwhile, EIS is a useful technique for characterization of interface property according to the information of impedance change. Therefore, the impedance changes of the different sensors (GO /GCE, Au NPs - GO /GCE) were recorded by EIS. The diameter of semicircle is directly proportional to the electro transfer resistance (Rct) in the Nyquist plots of EIS. As shown in Figure 3, GO /GCE had a large electron transfer resistance with a big semicircle domain, implying that GO had the larger obstruction effect. This should be ascribed to GO which was negatively charged. Also, the zeta potential of AuNPs /GO was found to be +24.6 mV due to the positive charged AuNPs. Thus, Au NPs -GO /GCE showed a relatively low Rct, due to the excellent electron transfer capability of AuNPs.

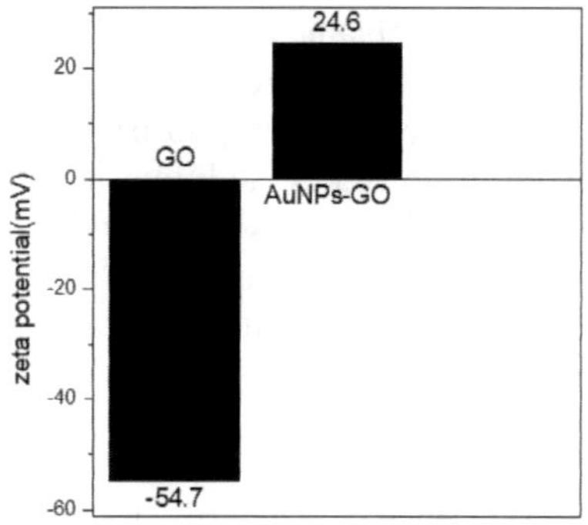

Figure 2. Zeta potentials of various materials.

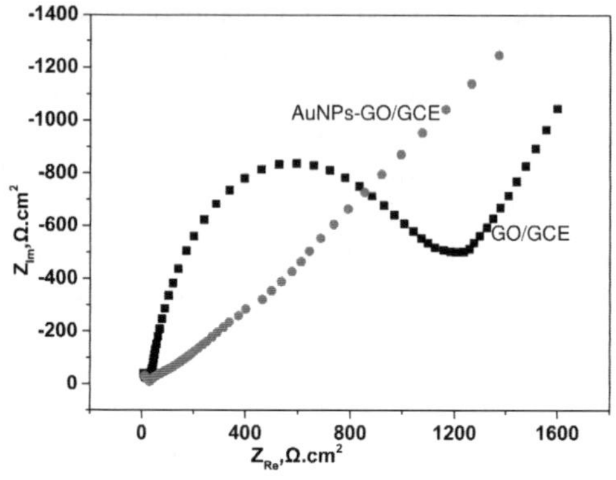

Figure 3. EIS of various sensors.

3.3. Optimization of Conditions for Sensor

Different influencing factors, including solution pH, and enrichment time of 5-HT would affect the properties of the sensor. DPV was employed to evaluate the electrochemical response of 5-HT under different conditions. In order to investigate the pH effect, 0.1 mol/L PBS was chosen as the buffer solution. As the pH of the PBS solution ranging from 5.2 to 8.1, the peak current response increased gradually and then decreased rapidly when increasing pH, and the current response reached maximum when the pH reached 6.2. Therefore, the PBS buffer solution with pH 6.2 was chosen as the supporting electrolyte.

The influence of accumulation time of 5-HT is also important for sensing performance which was evaluated from 30 to 240 s. The current of 5-HT increased while the accumulation time changing, maximum current was obtained at 210 s, which were adopted for optimal incubation time, implying the saturated sorption of 5-HT at the electrode surface.

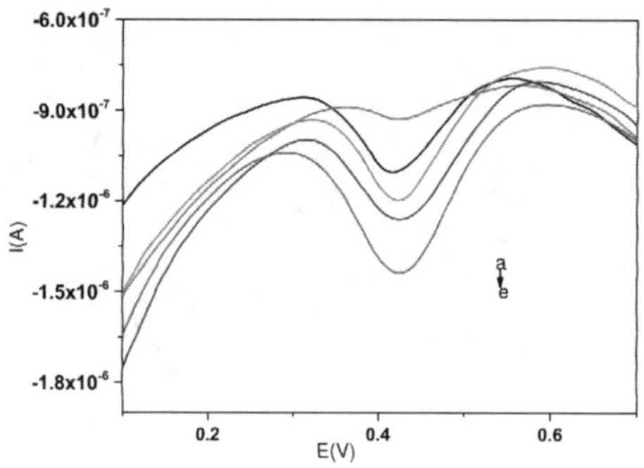

Figure 4. DPV responses for different concentrations of 5-HT on AuNPs-GO/GCE. (a–e) 0.06, 8, 13, 18, 30 μM, respectively.

Figure 4 showed amperometric responses gradually increased as the 5-HT concentration increasing from 0.06–30 μM. The calibration curve of the proposed sensor was I_P (μA)=-0.95-0.0169 C(μM), and a detection limit of 0.02 μM (S/N = 3).

3.4. Interferences and Applications

Structural analogues of 5-HT, including dopamine, epinephrine, norepinephrine, ascorbic acid, and various inorganic ions, were selected as competitors to examine the selectivity of the sensor. Based on the anti-interfered experimental results, 20-fold amounts of dopamine, epinephrine, norepinephrine, ascorbic acid exhibited interference in 5-HT sensing due to the structural analogues, but the peak current changes were all less than 5.2%. In addition, 50-fold amounts of Ca^{2+}, Mg^{2+}, Fe^{3+}, NO_3^-, SO_4^{2-} had little effect on the peak current of 5-HT. The reproducibility of AuNPs -GO/GCE was investigated. The measurements were performed at five different electrodes prepared under the same condition. The RSD of the current response for the individual measurement was 4.9%, demonstrating good repeatability of the fabricated sensor. According to the experimental results, the sensor had an acceptable selectivity and reproducibility for detecting 5-HT.

To evaluate the feasibility of AuNPs-GO/GCE sensor, the fabricated sensor was applied to detection of 5-HT in real samples. As 5-HT was not detected in the above samples, a recovery experiment was performed and the results are displayed in Table 1. The recoveries of this sensor were in the range of 94.0–102.0%, showing reliable detection of 5-HT in human serum samples.

Table 1. 5-HT sensing in serum samples (n=3) at AuNPs-GO/GCE.

Samples	Original	Added (μM)	Found (μM)	Recovery (%)
1	0	0.5	0.47	94.0%
2	0	5	4.74	94.8%
3	0	10	10.2	102%
4	0	20	19.7	98.5%

4. Conclusions

A new AuNPs-GO/GCE sensor was developed via in-situ synthetic method for quantitative analyses of 5-HT. In addition, AuNPs-GO nanocomposite is an excellent electrochemical sensing material which exhibited high sensitivity in detecting 5-HT. What is more, this sensor has been utilized to determine 5-HT in real samples, which is predicted to have wide applications in detecting 5-HT.

Acknowledgements

This work was supported by United Science Research Project of Guizhou Provincial Department of science and technology and Zunyi Normal College ([2017]7090 and [2014]17), Natural Science Research Project of Guizhou Provincial Department of Education (Qianjiaohe KY Zi [2018]004).

References

[1] Ressler KJ, Nemeroff CB (2000) Role of serotonergic and noradrenergic systems in the pathophysiology of depression and anxiety disorders. Depress Anxiety 12:2–19.

[2] Yang, Y. , Zeng, Y. , Tang, C. , Zhu, X. , Lu, X. , & Liu, L. , et al. (2018). Voltammetric determination of 5-hydroxytryptamine based on the use of platinum nanoparticles coated with molecularly imprinted silica. Microchimica Acta, 185(4), 219.

[3] Qaddare SH, Salimi A (2017) Amplified fluorescent sensing of DNA using luminescent carbon dots and AuNPs/GO as a sensing platform: A novel coupling of FRET and DNA hybridization for homogeneous HIV-1 gene detection at femtomolar level. Biosensors & Bioelectronics 89:773-780.

[4] Jian W, Li YF, Huang CZ, Wu T (2008) Rapid and selective detection of cysteine based on its induced aggregates of cetyltrimethylammonium bromide capped gold nanoparticles. Analytica Chimica Acta 626(1):37-43.

[5] Luo SX, Wu YH, Mou QS(2019) A thio-b-cyclodextrin functionalized graphene/gold nanoparticle electrochemical sensor: a study of the size effect of the gold nanoparticles and the determination of tetrabromobisphenol A. RSC Adv., 2019, 9, 17897

IOP Conf. Series: Materials Science and Engineering **964** (2020) 012017 doi:10.1088/1757-899X/964/1/012017

Solvothermal Synthesis and Characterization of Peanut-like $BiVO_4$

Jiangang Wei, Xiaoyu Wang, Hui Yang, Yuqin Mao and Luping Zhu*

School of Environmental and Materials Engineering, Research Center of Resource Recycling Science and Engineering, Shanghai Polytechnic University, Shanghai 201209, China

*Corresponding author email: lpzhu@sspu.edu.cn

Abstract. Peanut-like $BiVO_4$ were synthesized through a simple solvothermal method under the premise that the ratio of ethylene glycol (EG) to deionized water (DI) is 1:5, X-ray diffractometer was used to characterize the structure, scanning electron microscope was employed to characterize the morphology, and ultraviolet-visible diffuse reflectance spectroscopy were used to research optical properties of peanut-like $BiVO_4$. SEM images revealed that the as-synthesized $BiVO_4$ sample owned peanut-like and rods structures. XRD results revealed synthesized sample was monoclinic scheelite $BiVO_4$. The morphology of as-synthesized sample showed a shape of peanut-like. The strip width (Eg) of the $BiVO_4$ sample was 2.46 eV.
Keywords: Peanut-like; $BiVO_4$; Morphology.

1. Introduction

In the past decades, semiconductor nanosamples has gained widespread interest from the perspective of solving energy crisis and environmental remediation. People have made a lot of efforts in otocatalyst materials, such as cadmium sulfide (CdS)[1,2], tungsten oxide (WO_3)[3], zinc sulfide (ZnS)[4-8] and bismuth vanadate($BiVO_4$)[9,10]. Among these sampless, $BiVO_4$ has a moderate forbidden band width (2.4 eV), corresponds to the 516 nm wavelength in the visible region, where is close to the center of the solar spectrum. With a series of properties like non-toxic, good stability, low cost, and good photocatalytic performance, $BiVO_4$ has become a substitute for traditional sampless, and one of the most promising sampless.

In this work, under the premise that the total volume of EG and DI is unchanged, Herein, we have successfully prepared peanut-like $BiVO_4$ nanorods through a simple solvothermal method.

2. Experimental Sections

The peanut-like $BiVO_4$ were synthesized via a simple solvothermal method reported elsewhere [11]. Bismuth nitrate pentahydrate ($BiVO_4 \cdot 5H_2O$) and ammonium metavanadate (NH_4VO_3) were used as bismuth source and vanadium source, respectively. In this typical synthesis, 1.94 g $Bi(NO_3)_3 \cdot 5H_2O$ was added to 5 mL EG, and 0.47 g NH_4VO_3 was dissolved in 25 mL DI, relatively, and named them solution A and solution B, separately. After full magnetic stirring, solution B was added to solution A and stirred for 30 min. Then the mixed solution was transferred into a 100 mL Teflon-lined hydrothermal reactor and heated to 180℃ for 24 h. Ultimately, the yellow powder was filtered and washed with ultra-pure water and anhydrous ethanol three times, separately. Finally, the product was dried at 80℃ for 12 h.

The final product was characterized by XRD (Rigaku D/max 2500), SEM (Hitachi S-4800) and UV-vis DRS (Shimadzu 2550).

3. Result and Discussion

XRD pattern was used to determine the crystallinity and structural phase of the sample. As shown in Figure 1, all peaks of the $BiVO_4$ product's XRD pattern can be indexed to the monoclinic $BiVO_4$ (JCPDS no 14-0688). No extra peaks were observed, this reveals actuality that the conformation of monoclinic $BiVO_4$.

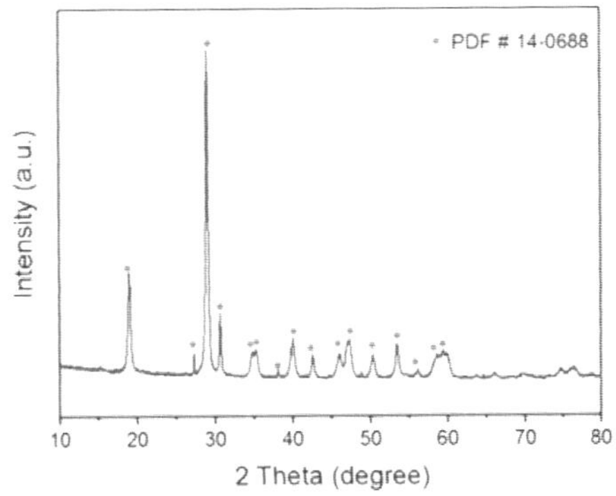

Figure 1. XRD pattern of BVO_4 product.

The morphology of product was shown in Figure 2. Fig 2a shows that as-prepared sample was made up of a large number of structures with good dispersion. Medium magnification picture (Figure 2b) displays obtained sample has a peanut-like architecture. Figure 3c displays that the average diameter of $BiVO_4$ sample is about 2.8 μm.

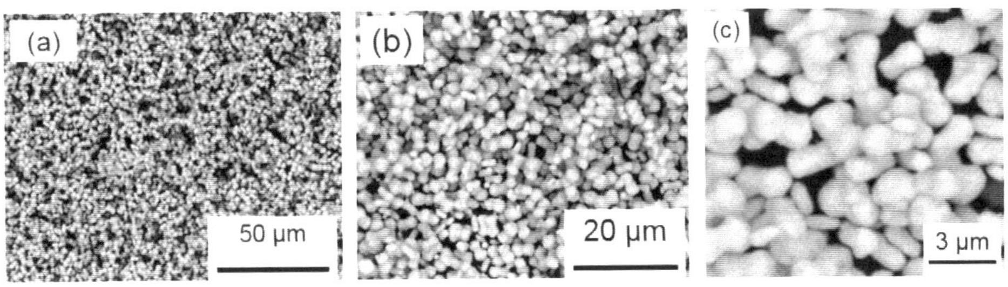

Figure 2. SEM image of prouct: (a-c) Low, medium and high magnification.

Figure 3 shows the sample's UV-vis DRS. Figure 3a presents that the as-synthesized $BiVO_4$ show great absorption in the visible light range of 450-550 nm. It proves that the $BiVO_4$ has excellent adsorption of visible light, causing the separation of photogenerated carriers. Figure 3b displays the plots of $(\alpha h\nu)^2$ versus photon energy ($h\nu$), where ν, h and α correspond to the frequency, Planck constan and tabsorption coefficient of the sample, separately[12]. Eg value of obtained sample can be evaluated to be about 2.46 eV.

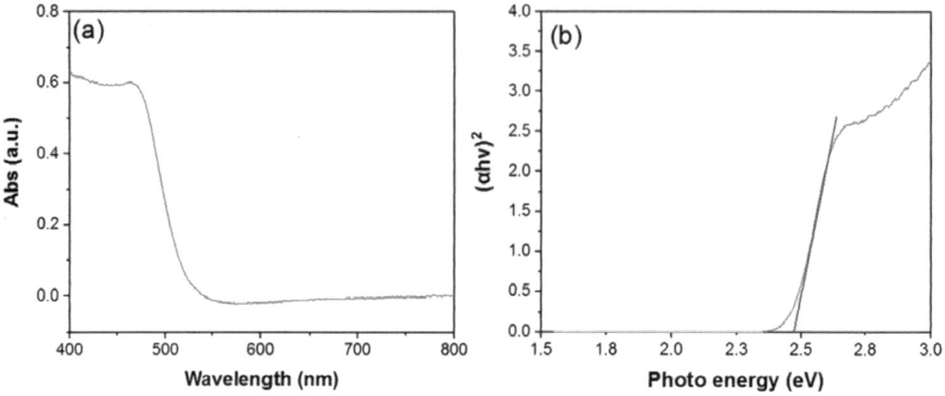

Figure 3. (a) UV-visible spectra of the obtained $BiVO_4$ sample, (b) the Taco plot for the sample.

4. Conclusion

In summary, peanut-like $BiVO_4$ was obtained successfully through a facile solvothermal route. The obtained $BiVO_4$ displays a nice peanut morphology, and a narrow bandgap. Furthermore, the preparation method reported in this study can also be applied to other vanadates' preparation by regulating the relevant synthetic conditions.

Acknowledgements

This study was co-sponsored by the following four institutions: one of them is Shanghai Municipal Natural Science Foundation (No: 18ZR1415700), the second is the Postgraduate fund (Shanghai Polytechnic University) (Nos: EGD18XQD26 and EGD18YJ0049), and the other one is key subject of Shanghai Polytechnic University (No. 4: Material Science and Engineering, XXKZD1601), the Gaoyuan Discipline of Shanghai-Environmental Science and Engineering (Resource Recycling Science and Engineering) is the last one.

References

[1] A.F. Qasrawi, and T.Y. Abed, Structural and optoelectronic properties of CdS/Y/CdS thin films[J]. Thin Solid Films 679. JUN.1(2019) 72-78.

[2] L.I. Bendavid, A.O. Atsango, and R.W. Smith. Interfacial properties of pure and doped CdS/graphene composites: CdS(0001)/graphene and a CdS/graphene bilayer[J]. Computational Materials ence. 177 (2020) 109537.

[3] M. Yang, J. Li, G. Ke, et al. WO_3 homojunction photoanode: Integrating the advantages of WO_3 different facets for efficient water oxidation[J]. Journal of Energy Chemistry. 56 (2021) 37-45.

[4] G. Zhu, J. Yang, C. Bao, et al. Organic–inorganic hybrid ZnS(butylamine) nanosheets and their transformation to porous ZnS[J]. Journal of Colloid & Interface ence. (2016) 136-144.

[5] F. Li, Z. Xi, Q. Liu, Controllable Synthesis and Optical Properties of ZnS: Mn^{+}/ZnS/ZnS: Cu^{2+}/ZnS Core/Multishell Quantum Dots toward Efficient White Light Emission[J]. ACS Applied Materials & Interfaces. (2017).

[6] X. Li, X. Li, B. Zhu, et al. Synthesis of porous ZnS, ZnO and ZnS/ZnO nanosheets and their photocatalytic properties[J]. Rsc Advances. 7 (2017) 30956-30962.

[7] H. Yang, D. Li, L. Jiang, et al. Preparation and Properties of ZnSlneinforced ZnS/ZnSe Composites[J]. Crystal Research and Technology. (2020).

[8] H. Tao, P. Choi, Q. Liu, et al. Chemical Functionalization of ZnS: A Perspective from the Ligand–ZnS Bond Character[J]. Journal of Physical Chemistry C. (2019).

[9] T. Ahmed, H. Zhang, Y. Gao, et al. Surfactant-free synthesis of m-$BiVO_4$ nanoribbons and enhanced visible-light photocatalytic properties[J]. Materials Research Bulletin. 99(MAR.) (2018) 298-305.

[10] G. Zhan, Z. Yao, W. Ng, et al. Economic production of monoclinic bismuth vanadate from waste vanadium ions: Process design and cost-benefit analysis[J]. Journal of Cleaner Production.

240(Dec.10) (2019) 118188.1-118188.11.

[11] A. Tayyebi, T. Soltani, H. Hong, et al. Improved photocatalytic and photoelectrochemical performance of monoclinic bismuth vanadate by surface defect states ($Bi_{1-x}VO_4$)[J]. Journal of Colloid and Interface ence. (2017) 514-565.

[12] A. Tayyebi, T. Soltani, and B.K. Lee, Effect of pH on photocatalytic and photoelectrochemical (PEC) properties of monoclinic bismuth vanadate[J]. Journal of Colloid and Interface ence. 534 (2019) 37-46.

Study on Properties of Waterborne Polyurethane Acrylate Modified Epoxy Acrylate Resin

Hongbo Liu, Xiaowu Xue, Jingjing Wei, Wangdong Xiao and Jing Bai*

School of Applied Chemistry and Biological Technology, Shenzhen Polytechnic, Shenzhen 518055, People's Republic of China

*Corresponding author email: baijing@szpt.edu.cn

Abstract. Polyurethane acrylate modified epoxy acrylate resin with high solid content and great mechanical properties was synthesized by addition of polyethylene glycol in this work. The conversion process from liquid to solid was confirmed by comparing the FT-IR of oligomer and film. The disappearance of the carbon-carbon double bond absorption peaks at 1639 cm^{-1} and 819 cm^{-1} indicates that the oligomer is cured completely. In addition, the thermal analysis of the resin was characterized by differential scanning calorimetry (DSC) and thermogravimetric analysis (TGA). According to the experimental data, it can be seen from that the resin started to decompose at 293.31 until it is almost completely decomposed at 481.73°C which indicated that the stability of the film was better and the decomposition temperature was higher. In addition: Compared with the thermosetting resin E51 and the modified resin, the glass transition temperature is reduced by about 31 °C, which solves the problem of brittleness and hardness.
Keyword: Polyurethane; Modified; Waterborne epoxy acrylate; UV-curable; Properties.

1. Introduction

Epoxy resin is thermosetting resin material with some characteristics like high hardness, good adhesion and good chemical resistance. It was synthesized in 1947 and has been extensively used in various fields until now. It is commonly used in composite materials, adhesives, steel anticorrosion and other industries. Solvent-based epoxy resin emulsion refers to a stable system constituted by the uniform distribution of epoxy resin in organic solvents. A large number of organic compounds in this emulsion are easily volatile into the atmosphere which will cause atmospheric pollution. With the strengthening of people's environmental awareness and national policies, solvents Type epoxy resin can no longer meet the current need, at this time, waterborne epoxy resin came into modern society [1-6].

Aqueous epoxy resin refers to a stable state obtained by uniformly dispersing epoxy resin in the form of droplets or particles in a system where the dispersion medium is water. Waterborne epoxy resin is a non-toxic, harmless and pollution-free environmentally friendly coating. Depending on different preparation methods, the preparation methods of waterborne epoxy resin are mainly divided into mechanical method, chemical modification method and reverse conversion method. The advantages of mechanically preparing water-based epoxy resin emulsion are simple process and low emulsifier content, but the disadvantages of emulsion are also obvious, such as a large particle size, irregular particle shape, poor stability, and poor film-forming performance [6-8]. The size of the dispersed phase particles in the aqueous epoxy resin emulsion prepared by the chemical modification method is unimportant, but the disadvantage is that the preparation steps are difficult to control and the cost is high. With the improvement of environmental awareness and national environmental regulations, ordinary coatings can no longer meet people's needs and special coatings gradually appear in people's field of vision [2,3].

UV-curing technology has been widely used in the coatings industry because of its advantages of energy saving, environmental protection and high efficiency [4-8]. As a special coating, UV-curable waterborne resin and its coatings are gradually appearing in people's lives. Considering the issues of energy conservation, environmental protection, UV-curing conductive coatings have received extensive attention from scientists and become one of the hot research topics in the coatings industry [7-11].
In this paper, epoxy resin E51 is used as the main body, and it is modified by polyurethane acrylate with a hydrophilic segment to obtain an environmentally friendly water-based resin with a lower glass transition temperature and good thermal stability.

2. Experimental

2.1. Materials
Epoxy resin (EP-E51, g/eq 210-244), Baling Petrochemical Branch of Sinopec Group Asset Management Co., Ltd. N, N-dimethylbenzylamine (DMBA), DaMao Chemical Reagent Factory.polyethylene glycol (PEG), Sinopharm Chemical Reagent Co., Ltd. Diethanolamine (DEA), polyethylene glycol (PEG-400), 2-hydroxyethyl acrylate (HEA), acrylic acid (AA), Shanghai Maclean Biochemical Technology Co., Ltd. Dibutyltin dilaurate (DBTDL), Dongguan Sparta Chemical Co., Ltd. 2,4-toluene diisocyanate (TDI), photoinitiators (Darocur 1173) IRGACURE BASF, 1-hydroxycyclohexyl phenyl ketone (photoinitiators 184) IRGACURE BASF. Acetic acid (HAc), phosphoric acid, Nanjing Chemical Reagent Co., Ltd. 4A Molecular Sieves were used to remove water from polyethylene glycol in this study.

2.2. Synthesis of Waterborne Resin
Initially, PEG 400 (1mol) and 2,4-TDI (2 mol) reacted using dibutyltin oxide as catalysis at room temperature under nitrogen atmosphere without any solvent. The content of -NCO during the reaction of PEG and TDI was measured through anti-titration, which could help us know that the help us to know the advancement of the entire experiment. In the first stage we get double-ended polyurethane(T1-P4). The 1:1mole ratio of HEA/ T1-P4 was used to obtain mono-terminated line like mono-isocyanate (T-P4) and monitor the progress of reaction using the same method as when synthesizing T1-P4. Lastly, the reaction mixture with T1-P4 and EAM (Obtained by EP-E51 and acrylic to 1:1) in the molar ratio of 1.01:1, dibutyltindilaurate as the catalyst for this step. The mixture was kept at 75°C for 1.5~2.0 h until the -NCO cannot be found and the nonionic waterborne resin (TP-EP) was obtained. However, Due to the poor hydrophilicity of TP-EP, further operation to be made by mixed with DEA in 80°C (0.5-1h) and HAc in 60 °C (0.5h) in sequence, the target resin of waterborne polyurethane acrylate modified epoxy acrylate resin (WTP-EP) was obtained[12,13]. Theoretically, the solid content of the resin reaches 80% before being water-based and the solid content reaches 65% after using water as a solvent to reduce viscosity.

2.3. Preparation of UV-cured Films
The waterborne resin WTP-EP was synthesized through the previous section 2.2. Mix the synthetic resin and photoinitiator according to Table 1 and add water to reduce viscosity. Apply the waterborne resin mixture to the surface of the polytetrafluoroethylene plate, remove water from the resin after flashing, and then irradiate the resin on the polytetrafluoroethylene plate for 8-10s under ultraviolet light [13].

Table 1. The composition of UV-curable liquid formulations (wt.%).

Sample	WTP-EP	H_2O	Darocur 1173	184
E51-400	65	30	4	1

2.4. Infrared Spectra
The blank background was prepared by KBr tableting method and scanned as a background file, the sample was coated evenly on the KBr wafer and dried under dried at low temperature environment. The Nicolet 6700 Fourier Infrared Spectrometer (FT-IR) was used for testing, with a scanning range of 4000 cm^{-1} to 400 cm^{-1}.

2.5. Gel Content

The cured film is made into a 3cm*3cm*1mm size and record the mass of the cured film. It was immersed in acetone for 48h and then dried in a vacuum oven at 90°C for 6h, and weighed the quality. Calculate the resin gel rate according to calculation formula 1.

$$\omega(\%) = \frac{m_1}{m_0} \times 100\%$$

ω the gel content
m_0 the resin weight before experiment
m_1 film weight after extraction and drying

2.6. Thermogravimetry (TG)

The thermodynamics of the coating film were analysed using a TA-Q50. Take about 7-13 mg of sample in a platinum crucible and heat to 480°C at a rate of 8°C/min under nitrogen environment.

2.7. Differential Scanning Calorimeter (DSC)

The glass transition temperature of the film is analysed by TA-Q200. Under a nitrogen atmosphere at a heating rate of 5°C/min, the temperature is increased from -20°C to 300°C, and the cycle is repeated twice.

3. Results and Discussion

3.1. Infrared Spectroscopy

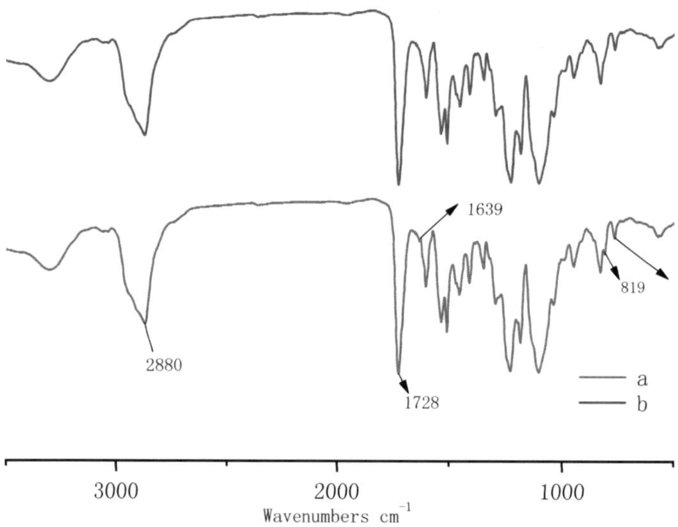

Figure 1. The FTIR spectra of oligomer (a), UV-cured film (b).

The FTIR spectra of the oligomer before UV-cured (a) and UV-cured film after radical UV-cured (b) are shown in Figure 1 respectively. After the addition of the free radical photoinitiators Darocur 1173 and 184, the unsaturated double bond radical polymerization of the epoxy acrylate is initiated after brief UV irradiation. In the final product synthesis, an epoxy bond is subjected to a ring-opening reaction with a diethanolamine epoxy group to introduce a hydrophilic group (hydroxyl group, amino group), and to retain as much epoxy group as possible in each resin molecule. It can be seen from Figure 1 that the prepolymer has a characteristic peak of epoxy at 981 cm^{-1} and 773 cm^{-1}, which indicates that DEA retains a small amount of epoxy bond after reacting with the remaining epoxy bond, avoiding the occurrence of DEA excess phenomenon. It can be seen from the comparison of the infrared spectrum of the prepolymer and the photocurable coating film that the characteristic absorption peak of the double bond of the cured film at 1639cm^{-1} and 819cm^{-1} disappears obviously, which indicates that the free radical photoinitiator initiates the prepolymer.

After the polymerization, the prepolymer is sufficiently polymerized; after the addition of the radical photoinitiator 184, the ring-opening polymerization of the epoxy group occurs in the action of the epoxy group, which also indicates the synergistic effect of the two photoinitiators. Since the polymerization is enough, the curing speed is fast and sufficient under the action of the two photoinitiators. The characteristic absorption peak of the double bond disappeared at 1639 cm^{-1} and 819 cm^{-1}, and the characteristic absorption peak of the epoxy group at 981 cm^{-1} disappeared, indicating that the curing was complete.

3.2. Thermal Decomposition

Figure 2 shows the results of thermal decomposition test of the aqueous resin cured film. At 273.31°C to 315 °C, the weight loss caused by the residual moisture and volatilization of small molecules in this stage. The quality decrease between 314 and 400°C is mainly caused by the decomposition of the hard segment in the polymer side chain and the range of quality decrease is caused by the decomposition of the soft segment in the polyether polyol and the breakage of the molecular main chain; after 400°C The thermal weight loss is mainly the decomposition of crosslinks. At a temperature of 481.73, the cured film is almost completely decomposed. The film shows better thermal stability because the maximum weight loss temperature of the cured film is higher than 370°C.

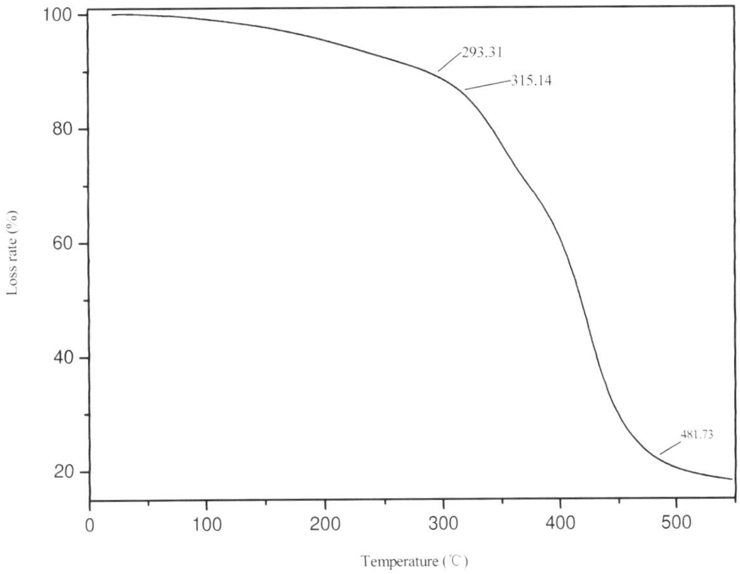

Figure 2. Thermal decomposition temperature curve of waterborne resin.

3.3. Glass Transition Temperature

Figure 3 show the differential scanning calorimetry (DSC) of cured films of the modified epoxy resin WTP-EP and EP-51. We can explain as follows: Due to the flexible urethane acrylate segment on the side link of the rigid epoxy acrylate monoester, compared with the rigid epoxy acrylate cured film, the aqueous resin cured film of this test has reduced brittleness, enhanced flexibility, and cured film. The glass transition temperature also decreased and the cured film had a glass transition temperature of 58.74 °C.

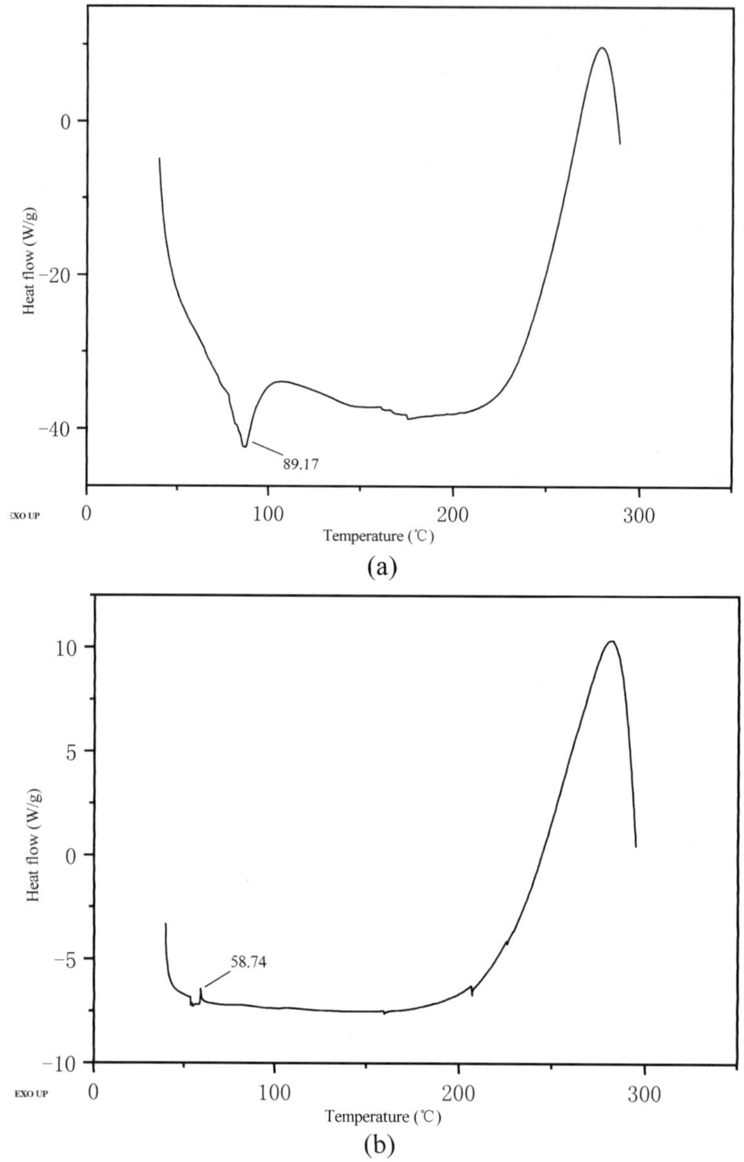

Figure 3. Glass transition temperature curve of cured films of EP-E51(a) and WTP-EP (b).

3.4. Gel Fraction

The photoinitiator was added to the water-based resin, and a solid film was formed after irradiation under ultraviolet light. The gelation rate of the cured film reached 96.65%, which indicates that the aqueous resin can be effectively crosslinked under the action of the photoinitiator, forming a relatively complete Network cross-linking structure.

4. Conclusion

The polyurethane acrylate modified epoxy acrylate waterborne resin with great physical properties was successfully synthesized from EP-E51, 2,4-TDI, PEG400 and HEA in this study. The pre-polymer was utilized in preparation of green environment-friendly UV/energy-saving quickly curing with great properties. After compared the cured film conforms with intended design, the resin is in line with the expected results in the FT-IR. The cured coating film has a great thermal decomposition temperature, a lower glass transition temperature compared with the EP-E51.

Acknowledgements
This paper was financially supported by the Shenzhen Basic Research Project (JCYJ20170818 114324998).

References
[1] Yuan X, Zhu B, Cai X, et al. 2017 *Journal of Applied Polymer Science* **17** 134
[2] Sheng J G, Zeng P, Shan Y D. 2013 *Advanced Materials Research* **815** 547
[3] Qiu S, Cheng C, Cui M, et al. 2017 *Applied Surface Science* **407** 213
[4] Liu T, Li J Y, Li X Y, et al. 2019 *Progress in Organic Coatings* **128** 137
[5] Gu Y, Huang Y, Liao B, et al. 2000 *Journal of Applied Polymer Science* **76** 690
[6] Zhang J, Xu H, Hu L, et al. 2017 *ACS Omega* **2** 7546
[7] Krylova I 2001 *Progress in Organic Coatings* **42** 119
[8] El-Ghaffar M, Kantouch F A, Hashem A I, et al. 2014 *Pigment and Resin Technology* **43** 84
[9] Krylova I. 2001 *Progress in Organic Coatings* **42** 119
[10] Wang Y, Liu F, Xue X. 2015, *Progress in Organic Coating* **78** 404
[11] Arsu N, Davidson R S, Holman R.1995 *Journal of Photochemistry and Photobiology A Chemistry* **87** 169
[12] Liu H B, Zhang Y Y, Lin f, et al. 2013 *Applied Mechanics & Materials* **447** 1169
[13] Liu, H B, Xue X W, Zhang W Y, et al. 2020 *Materials Science and Engineering* **770** 012062

IOP Conf. Series: Materials Science and Engineering **964** (2020) 012019 doi:10.1088/1757-899X/964/1/012019

A Study on Colour Print Quality Attributes Variance of OMR between PET and ABS by Digital Inkjet Printing

Xiaoling Lu[1,*] and Changlang Chen[2]

[1]Art and Design College of Jiangsu University of Technology, China
[2]Department of Graphic Communication Arts, National Taiwan University of Arts, Taiwan

*Corresponding author email: 2002500042@jstu.edu.cn

Abstract. In-mould transfer printings are limited, such as surface bending, only on plastic parts, cost higher and print for a small run length of productions. OMR transfer can be transferred to plastic or metal, transfer process accurate register, high quality and tolerance. The current printing market tend to develop personalized, so this study to use digital inkjet printing, through rigorous experimental process and quantitative measurement and comprehensive analysis, summarized the most suitable digital inkjet printing methods and the best quality attributes. In each combination, the samples will be selected to transfer printed products which will be examined for print quality attributes (SID, TVI, PC, colour difference and colour gamut). This study results found that when the higher coverage and the lower pass of inkjet printing, the SIDs will become higher as well. When transferred to ABS, SIDs in each combination will increase. The higher coverage printed, the higher TVI. After transferred into ABS, each combination of PC is worst. When transferred to ABS, colour variances of each combination are different from orange and cyan colour. Through this study, we can get the colour correspondence between the original and the finished product ABS, to establish the personalized service process, to improve the digital inkjet printing and technology application to enhance its product added value, and to provide research results to give back to the printing market. It can bring new opportunities for the industry.

1. Introduction

With the advancement of society (especially the rapid development of 3C and the automotive decoration industry), the use and function of plastics are becoming more diversified, consumers are constantly demanding the pattern and texture of the surface of the products, and various production technologies and equipment are changing with each passing day. Under the pressure of survival competition, the production technology, cost and production cycle are facing serious challenges, such as improving product quality, reducing defective products, reducing processing, assembly time and cost. The printing method that can print the most texture and the best effect on irregular objects has been widely used in industrial production. It can make up for the lack of in mould roller (IMR) production process. Due to the wide application field of the Outside mould (OMD) transfer technology, the process stable is better than the water transfer printing, the beautiful appearance of the product, and the good economic benefits, it has gradually been favored by many manufacturers.

OMR is now using Gravure and Screen Printing to print Polyethylene terephthalate (PET) film, however, there are still some problems with these two technologies. For example, the process of Gravure Printing is more complex and expensive. Screen Printing is cheaper but the process is also complex. These two kinds of printings are both unable to create personalized work, while Digital Inkjet Printing can.

Therefore, this research is to investigate the potential use of combining OMR and Inkjet Printer in 3D decoration. Digital Inkjet Printing can print precision colour products without the need for platemaking, compatibility, ease of operation [1]. It is the trend of the printing [2]. This research is mainly to expand the application of the results of the in mould transfer, through the digital inkjet printing combined with the process of the mould transfer, the process is standardized, and the literature analysis and quantitative experiments are supported to derive the best. It's bringing about the advancement of digital inkjet printing and mould transfer technology, the improvement of printing quality and customer satisfaction.

2. Literature

2.1. Digital UV Inkjet Printing

Digital Printing uses digital technology to personalize various documents and materials, and uses the prepress system to transmit various types of graphic information directly to the digital printing machine through the network, and prints a printing of the product. The computer uses the digital format to print the information of the archived content on the printed material. Digital printing is mainly referred to as Non-impact printing (NIP) [3]. Its turnover has gradually increased and it has become a printing in recent years. It's the new trends in the media.

Digital UV printing is a very potential printing method. It is widely used for printing. UV-dried inks can be used in a wide variety of different printed materials. This ink enables faster production because it is polymerized using ultraviolet light for instant drying. It is different from traditional ink structures and is mainly used for the printing of non-absorbent substrates, such as plastic and metal sheets. At present, with the advancement of technology, LED-UV light sources have been successively adopted.

2.2. Type of Transfer Printing

In printing technology applications, transfer printing technology, such as non-planar products, special materials, is required. This study is one of transfer printing, Transfer printing decoration technologies had including thermal transfer printing, water transfer printing, in mould roller /reprint (IMR) / outside mould roller (OMR), in mould label (IML), in mould film/forming (IMF) [4].

2.3. Print Quality Attributes

The print quality attributes are a general term for quantitative analysis of images, and may include colour print quality attributes and physical print quality attributes. Physical print quality attributes are physical testing of products such as weather resistance, water resistance, wear resistance...etc. Colour print quality attributes are print colour testing of products such as solid ink density(SID) of cyan(C), magenta(M), yellow(Y) and black(K), tone value increase (TVI), print contrast (PC), colour difference (ΔE), colour gamut, etc. [5].

3. Research Methods

In this study, the True Experimental Method was used to investigate the suitability of the digital inkjet printing process for the combination of OMD printing. After the inkjet printing of the PET transfer film, the printed colour quality attributes exhibited by the OMD transfer to the printed material -Acrylonitrile butadiene styrene (ABS) are transferred. Digital inkjet equipment adopts Mimaky's roll to roll flat-bed UV inkjet, which is one of the main equipment that can simulate the beautiful quality of printing [6]. The experiment is to continue the optimum combination mode of the film coating release agent and the hardened layer, using two resolutions (720 dpi, 1440 dpi), three kinds of printing passes (8 pass, 16 pass, 32 pass), two kinds Ink coverages (100%, 120%) combination, in addition to analysing the colour quality attributes exhibited after inkjet printing, and examining the printing colour quality performance and physical attributes test. They are 12 experimental combinations from A to L, such as A (720dpi, 8pass, 100coverage), B (720dpi, 8pass, 120coverage), G (1440dpi, 8pass, 100coverage), L (1440dpi, 32pass, 120coverage) ...etc.

4. Research Results and Discussion

The experimental results obtained are shown and discussed below.

4.1. Solid Ink Density (SID)

Density is the ability of the printed material to absorb light, and SID refers to the highest density of colour printed on the printed material. This study refers to the highest density of each colour of the printed PET after different inkjet printing conditions and after transfer to ABS. As shown in Figure 1 and 2, the coverage is high, and the SID is relatively high, in the case of low coverage, SID after transfer is generally higher than that before. After the ink is high temperature, it is easy to carbonize and blacken, causing the SID to rise, but when the coverage is increased, it is not significant.

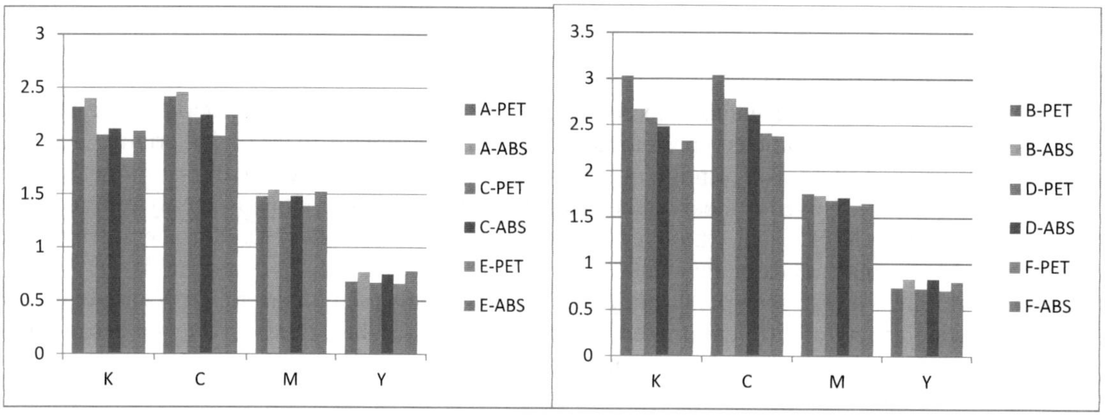

Figure 1. The SID of different printing pass (8.16.32), same resolution (720) and coverage 100% (ACE) (left) and different printing pass (8.16.32), same resolution (720) and coverage 120% (BDF) (right).

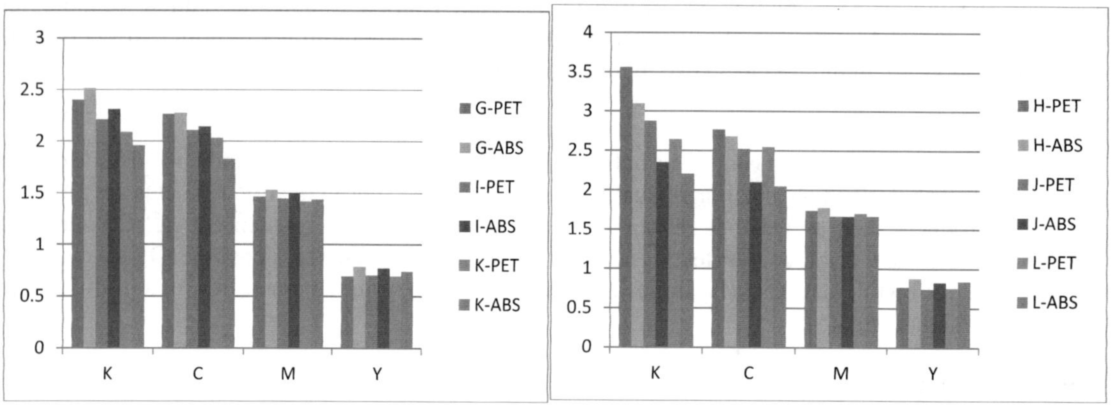

Figure 2. The SID of different printing pass (8.16.32), same resolution (1440) and coverage 100% (GIK) (left) and different printing pass (8.16.32), same resolution (1440) and coverage 120% (HJL) (right).

4.2. Tone Value (TV) and Tone Value Increase (TVI)

Before transfer (PET) and after transfer (ABS), the reproduction curve is adjusted by A (720dpi, 8pass, 100coverage) and G (1440dpi, 8pass, 100coverage). After the transfer is completed, the dot area is increased in the highlight area, But the middle to shadow tone area are not significant, as shown in Figure 3 and 4.

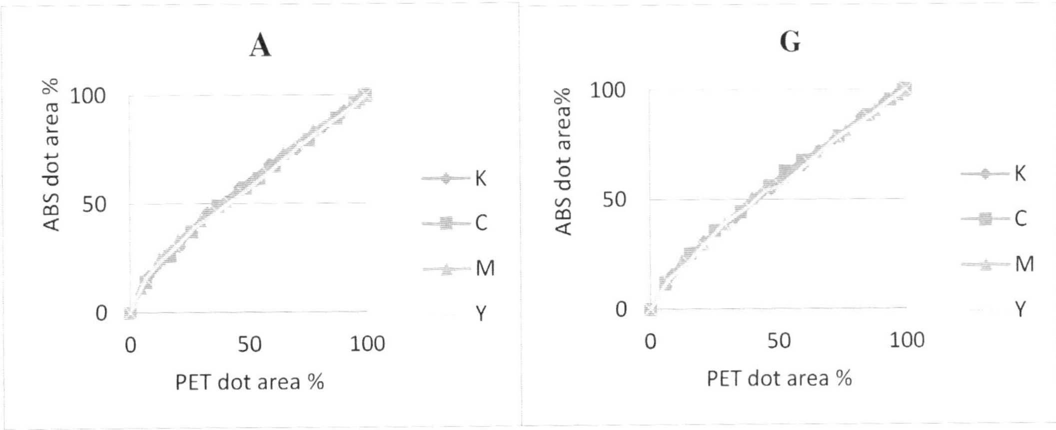

Figure 3. Comparison A & G of the reproduction curve before and after transfer.

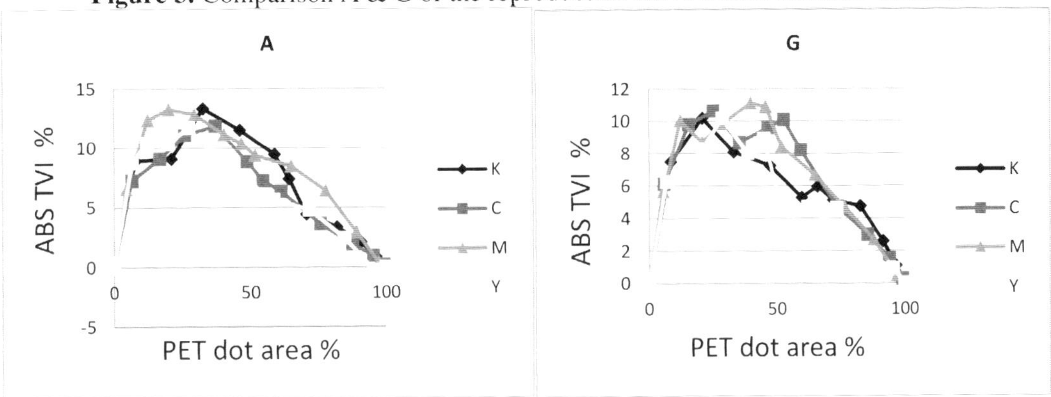

Figure 4. Relative percentage tone value increase (TVI) of dots after transfer.

4.3. Print Contrast (PC)

Before (PET) and after (ABS) printing comparison of the experimental combination A (720dpi, 8pass, 100coverage) and G (1440dpi, 8pass, 100coverage) after the transfer, the PC is lower than before, the high temperature transfer leads to the shadow area. The level is getting worse. In this study, a comparison of the 75% dot area was used for comparative analysis. The larger the PC, the richer the tone, as shown in Figure 5.

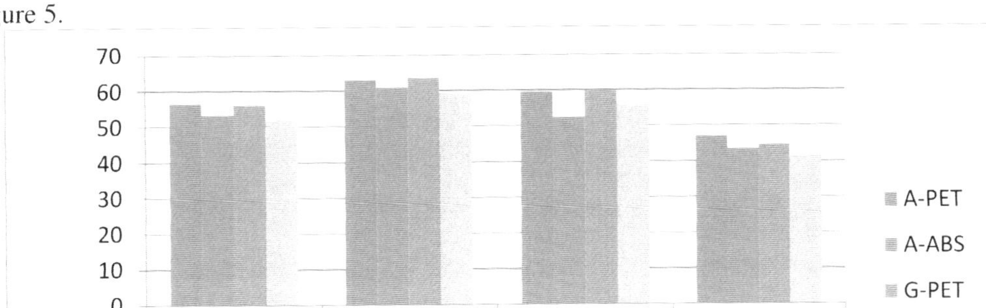

	K	C	M	Y
A-PET	56.39	63.01	59.51	46.93
A-ABS	53.35	60.98	52.42	43.52
G-PET	56.13	63.6	60	44.64
G-ABS	51.9	59.35	55.74	41.4

Figure 5. Pre-transfer (PET) versus post-transfer (ABS) comparison print contrast (PC) values.

4.4. Colour Gamut Analysis

Based on the Lab colour space, the colour gamut of each experimental combination before and after the transfer are depicted. As shown in Figure 6, after the OMD transfer to the ABS material, the brightness value (L) of all products decreased. After the transfer, the ink is slightly burnt, in the colour and colour domain, it can be known that the first and third quadrants change greatly after the transfer.

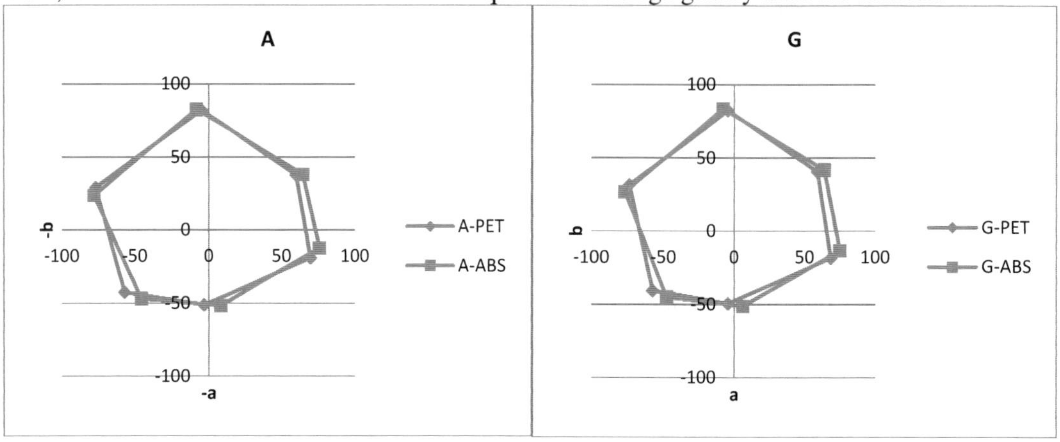

Figure 6. Comparison of pre-transfer and post-transfer colour gamut.

4.5. Colour Difference(ΔE)

Comparing the colour difference between the pre-transfer (PET) and post-transfer (ABS) colours, the colour differenceΔE of the K colour is below 1-3.2, and the C colour is 6-15, the M colour is 5-11 and the Y colour is 1-7. The K colour has better consistency in colour performance, and the Y colour and the M colour are the second, and the C colour is inferior. It should be that the high temperature in the transfer process has a great influence on the ink of the Cyan colour.

4.6. Dot Characteristics Reproduction Curve

As shown in Figure 7, the dot characteristics reproduction curve of the different combinations (A (720dpi, 8pass, 100coverage) and H (1440dpi, 8pass, 120coverage)) are related to the calibration curve, and the first quadrant of the figure is the reproduction curve between the original and the finished product (ABS), the second quadrant is the curve of PET before transfer to after the transfer ABS. the third quadrant is PET 45° corresponding straight line, the fourth quadrant is the curve of original to PET, from this figure can be obtained from the original to the finished product ABS between the dot change.

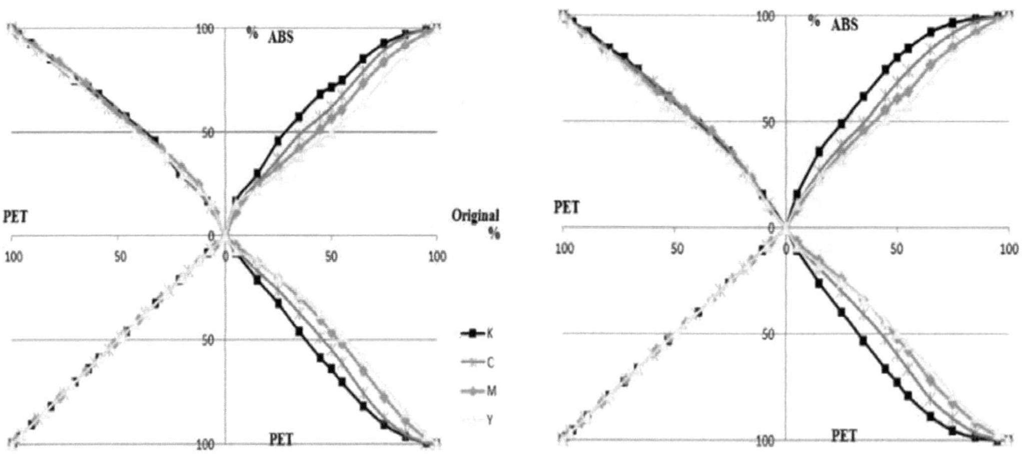

Figure 7. The dot characteristics reproduction curve of A (left) and H (right).

5. Conclusion

Following conclusions are found in this study below.

- When the coverage is low, the SID is higher than that before the transfer.
- After the transfer, the dot area is increased in the highlight, which makes the tone reproducibility better.
- The experimental combination of A (720dpi, 8pass, 100coverage) and G (1440dpi, 8pass, 100coverage) after transfer, the maximum TVI of each colour occurs in the highlight area.
- All experimental combinations after the transfer, the PC is lower than before the transfer.
- After the OMD transfer to the ABS material, the brightness value (L) decreased, mainly due to carbonization and charring after transfer. It can be known that the change of orange and cyan after transfer is significant.
- Comparison of colour difference(ΔE) before and after transfer: K colour has better consistency in colour performance before and after transfer, Y colour and M colour second, C colour is poor. It should be that the high temperature in the transfer process has a great influence on the ink of the Cyan.
- Through the integration before and after the transfer, the correspondence of the dots presented from the original to PET to ABS product can be obtained. This allows us to understand the changes in the dot area between the original electronic file and the finished product ABS. It can control the colour at production process.

References

[1] Chen C L and Lo M C and Su Y T and Chang Y T 2012 *The study on colour print quality attributes of In-Mold Roller using digital inkjet printing* Applied Mechanics and Materials Vol 262 pp 340-344

[2] Chen C L 2020 *A Study on tone reproduction integration curves of injection moulding using digital transfer printing* Chinese Association Graphics Science and Technology Journal Chinese Association Graphics Science and Technology Taiwan pp 3-12

[3] Kipphan Helmut Ed 2001 *Handbook of Print Media* Berlin Germany Springer pp677-687

[4] Chen C L and Chang Y T and Taso S H and Hsu W C 2014 *The study on colour quality attributes of IMR with PET on white ink by UV inkjet* Advanced Materials Research Vols 1004-1005 pp799-802

[5] Chen C L and Wang Y L 2009 *The study on colour print quality attributes of Water Transfer by Digital Printing* Chinese Association Graphics Science and Technology Journal Chinese Association Graphics Science and Technology Taiwan pp 169-190

[6] Lo M C and Chen C L and Perng R K and Hsieh T H 2006 *The Characterisation of Colour Printing Devices via Physical, Numerical and LUT Models,* CGIV 2006 IS&T's Third European Conference on Colour in Graphics Imaging and Vision University of Leeds UK pp 95-99

IOP Conf. Series: Materials Science and Engineering **964** (2020) 012020 doi:10.1088/1757-899X/964/1/012020

Investigation of Explosion Simulation on Jet Forming of ZrCuNiAlAg Amorphous Alloy Liner with Eccentric Sub-hemisphere Structure

Ping Cui[1,2], Dongmei Shi[2], Jingqing Xu[2,*] and Deshi Wang[1]

[1]College of Weaponry Engineering, Naval University of Engineering, Wuhan, China
[2]Department of Ammunition Engineering of Shijiazhuang Campus, Army Engineering University of PLA, Shijiazhuang, China

*Corresponding author email: dygcx12324@163.com

Abstract. Structure and material are the two key factors affecting the jet forming of the liner. To investigate the influence of new structure and new materials on the jet forming of the liner, a shaped charge model with eccentric sub-hemisphere structure based on ZrCuNiAlAg amorphous energetic alloy was proposed. The jet forming law of eccentric sub-hemispherical liner is studied by fixing other parameters and changing the structure parameters of liner in turn by means of AUTODYN numerical simulation software. The simulation results show that the eccentric sub-hemispherical liner of ZrCuNiAlAg amorphous alloy can form a shaped jet very well by using gathering energy effect. The speed of the shaped jet decreases gradually with time, and the shape of the jet changes from short and thick to long and thin. The curvature radius, eccentricity and wall thickness have a significant effect on the jet forming performance of ZrCuNiAlAg amorphous alloy. As the curvature radius or wall thickness increase, the tip velocity and the length diameter ratio (LDR) of the jet will decrease gradually, while with the increase of the eccentricity of the liner, the jet tip velocity and the LDR will increase gradually.The analysis of simulation data shows that the change of eccentricity has the greatest influence on the jet LDR. When the eccentricity increases from 25 mm to 39 mm, the jet LDR increases by 74% at 60μs. The wall thickness has the greatest influence on the tip velocity of the jet. When the wall thickness increases from 1.4 mm to 3.8 mm, the tip velocity of the jet at 60μs decreases by 29.77%. The results of this paper can be used for scientific reference in the structural design of this kind of liner.

1. Introduction

The liner is the core component of shaped charge. The structure and material are the two key factors affecting the performance of the liner. For more than 100 years, scholars have been carrying out exploration and research on the selection of the liner materials. The liner materials have experienced the changes of high density single metal materials, non-metal materials with low density, composite materials and energetic materials. Especially in the past 20 years, the application of energetic reactive materials in liner has become a research hotspot in the field of shaped charge. Some scholars have carried out theoretical analysis and experimental research on the formulation, manufacturing process, mechanical properties, energy release and composite damage mechanism of metal/fluoropolymer energetic materials[1-5] (such as PTFE/Al) and metal-based energetic materials[6-8] (such as Al/Ni). Since Duwez[9] first prepared the world Au-Si amorphous alloy by means of rapid cooling of molten metal in 1960, the amorphous alloy has shown a broad application prospect for more than half a century because of its unique atomic structure and excellent performance. Zirconium based (Zr-based) amorphous alloy is a newly developed energetic material, which has better mechanical properties and good energy release properties than metal/fluoropolymer and metal-based active energetic materials. It has great application potential in energetic fragments [10-11], armour piercing core[12-15], and especially in the design of new shaped charge jet materials[16-20].

The shape of the liner is also a long-term concerned focus of the scholars in the field of shaped charge. Different liner shapes have a crucial influence on jet morphology, velocity gradient and maximum stretch length[21]. At present, the conical liner is widely used in shaped charge warhead. Relevant research shows that the eccentric sub-hemispherical liner has a unique advantage in the forming of shaped jet due to its large curvature radius and long generatrix. Curvature radius, eccentricity and wall thickness are the three most important factors that restrict the jet forming performance of eccentric sub- hemispherical liners. Yao[22-23] and Yi[24] respectively studied the jet forming performance of eccentric sub- hemispherical copper and molybdenum liner, which has a good guiding significance for the structural design of such liners.

In this paper, a new Zr-based amorphous alloy energetic material and a good liner geometry are combined by means of numerical simulation. That is, the explosive forming process of the eccentric sub-hemispherical liner of Zr-based amorphous alloy is studied, and the influence of various factors on the jet forming effect is explored, which could provide a scientific reference for the liner structure design of this material.

2. Model of Simulation

2.1. Structure Model of Shaped Charge (SC)

The structure of Zr-based amorphous alloy eccentric sub-hemispherical SC is shown in Figure 1, which is mainly composed of explosive charge and the liner. The charge diameter (D)is 80mm, and the height of the charge (H) is 96mm. The geometrical parameters of the liner with an eccentric sub-hemispherical structure and constant wall thickness include the radius of arc curvature (r), the eccentricity (e) and the wall thickness (δ). The SC is detonated by central point initiation mode. Different simulation schemes are designed by changing the curvature radius, eccentricity and wall thickness of the liner structure, and the influence of them on the jet forming is studied.

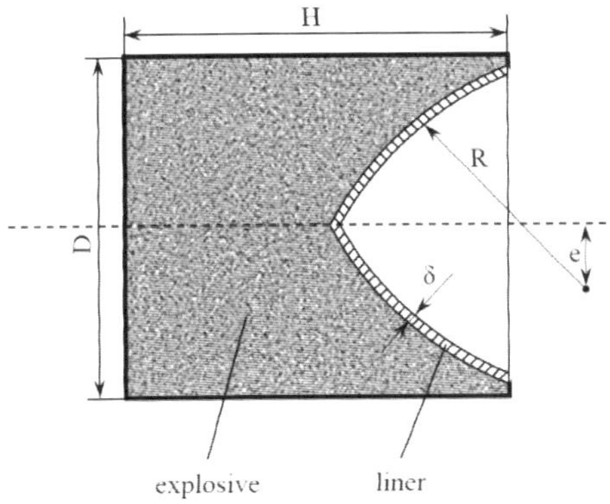

Figure 1. The shaped charge structure.

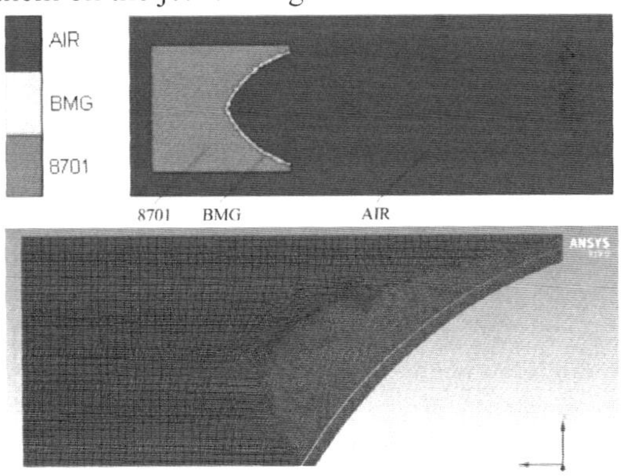

Figure 2. The finite element model.

2.2. Finite Element (FE) Calculation Model

The FE calculation model of eccentric sub-hemispherical SC established by AUTODYN software is shown in Figure 2. It is assumed that the explosive, liner and air are homogeneous continuous media, and the whole explosion process is adiabatic. The effect of gravity on the jet is ignored. Considering the axisymmetry of the entireFE model and the explosion load, in order to save the calculation time, a two-dimensional axisymmetric model is used in the simulation for computational efficiency. A total of 63650 elements were createdwithin the numerical analysis of the ANSYS which provided a sufficient mesh density for the calculation.Because of the high strain rate and high overload process in the formation of damage element of SC, Arbitrary Lagrange Euler (ALE) algorithm is used in numerical simulation to calculate the formation process of shaped jet involving large grid deformation and material flow. Euler algorithm is used for the explosive, the liner and the air. Mm-mg-ms unit system is used in the modeling process.

2.3. Material Model and Equation of State(EOS)

8701 explosive is used as the main charge, with a density of 1.69g/cm^3 and a detonation speed of 8425m/s. The constitutive relation is described in the form of High-Explosive-Burn. The EOS of the explosive is chosen as JWL and the material parameters[4-5] are shown in Table 1. The liner material is ZrCuNiAlAg amorphous alloy, the EOS of which is polynomial in JH-2 model. The material parameters of the liner[25]are shown in Table 2. The Euler domain is filled by air, and the EOS of which is Ideal Gas.The air material parameters[5] are shown in Table 3. The boundary condition of the air (Euler) domain is set as "Flow_out (ALL EQUAL)" to eliminate the boundary effect.

Table 1. Parameters of the 8701 explosive[4-5].

Material	ρ/(g·cm^{-3})	D/(m·s^{-1})	P_{CJ}/(GPa)	E/(GPa)	A/(GPa)	B/(GPa)	R_1	R_2	ω	V_0
8701 explosive	1.69	8425	29.6	8.499	524.23	7.678	4.2	1.1	0.34	1.00

Table 2.JH-2 material parameters of ZrCuNiAlAgamorphous alloy[25].

Parameter	P/(g/cm^3)	A_1/(kPa)	A_3/(kPa)	G/kPa	B	M	D_1	σ_{HEL}/(kPa)	A_2/(kPa)	T_1/(kPa)	A	C	N	D_2
Value	6.581	1.117×10^8	8.044×10^9	3.704×10^6	0.258	0.59	0.005	3.467×10^6	4.037×10^8	1.117×10^8	1.397	0.0044	2.432	1

Table 3. Material parameters of air[5].

Material	P/(g/cm^3)	γ	C_P(kJ/kg·K)	C_V(kJ/kg·K)	T(K)	E_0(kJ/kg^{-1})
Air	1.225	1.4	1.005	0.718	288.2	2.068×10^5

3. Analysis of Influencing Factors of Jet Forming

3.1. Radius of Curvature

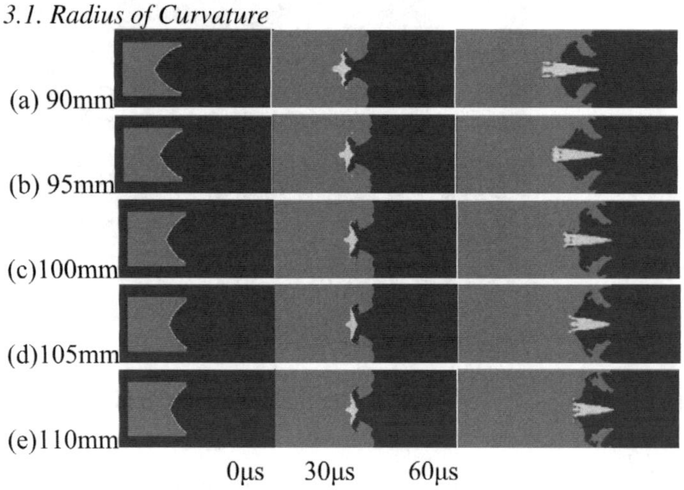

Figure 3. The jet forming process with different curvature radius.

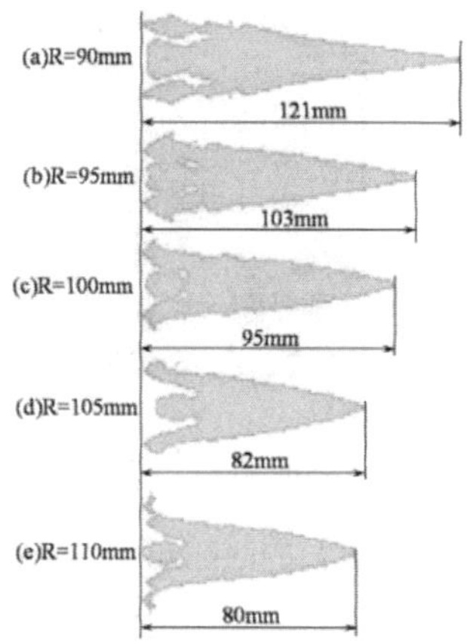

Figure 4. The jet morphology of different curvature radius.

The wall thickness and eccentricity of the liner are fixed to 2.6mm and 37mm, respectively. The jet forming process of ZrCuNiAlAg amorphous alloy liner is simulated when the curvature radius of the liner is 90mm, 95mm, 100mm, 105mm and 110mm, respectively. The jet forming process corresponding to the aforementioned curvature radius are shown in Figure 3. The shaped jets ofZrCuNiAlAg amorphous alloy liners with different curvature radius at 60 μs are shown in Figure 4. It can be seen from Figure. 4that the eccentric sub-hemispherical liner of ZrCuNiAlAg amorphous alloy can form a shaped jet with good morphology at 60μs.And the jet length decreases as the curvature radius increases. The jet length when the curvature radius is 110 mm is about 66.1% of that when the curvature radius is 90 mm. The tip velocity and the length diameterratio (LDR) of the jet under different curvature radiuswith time are shown in Figure 5 and Figure 6.

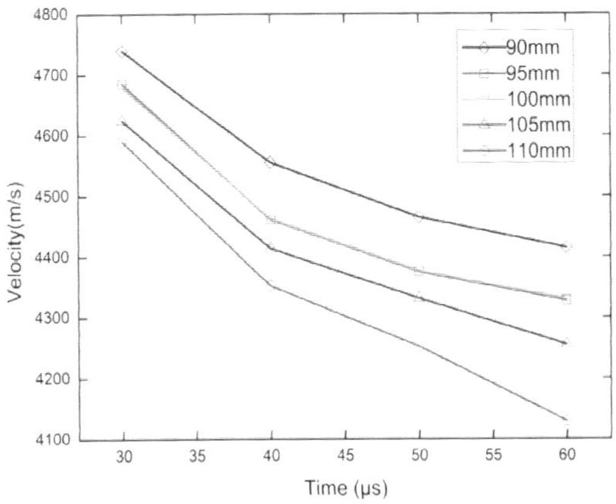

Figure 5. The jet tip velocity with time under different curvature radius.

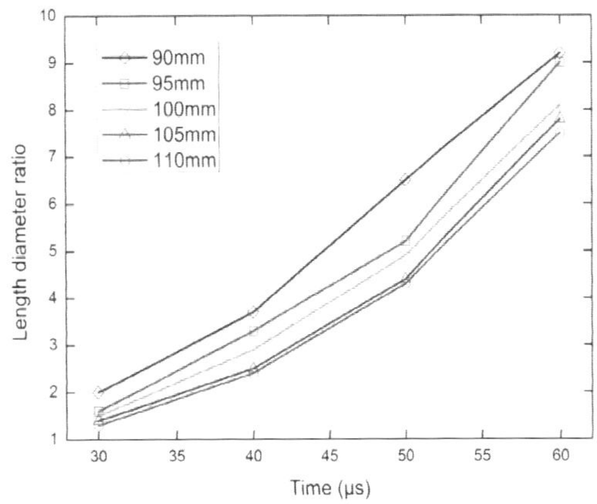

Figure 6. The jetLDRwith time under different curvature radius.

Figure. 5 and Figure. 6 show that during the process from collapse to the formation of the jet, the tip velocity decreases while the LDR increases gradually with time. Taking the curvature radius of 100 mm as an example, the tip velocity of the shaped jet is 4330 m/s at 60 μs, which is about 7.5% lower than that of 4680 m/s at 30 s, while the LDR increases from 1.5 to 8.1, which is about 4.4 times higher. Both the tip velocity and the LDR of the shaped jet are affected by the curvature radius of the liner. With the increase of the curvatureradius, both the two parameters of the shaped jet decrease gradually. This trend can also be seen from the shaped jet parameters of the eccentric sub-hemispherical liner (see Table 4) under different curvature radiusat 60μs. The jet tip velocitydecreases by 6.48% and the LDR decreases by 18.21% at 60 μs when the curvature radius increases from 90 mm to 110 mm.

Table 4. The shaped jet parameters under different curvature radius (60μs).

Curvature radius / (mm)	90	95	100	105	110
Tip velocity /(m/s)	4415	4327	4330	4255	4129
LDR	9.17	9	8.1	7.8	7.5

3.2. Eccentricity

The wall thickness and curvature radius of the liner are fixed to 2.6mm and 100mm, respectively. The jet forming process of ZrCuNiAlAg amorphous alloy liner is simulated when the eccentricity of the liner is 25mm, 30mm, 30mm, 37mm and 39mm, respectively. The jet forming process corresponding to the aforementioned eccentricity are shown in Figure 7. The shaped jets ofZrCuNiAlAg amorphous alloy liners with different eccentricityat 60 μs are shown in Figure 8.

IOP Conf. Series: Materials Science and Engineering **964** (2020) 012020 doi:10.1088/1757-899X/964/1/012020

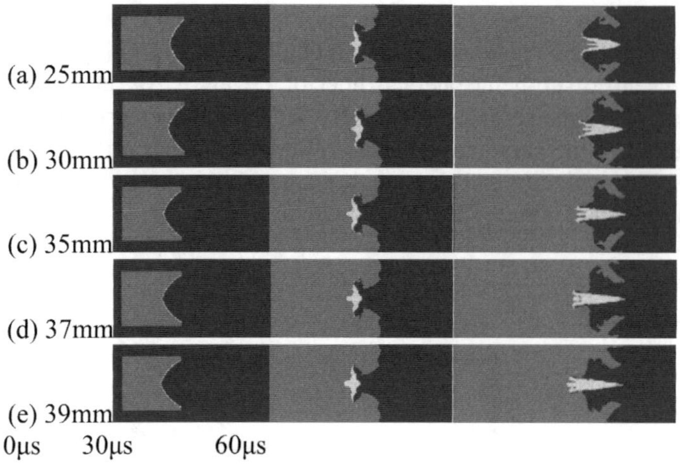

Figure 7. The jet forming process with different eccentricity.

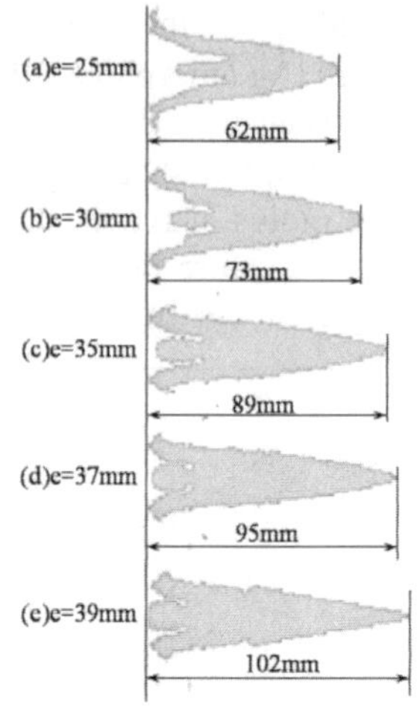

Figure 8. The jet morphology of different eccentricity.

As we can seefrom Figure. 8, the jet length increases gradually with the increase of the eccentricity. By 60 μs,the jet length when the eccentricity is 39 mm is about 1.65 times of that when the eccentricity is 25 mm. The tip velocity and the LDR of the jet under different eccentricity with time are shown in Figure 9 and Figure 10.

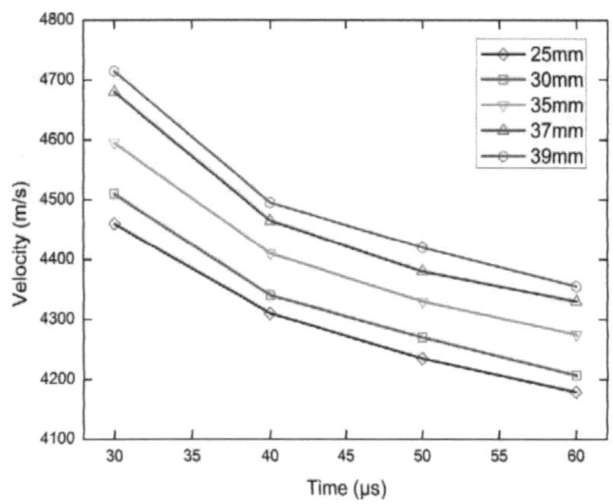

Figure 9. The jet tip velocity with time under different eccentricity.

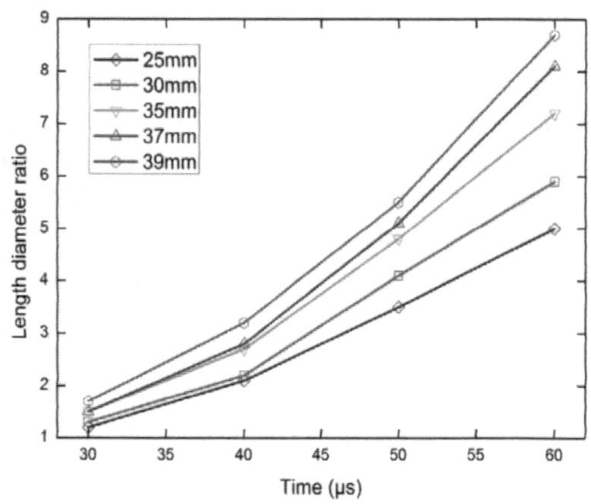

Figure 10. The jetLDRwith time under different eccentricity.

Figure. 9 and Figure. 10point out that during the process from collapse to the formation of the jet, the tip velocity decreases while the LDR increases gradually with time. Taking the eccentricity of 35 mm as an example, the tip velocity of the shaped jet is 4275m/s at 60μs, which is about 6.86% lower than that of 4590 m/s at 30 μs, while the LDR increases from 1.5 to 7.2, which is about 3.8 times higher. Both the tip velocity and the LDR of the shaped jet are affected by the eccentricity of the liner. With the increase of the eccentricity, both the two parameters of the shaped jet increase gradually. This trend can also be seen from the shaped jet parameters of the eccentric sub-hemispherical liner (see Table 5) under different eccentricityat 60μs. The jet tip velocityincreases by 4.21% and the LDR increases by 74% at 60 μs when the eccentricity increases from 25 mm to 39 mm.

Table 5. The shaped jet parameters under different eccentricity (60μs).

Eccentricity / (mm)	25	30	35	37	39
Tip velocity /(m/s)	4179	4207	4275	4330	4355
LDR	5	5.9	7.2	8.1	8.7

IOP Conf. Series: Materials Science and Engineering **964** (2020) 012020 doi:10.1088/1757-899X/964/1/012020

3.3. Wall Thickness

The curvature radius and eccentricity of the liner are fixed to 100 mm and 37mm, respectively. The jet forming process of ZrCuNiAlAg amorphous alloy liner is simulated when the wall thickness of the liner is 1.4mm, 2.0mm, 2.6mm, 3.2mm and 3.8mm, respectively. The jet forming process corresponding to the aforementioned wall thickness are shown in Figure 11. The shaped jets ofZrCuNiAlAg amorphous alloy liners with different wall thickness at 60 μs are shown in Figure 12.

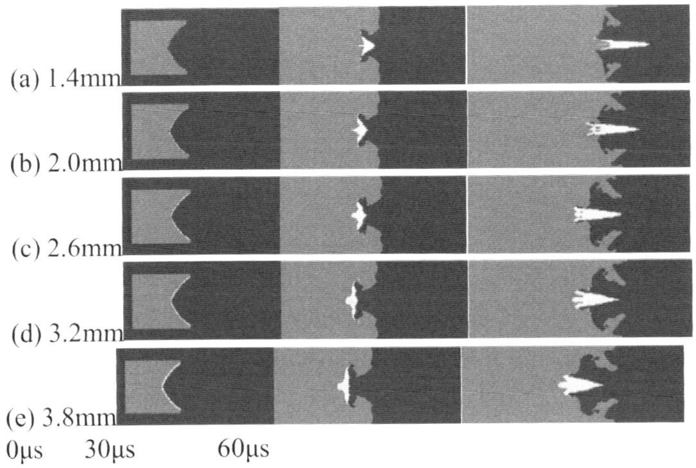

Figure 11. The jet forming process with different wall thickness.

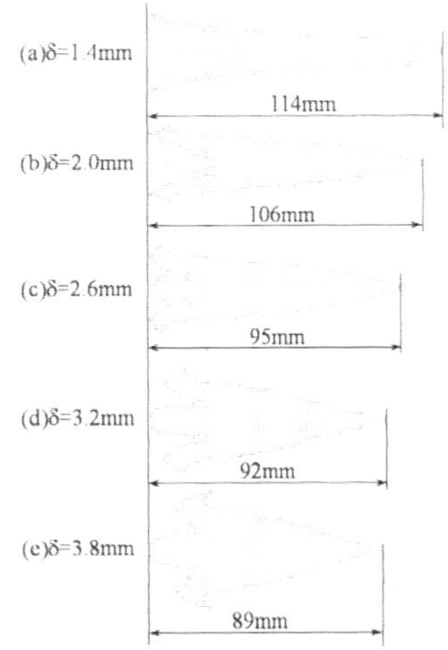

Figure 12. The jet morphology of different wall thickness.

It can be seen from Figure. 12 that the length and diameter of the jet become shorter and larger gradually with the increase of the wall thickness. By 60 μs,the jet length when the wall thickness is 3.8 mm is about 78.07% of that when the wall thickness is 1.4 mm. The tip velocity and the LDR of the jet under different wall thickness with time are shown in Figure 13 and Figure 14.

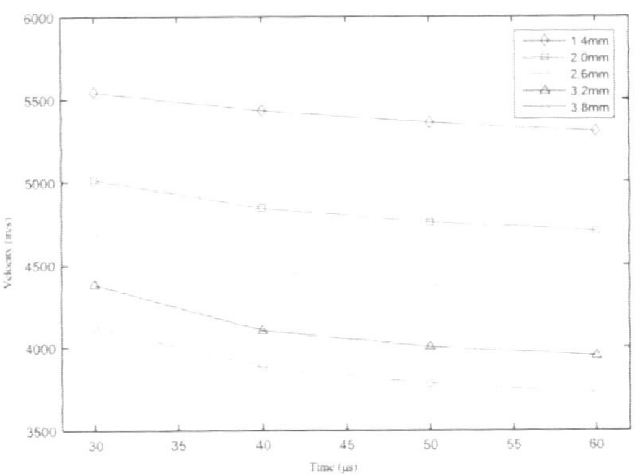

Figure 13. The jet tip velocity with time under different wall thickness.

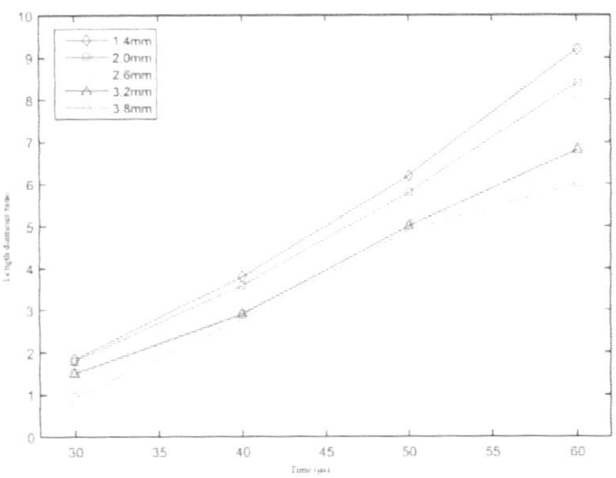

Figure 14. The jetLDRwith time under different wall thickness.

Figure. 13 and Figure. 14illustrate that during the process from collapse to the formation of the jet, the tip velocity decreases while the LDR increases gradually with time. Taking the wall thickness of 2.6 mm as an example, the tip velocity of the shaped jet is 4330m/s at 60μs, which is about 7.62% lower than that of 4660 m/s at 30 μs, while the LDR increases from 1.5 to 8.1, which is about 4.4 times higher. Both the tip velocity and the LDR of the shaped jet are affected by the wall thickness of the liner. As the wall thickness increases, both the two parameters of the shaped jet decrease gradually. This trend can also be seen from the shaped jet parameters of the eccentric sub-hemispherical liner(see

Table 6) under different wall thickness at 60μs. The jet tip velocity decreases by 29.77% and the LDRdecreases by 34.78% at 60 μs when the wall thickness increases from 1.4 mm to 3.8 mm.

Table 6. The shaped jet parameters under different wall thickness (60μs).

Wall thickness / (mm)	1.4	2.0	2.6	3.2	3.8
Tip velocity /(m/s)	5304	4704	4330	3950	3725
LDR	9.2	8.4	8.1	6.8	6

4. Conclusion

(1) The eccentric sub-hemispherical liner of ZrCuNiAlAg amorphous alloy can form a shaped jet very wellby using the gathering energy effect. As a whole, the tip velocity of the shaped jet decreases graduallywith the simulation time, and the jet shape changes from short and thick to long and thin, that is, the jet LDR increases. From 30μs to 60μs, the tip velocity of the shaped jet decreases by 7.33% on average, and the LDR increases by 4.2 times on average.

(2) The curvature radius, eccentricity and wall thickness have a significant effect on the jet forming performance of ZrCuNiAlAg amorphous alloy. Simulation results show that the tip velocity and the LDR of the jet will decrease gradually with the increase of curvature radius or wall thickness, while the jet tip velocity and the LDR will increase graduallywith the increase of the eccentricity of the liner.

(3) The eccentricity has the greatest influence on the jet LDR. When the eccentricity increases from 25 mm to 39 mm, the jet LDR increases by 74% at 60μs. The wall thickness has the greatest influence on the tip velocity of the jet. When the wall thickness increases from 1.4 mm to 3.8 mm, the tip velocity of the jet at 60μs decreases by 29.77%.

Acknowledgments

The authors were appreciated for the precious comments and suggestions of the editors and reviewers.

References

[1] J. G. Xiao, X. P. Zhang, Y. Z. Wang, F. Y. Xu and H. F. Wang. "Demolition Mechanism and Behavior of Shaped Charge with Reactive Liner,"*Propellants, Explosives, Pyrotechnics*, vol. 41,no. 4, 2016, pp. 612-617.

[2] F.Y. Xu, Y. F. Zheng, Q. B. Yu, Y. Z. Wang and H. F. Wang. "Experimental Study on Penetration Behavior of Reactive Material Projectile Impacting Aluminium Plate,"*International Journal of Impact Engineering*, vol. 95, no. 9, 2016, pp. 125-132.

[3] Y. F. Zheng, C. H. Su, H. G. Guo, Q. B. Yu and H. F. Wang."Behind-target Rupturing Effects of Sandwich-like Plates byReactive Liner Shaped Charge Jet,"*Propellants, Explosives.Pyrotechnics*,vol. 44, no. 11, 2019, pp. 1400-1409.

[4] H. F. Wang, H. G. Guo, B. Q. Geng, Q. B. Yu and Y. F. Zheng. "Application of PTFE/Al Reactive Materials for Double-Layered Liner Shaped Charge," *Materials*, vol. 12, no. 16, 2019,pp. 2768.

[5] H. G. Guo, Y. F. Zheng,Q. B. Yu, C. Ge and H. F. Wang."Penetration Behavior of Reactive Liner Shaped Charge Jet Impacting Thick Steel Plates,"*International Journal of Impact Engineering*, vol. 126, no. 4, 2019, pp. 76-84.

[6] C. Ji, Y. He, C. T. Wang, Y. He,X. C. Pan, J. J. Jiao, L. Guo."Investigation on Shock-induced Reaction Characteristics of an Al/Ni Composite Processed via Accumulative Roll-bonding,"*Materials& Design*, vol. 116,2017, pp. 591-598.

[7] W.Xiong, X. F.Zhang andY. Wu."Influence of Additives on Microstructures, Mechanical Properties and ShockInduced Reaction Characteristics of Al/Ni Composites,"*Journal of Alloys Compound*. vol. 648,2015, pp. 540-549.

[8] M. Sun, C. Li, X. G. Zhang, X. M. Hu, X. Y. Hu and Y. B. Liu. "Reactivity and Penetration Performance Ni-Al and Cu-Ni-Al Mixtures as Shaped Charge Liner Materials,"*Materials*, no.11,2018,2267.

[9] W. Klement, R. H. Willens and P. Duwez. "Non-crystalline Structure in Solidified Gold-silicon Alloy,"*Nature*, no. 187, 1960, pp. 869-870.

[10] C. T. Wang, Y. He, C.Ji, Y. He, W. Han and X. C. Pan. "Investigation on Shock-induced Reaction Characteristics of a Zr-based Metallic Glass,"*Intermetallics*, 2017,http://dx.doi.org/10.1016/j.intermet.2017.11.004.
[11] Y.F.Zhang,X.B.Luo and G.Q. Liu."Penetration and Energy Release Effect of W/ZrNiAlCuMetastable Reactive Alloy Composite Fragment Against RHA Target,"*Explosion and Shock Waves*, July 16th, 2019, Web-start, DOI: 10.11883/BZYCJ-2009-0065.
[12] Z. H. Du, C. X. Du, Z. W. Zhu andL. X. Xia."An Experimental Study on Perforation Behaviorof SegmentedW_f/Zr-Based Bulk Metallic Glass Matrix Composite,"*Rare Metal Materials and Engineering*, vol.45, no.9, 2016, pp.2359-2365.
[13] Z. H. Du, C. X. Du and Z. W. Zhu. "An Experimental Study on Perforation Behavior of Pole Penetrator Preparedfrom W_F/Zr-Based Bulk Metallic Glass Matrix Composite,"*Rare Metal Materials and Engineerin*, vol.45, no.5, 2016, pp. 1308-1313.
[14] C. X. Du, Z. H. Du andZ. W. Zhu."Effect of Impact Velocity and Diameter of Tungsten Fiber on Penetration Ability ofW_f/Zr-based Metallic Glass Composite Penetrator,"*Rare Metal Materials and Engineering*, vol.46, no.6, 2017, pp.1632-1637.
[15] C. X. Du, Z. H. Du andZ. W. Zhu."Effect of Tungsten Fiber Diameter on Penetration Ability of Zr-based Metallic GlassComposites,"*Rare Metal Materials and Engineering*, vol.46, no.4, 2017, pp.1080-1085.
[16] J.K.Laszlo, P.W.William. "Investigation of a Bulk Metallic Glass as a Shaped Charge Liner Material,"in *Proceedings of the 23rd International Symposium on Ballistics*, Tarragona, Spain,Apr. 2007, pp.269-274.
[17] P.W.William, J.K.Laszlo and E. P.Justin. "Investigation of a Bulk Metallic Glass as a Shaped Charge Liner Material," US: Aberdeen Proving Ground,US Army Research Laboratory, ARL-TR-3864, Aug. 2006.
[18] J. L. Han, X. Chen,Z. H. Du, H. M. Fu, Y. D. Jing, T. Yuan, C. Cheng, C. X. Du and L. Z. Xu."Application of W/ZrAmorphous Alloy for Shaped Charge Liner,"*Materials Research Express*,vol. 6, no. 11, 2019, pp. 115209.
[19] J. L. Han, X. Chen,Z. H. Du, H. M. Fu, C. Cheng, L. Z. Xu,J. B. Wang and C. X. Du."Compressive Mechanical Properties of W-particle/Zr-based Amorphous Alloy Composites,"*Materials Research Express*, vol. 6, no. 8, 2019, pp. 085206.
[20] P. Cui, D. S. Wang, D. M. Shi, J. Q. Xu, and J. W. Zhen. "Investigation of Penetration Performance of Zr-based Amorphous Alloy Liner Compared with Copper," *Materials*, vol. 13,no. 4, 2020,pp. 912.
[21] T. CAO, W. B. GU andJ. Q. LIU."The Effects of Liner's Shape on Performance of Lateral Annular Jet,"*Initiators & Pyrotechnics*, no. 6, 2017, pp.14-18.
[22] Z. H. Yao, D. Z. Li, Q. H. Fu and C. Wu."Numerical Simulation of Shaped Charge with an Eccentric Sub-hemisphere Liner,"*Explosive Materials*, vol. 42, no.2, 2013, pp. 17-20.
[23] Y. Y. Han, J. P. Yin, C. Chen andF. D. Dong."Analysis on the Influence of Liner Structural Parameters on the Formation of K-shaped Charge," ACTA ARMAMENTARII, vol. 39, no. suppl.1, 2018, pp. 57-61.
[24] J. P. Yi, J. P. Yin, Z. J. Wang andL. K. Ma."Rod-like Jet Properties of Eccentric Sub-hemisphere Shaped Charge for Molybdenum Liner,"*Ordnance Material Science and Engineering*, vol. 38, no. 2, 2015, pp. 53-57.
[25] Y.X.Shi, D.M.Shi, W.Z.Li,Z. T. Yu and C. M. Shang."Study on JH-2 Model of the ZrCuNiAlAgBulk Amorphous Alloy,"*Explosion and Shock Waves*, vol. 39, no. 9, 2019, pp. 093104-1-093104-8.

IOP Conf. Series: Materials Science and Engineering **964** (2020) 012021 doi:10.1088/1757-899X/964/1/012021

Abnormal Magnetic Signals Characterization of Fatigue Crack Propagation Life

Changliang Shi[1,*], Xin Zhang[2], Yimin Lin[1], Rutao Guan[2] and Hongyan Yang[2]

[1]Center for Industrial Analysis and Testing, Guangdong Academy of Sciences, Guangzhou, China
[2]CSSC Huangpu Wenchong Shipbuilding Company Limited, Guangzhou, China

*Corresponding author email: barrystone@163.com

Abstract. In this research, the abnormal magnetic signals were detected in dynamic tension tests on the ferromagnetic material specimens, in order to study the characterization method of abnormal magnetic signals for the crack propagation life. It is indicated that fatigue crack length a is linear with the peak-peak value $\Delta H_{pa}(y)$ of abnormal magnetic signals. The $\Delta H_{pa}(y)$ of fatigue crack and the gradient of magnetic field intensity k increase with the increase of fatigue cycle number. Therefore, $\Delta H_{pa}(y)$ and k are confirmed as the key parameters of characterizing crack propagation life. Then, a life prediction model of abnormal magnetic signals for the crack propagation life is founded based on the Paris equation. The maximum error between measured values and calculated values is less than 15%.

1. Introduction

At present, most key components of machinery and equipment are made of ferromagnetic materials[1]. Because the components are subjected to repeated loads, fatigue fracture is prone to occur. Fatigue failure is one of the main failure modes for the components. Because the fatigue failure has a certain suddenness and contingency, it often leads to major accidents. Therefore, the fatigue life prediction is becoming more and more important[2]. The various defects in the ferromagnetic materials will be produced in the process of manufacturing. Under the action of fatigue load, these defects are easily becoming the source of crack, which leads to the fatigue fracture of the component. However, according to the fracture mechanics, it is found that even if there are cracks in some components, as long as we can monitor the crack growth and ensure that the crack will not expand to the critical size, the component can also be safely applied. Therefore, it is of great significance to study the fatigue crack propagation life for the components. In the field of engineering, conventional mechanical testing methods are very difficult to complete the monitoring tasks of the fatigue crack propagation. The nondestructive testing method is very suitable for monitoring the fatigue crack propagation, due to its sensitivity and diversity[3-5].

The metal magnetic memory testing is promoted by the Russian scholar Dubov in 1997 for the first time. The abnormal magnetic signals, which are affected by the geomagnetic field and stress, could be detected by the method. This novel nondestructive testing technology could be applied to detect the stress concentration and surface defect of ferromagnetic material[6,7]. Compared with the traditional nondestructive testing technology, the metal magnetic memory testing does not need to pretreat specimen, and it can be carried out when the specimen is in service. Meanwhile, the stress concentration area or micro defects can be detected in order to predict early warning of fatigue failure, unlike other traditional nondestructive testing methods only have the ability to detect macro defects[8,9].

Using metal magnetic memory testing technology, the loading history of ferromagnetic parts can be detected, and the stress concentration position can be measured. It is possible to characterize the stress concentration degree, damage degree and elastic-plastic state of the material. Therefore, metal magnetic memory testing technology has great potential in life prediction of ferromagnetic components. In this paper, the abnormal magnetic signals in metal magnetic memory testing were detected during dynamic tension tests on the ferromagnetic specimens with U notch. The relationship between abnormal magnetic signals and fatigue crack propagation was studied.

2. Experimental

2.1. Specimens

The base metal of specimens are 45 steel, which is a kind of medium carbon steel. The tensile strength of specimen is 679MPa, the yield strength is 386MPa and the elongation is 16.0%. The specimens were polished. Because the initial magnetic signals could be easily affected by the machining process and heat treatment condition, the specimens were demagnetized before testing in order to study the relationship between the stress, crack length and abnormal magnetic signals in the initial state.

2.2. Experimental Instruments

Instron-8801 hydraulic servo fatigue test machine were used, the dynamic load error was 1.0%. Tension–tension fatigue tests of sinusoidal waveform constant amplitude were carried out. The metal magnetic memory testing device is the EMS-2003 intelligent magnetic memory detector, and the detection probe is a two channel pen probe, the instruments are shown in figure 1. The normal component $H_p(y)$ of the scattering magnetic field intensity can be tested using this equipment. The magnetic memory detection probe is clamped on the three-dimensional electronic control displacement platform made of aluminium alloy, and the detection step distance is 2.5 m. The fatigue crack propagation was monitored by Olympus i-speed high speed camera, and the length of fatigue crack could be detected.

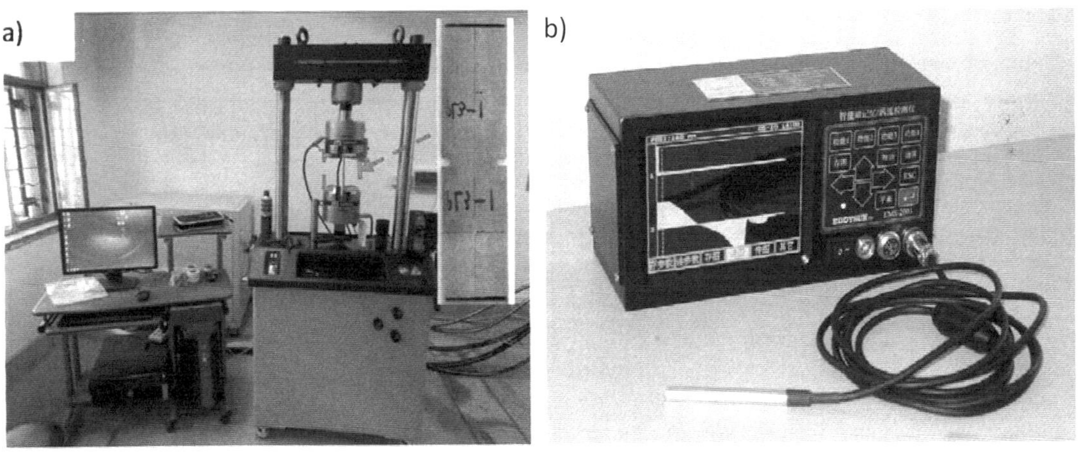

Figure 1. Testing instruments: a) Instron-8801 fatigue testing machine b) EMS-2003 detector.

2.3. Testing Procedure

The tension-tension fatigue test were carry out on the flat U notched specimens (stress concentration factor K_t was 5). Under the cyclic loading (maximum stress was 80MPa, sine wave, stress radio R was 0.1, loading frequency f was 10), the number of cyclic loading sequentially increased. The specimen was removed from the fatigue testing machine and placed in the north to south direction on the platform after loading to the predetermined number of cycles. In the middle part of the test piece, 5 parallel lines, whose interval was 5 mm and length was 60mm, was drawn as the detection paths, shown in figure 2. The probe was gripped on the scanning platform, the lift-off value was 0.5mm, the testing speed was 10mm/s, and the sampling time interval was 30ms. The specimen was loaded again to a higher preset circle value after metal magnetic memory testing, and the above procedure repeated until the specimen cracked.

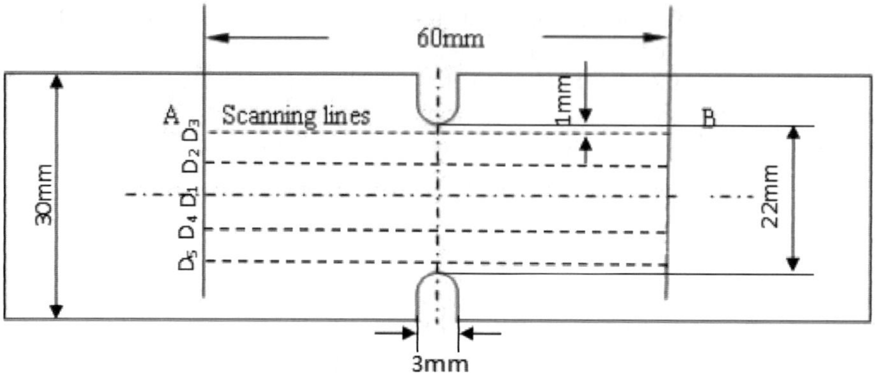

Figure 2. Scanning lines of the tension-tension fatigue testing specimen with U notch.

3. Results and Analysis

The experiments are carried out in laboratory condition, so the environmental magnetic field is dominated by the relatively constant geomagnetic field. The abnormal magnetic signals $H_p(y)$ in different detection lines are detected after 1000 fatigue cycle times, shown in figure 3.

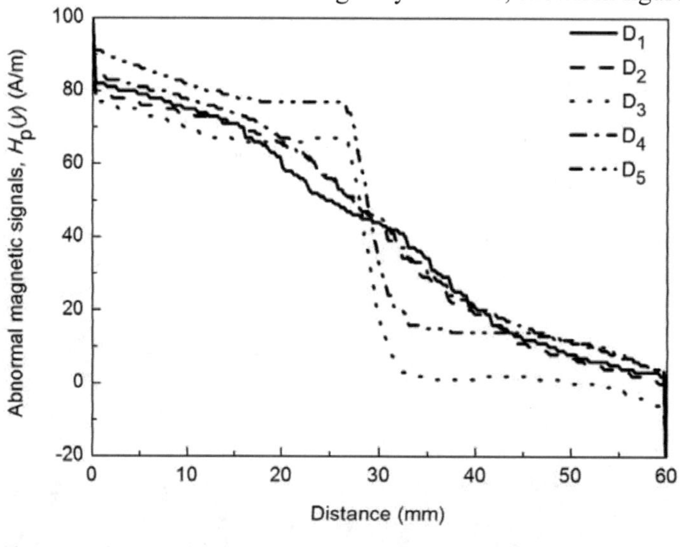

Figure 3. Abnormal magnetic signals $H_p(y)$ in different lines of the specimen after 1000 cycle times.

It is indicated that the gradient k of the abnormal magnetic signal $H_p(y)$ increases rapidly near the U notch in the specimen, and the gradient k of abnormal magnetic signals increases more when it is closer to the U notch. According to the material mechanics, the stress concentration degree increases nearby the U notch. The gradient k increases with the increase of stress concentration[8], so the gradient k dramatically increases near the U notch in the specimen and the stress concentration of the specimen can be qualitatively determined according to the gradient k of the H_p (y) detection curve. It can be seen that the change of abnormal magnetic signals $H_p(y)$ in D_5 detection path is the most obvious in figure 3. So the following analysis will be only around the results in D_5 detection path in order to facilitate the analysis and make it representative.

The abnormal magnetic signals $H_p(y)$ in D_5 detection path of the specimen after different fatigue cycle times is shown in figure 4. It is indicated that the abnormal magnetic signals $H_p(y)$ increase when the fatigue cycle times increase. The gradient k of the abnormal magnetic signal $H_p(y)$ curves increases dramatically near the U notch in every fatigue cycle time, and the peak of $H_p(y)$ initiates. Therefore, the maximum value k_{max} of the gradient and the peak-peak value $\Delta H_{pa}(y)$ of abnormal magnetic signals $H_p(y)$ could be the key parameters to reflect the relationship between the fatigue cycle time and the abnormal magnetic signals.

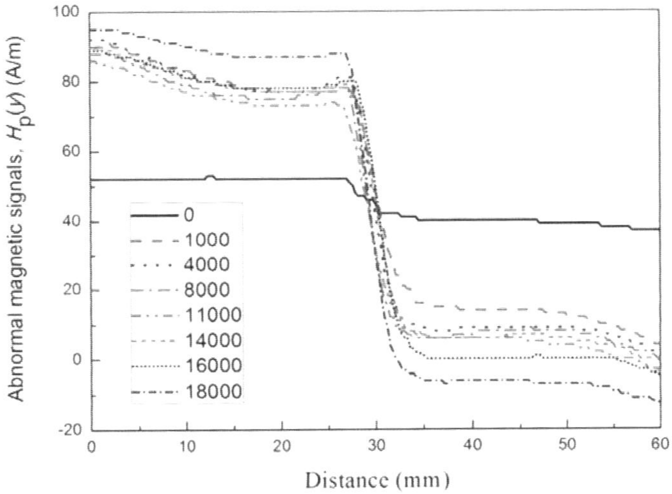

Figure 4. Abnormal magnetic signals $H_p(y)$ of specimen in D_5 path after different fatigue cycle times.

Firstly, the maximum value k_{max} of the gradient was studied. The relationship between k_{max} and fatigue cycle time N is shown in figure 5. It is indicated that the maximum value k_{max} of the gradient apparently increase at first when the fatigue cycle time increase, which is because the magnetic field strength will increase significantly with the increase of the load when the specimen is in the elastic region at the initial stage of fatigue. But when the cycle number increases to a certain extent, the maximum value k_{max} trends in the stable, which is because the influence of fatigue load on the magnetic domain structure of the material reaches saturation.

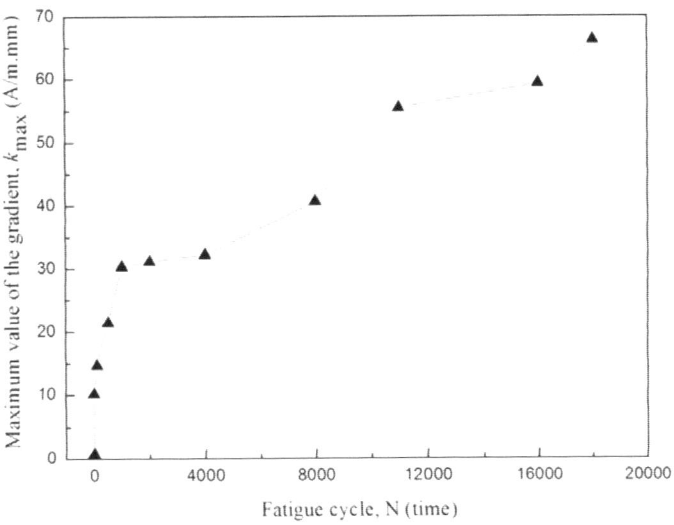

Figure 5. Relationship between k_{max} and fatigue cycle N.

With the increase of fatigue cycle time, the micro fatigue cracking is detected by the high speed camera at 8000 cycle times, then the maximum value k_{max} increases dramatically to the maximum value until the specimen cracked at 18000 cycle times. Meanwhile, there is also the same relationship between peak-peak value $\Delta H_{pa}(y)$ and the fatigue cycle time N. According to the fracture mechanics[4], fatigue crack length increases when the fatigue cycle time increases. At the same time, the preliminary results indicate that the maximum value k_{max} and the peak-peak value $\Delta H_{pa}(y)$ also increases with the increase of fatigue cycle time. Therefore, there must be some relationship between peak-peak value

$\Delta H_{pa}(y)$ and the length of fatigue crack a. In order to find the correlation, a fitted curve of peak-peak value $\Delta H_{pa}(y)$ and the length of fatigue crack a is shown in figure 6.

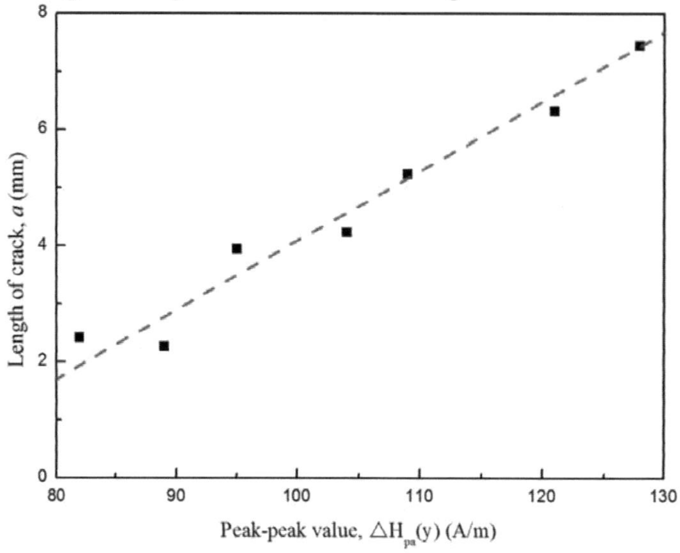

Figure 6. Fitted curve of $\Delta H_{pa}(y)$ and a.

It can be seen that the peak-peak value $\Delta H_{pa}(y)$ of abnormal magnetic signals gradually increases with the increase of length of fatigue crack a. Then, the experimental data are fitted, and the fitting equation is shown in equation (1) as following.

$$a = 0.12 \times \triangle H_{pa}(y) - 4.96 \tag{1}$$

According to the equation (1), there is a linear relationship between the peak-peak value $\Delta H_{pa}(y)$ and the length of fatigue crack a. In this research, the equation (1) is simplified to be equation (2) as following in order to be easy to analyze.

$$a = A \times \triangle H_{pa}(y) + B \tag{2}$$

Where A and B are constants, affected by material and stress.

At present, the fatigue crack growth life is mainly predicted by the fracture mechanics method[7]. In engineering practice, the method, which is most commonly used to the prediction of the fatigue crack propagation life, is the classical Paris formula, shown in equation (3) as following.

$$N = \frac{1}{C(f\Delta\sigma\sqrt{\pi})^m(0.5m-1)}\left(\frac{1}{a_0^{0.5m-1}} - \frac{1}{a_c^{0.5m-1}}\right) \qquad m \neq 2$$

$$N = \frac{1}{C(f\Delta\sigma\sqrt{\pi})^m}\ln\left(\frac{a_c}{a_0}\right) \qquad m = 2 \tag{3}$$

Where C and m are constants, calculated by fatigue crack propagation test. f is the geometric correction coefficient.

The fatigue test is carried out in constant amplitude condition, so the equation (2) could be used in equation (3), shown in equation (4) as following.

$$N = \frac{1}{C(f\Delta\sigma\sqrt{\pi})^m(0.5m-1)}\left(\frac{1}{(A\Delta H_{pa}(y)_0 + B)^{0.5m-1}} - \frac{1}{(A\Delta H_{pa}(y)_c + B)^{0.5m-1}}\right) \qquad m \neq 2$$

$$N = \frac{1}{C(f\Delta\sigma\sqrt{\pi})^m}\ln\left(\frac{(A\Delta H_{pa}(y)_c + B)}{(A\Delta H_{pa}(y)_0 + B)}\right) \qquad m = 2 \tag{4}$$

Where $H_{pa}(y)_0$ is the origin $H_{pa}(y)$, $H_{pa}(y)_c$ is the critical $H_{pa}(y)$.
In order to verify the accuracy of the equation (4), the test parameters are used as geometric correction coefficient *f* is 1.12, maximum stress is 80MPa, minimum stress is 8MPa, *m* is 2.32 and *C* is 2.47×10^{-12}. The test parameters are brought into the equation (4), and the calculated values of the fatigue crack propagation life could be obtained, which are compared with the measured values, shown in figure 7.

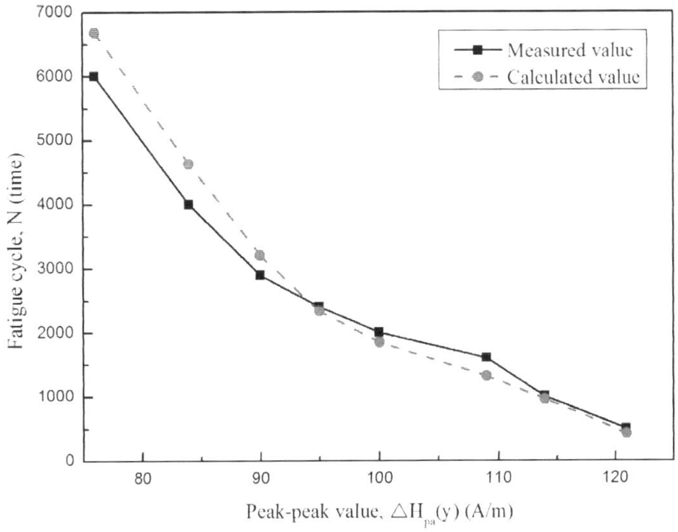

Figure 7. Measured values and calculated values for propagation life of the specimen.

Table 1. The relative error between measured values and calculated values shown in figure 7.

Peak-peak value $\Delta H_{pa}(y)$ (A/m)	121	114	109	100	95	90	84	76
Calculated value (time)	490	1000	1450	1860	2460	3200	4590	6720
Measured value (time)	500	1000	1700	2000	2500	2900	4000	6000
Relative error (%)	2.00	0.00	14.70	7.00	1.60	10.34	14.75	12.00

Validity analysis based on the actually measured value demonstrates that the measured values fit well with the calculated values, and the maximum error is less than 15%, shown in Table 1. The error is related to the constant A and B in equation (2). The better the linear fitting is, the smaller the error is. The equation (4) is obtained in the conditions of this paper, and there are more restrictive conditions. Although the method is obtained in laboratory conditions, it can direct the prediction for fatigue crack propagation life of ferromagnetic materials by the abnormal magnetic signals, and it can greatly promote the development and perfection of the components reliability.

4. Conclusions

In this research, the gradient *k* of the abnormal magnetic signals increases when the stress concentration increases. The gradient *k* and the peak-peak value $\Delta H_{pa}(y)$ of the abnormal magnetic signals increase with the increase of fatigue cycle times. The maximum value k_{max} of the gradient and the peak-peak value $\Delta H_{pa}(y)$ of abnormal magnetic signals are the key parameters to reflect the relationship between the fatigue cycle time and the abnormal magnetic signals. There is a linear relationship between the peak-peak value $\Delta H_{pa}(y)$ and the length of fatigue crack *a*. A predictive model for fatigue crack propagation life of ferromagnetic material is established, combined with the classical mechanics Paris formula. Validity analysis based on the actually measured value demonstrates that the measured values fit well with the calculated values and the maximum error is less than 15%.

Acknowledgments
This work is sponsored by Guangzhou Foreign Science and Technology Cooperation Project (201907010004), GDAS' Project of Science and Technology Development (No. 2018GDASCX-0114 and No. 2020GDASYL-20200302014).

References

[1] Gadallah R, Osawa N K and Tanaka S K 2017 Evaluation of Stress Intensity Factor for A Surface Cracked Butt Welded Joint Based on Real Welding Residual Stress *Ocean Engineering* **138** 123-139
[2] Pavan M, Furfari D, Ahmad B, et al. 2019 Fatigue crack growth in a laser shock peened residual stress field *International Journal of Fatigue* **123** 157-167
[3] Reza M N, Mahmoud S and Khalil F 2019 Prediction of fatigue crack propagation and fractography of rail steel *Theoretical and Applied Fracture Mechanics* **101** 320-331
[4] Branco R, Prates P A, Costa J D, et al. 2018 New methodology of fatigue life evaluation for multiaxially loaded notched components based on two uniaxial strain-controlled tests *International Journal of Fatigue* **111** 308-320
[5] Falkowska A, Seweryn A and Tomczyk A 2018 Fatigue life and strength of 316L sintered steel of varying porosity *International Journal of Fatigue* **111** 161-176
[6] Dubov A A 1997 A Study of Metal Properties using the Method of Magnetic Memory *Metal Science and Heat Treatment* **39** 401-402
[7] Liu B, Xue X M, LI J M, et al. 2019 Grain size effect on metal magnetic memory signal for stress damage evaluation of low carbon steel *Nondestructive Testing and Evaluation* **34(3)** 267-282
[8] Xing H Y, Dang Y B, Wang B, et al. 2016 Quantitative Metal Magnetic Memory Reliability Modeling for Welded Joints *Chinese Journal of Mechanical Engineering* **29(2)** 372-377
[9] Su S Q, Zhao X R, Wang W, et al. 2019 Metal Magnetic Memory Inspection of Q345 Steel Specimens with Butt Weld in Tensile and Bending Test *Journal of Nondestructive Evaluation* **38(3)** 1-17

IOP Conf. Series: Materials Science and Engineering **964** (2020) 012022 doi:10.1088/1757-899X/964/1/012022

Fatigue Damage of Crystalline Solar Cell under Cyclic Loadings

Zhiqi Li*

College of Civil Engineering, Hunan University, Changsha 410082, China

*Corresponding author email: zqlee@hun.edu.cn

Abstract. In application of building attached photovoltaic (PV) in China, the wind loads are the main mechanical loading for the modules of the PV system installed on the roofs of the bulidings. Due to the fluctuation of the wind load, the initial damages inside the cells will develop under this cyclic loading, as a result, the power loss of the solar cell can be observed due to the development of the fatigue damage. In this paper, the damage evolution equation for solar cell was derived based on the continuum damage mechanics, the fatigue damage evolution in the solar cell was simulated by FE and the results are in good agreement with that of experiment.

1. Introduction

In recent years, photovoltaic has been widely used in China. A considerable numbers of PV modules are installed on roofs or integrated in buildings for sufficient sunlight. In this case, wind pressure is the main load of modules. For PV modules installed at this position, wind load is the most common external load besides gravity. Due to the fluctuation of wind load, the long-term effect leads to the development of the initial damage in the crystalline solar cells of photovoltaic modules and the decay of some electrical properties or overall failure.

It is one of the effective methods to predict the fatigue life of the structures by applying the continuous damage mechanics in the fatigue problem to derive the reasonable fatigue damage evolution equation. Chaboche and Lesne (1988) first applied the continuous damage mechanics (CDM) method to the fatigue problems. Lemaitre (1996) proposed the fatigue damage evolution equation based on the CDM method and the model was improved for the low cycle fatigue damage prediction (Yang et al, 1997) and the high cycle problem (Xiao et al, 1998). Later, this model was extended to the anisotropic damage problem (Lemaitre et al, 2000). During past years, some models in CDM framework were developed and applied for brittle or quasi-brittle materials (Richard et al, 2010; Richard and Ragueneau, 2013; Yadav and Thapa, 2020; Yang and Liu, 2020).

Despite the fact that some efforts on calculation of the stress and strain of the solar cells in modules brought us the information, the investigations on the development of fatigue damage resulting from the cyclic loading and the corresponding power loss of the cells is, however, not available in literature so far. In this paper, the damage accumulation and the development in the solar cell were calculation based on the Lemaitre's model, then the ANSYS was used to simulate the evaluation of the fatigue damage evolution in the solar cell under cyclic loading and the results were compared to experiment.

2. Damage Evolution in Crystalline Solar Cell

Monocrystalline and polycrystalline silicon cells are mainly two types of the crystal solar cells in applications. In general, monocrystalline silicon is an anisotropic material (Li et al, 2008), its elastic modulus in (111) crystal surface (about 140GPa) is lower than that of other two crystal surfaces (about 180GPa) while the anisotropy of polycrystalline silicon is much less than that of monocrystalline silicon. In the present analysis, for simplicity, the solar cell is assumed to be isotropic material and the damage variable D is independent on the direction of cross section.

For isotropic damage, the effective stress tensor of the cell is expressed as

$$\tilde{\sigma}_{ij} = \frac{\sigma_{ij}}{1-D} \tag{1}$$

where σ_{ij} represents the Cauchy stress tensor. When the isotropic damage is considered, the damage evolution law reads (Lemaitre and Desmorat, 2005)

$$\dot{D} = \dot{\lambda}\frac{\partial \psi_D}{\partial Y} \tag{2}$$

where $\dot{\lambda}$ is the plastic multiplier, ψ_D denotes the dissipative damage potential, Y represents the strain energy density release rate

$$Y = \frac{\tilde{\sigma}_{\text{eq}}^2}{2E} R_{\text{v}} \tag{3}$$

$$\tilde{\sigma}_{\text{eq}} = \frac{\sigma_{\text{eq}}}{1-D} \tag{4}$$

$$R_{\text{v}} = \frac{2}{3}(1+\mu) + 3(1-2\mu)\left(\frac{\sigma_{\text{H}}}{\sigma_{\text{eq}}}\right)^2 \tag{5}$$

in which E and μ are the Young's modulus and Poison ratio of the material, respectively, R_{v} is the triaxiality function, $\tilde{\sigma}_{\text{eq}}$ is effective equivalent stress, σ_{eq} and σ_{H} are von Mises stress and hydrostatic stress, respectively:

$$\sigma_{\text{eq}} = \left(\frac{3}{2}s_{ij}s_{ij}\right)^{\frac{1}{2}} \tag{6}$$

$$s_{ij} = \sigma_{ij} - \sigma_H \delta_{ij} \tag{7}$$

$$\sigma_{\text{H}} = \frac{1}{3}\sigma_{kk} \tag{8}$$

where s_{ij} denotes the stress deviator, δ_{ij} is Kronecker symbol. Based on the experimental observations, Lemaitre and Desmorat (2005) suggested that the dissipative damage potential function ψ_D can be expressed in a simple form as

$$\psi_D = \frac{s_0}{(s_1+1)(1-D)}\left(\frac{Y}{s_0}\right)^{s_1+1} \tag{9}$$

where s_0 and s_1 are material parameters.

Crystalline silicon is a brittle material and the plastic deformation in macroscopic level or even in mesoscopic level can not be observed in the range of working temperature of the module, however, the micro-scale plastic strains always exist at the microcrack tips and the micro-scale defects.

If π denotes the microplastic equivalent strain, the plastic multiplier $\dot{\lambda}$ can be expressed in terms of the accumulated plastic strain rate $\dot{\pi}$ and the damage variable D as

$$\dot{\lambda} = (1 - D)\dot{\pi} \tag{10}$$

Assuming that the microplastic equivalent strain, π, for the cyclic loading can be expressed in the form of the Ramberg–Osgood relation

$$\pi = \left(\frac{\tilde{\sigma}_{\text{eq}}}{K}\right)^{\frac{1}{n}} \tag{11}$$

in which K is the cyclic strength coefficient, and n represents the cyclic strain hardening component. Substituting (3), (4), (5), (9), (10) and (11) into (2), yields

$$\dot{D} = \frac{1}{nK^{1/n}}\left(\frac{R_{\text{v}}}{2Es_0}\right)^{s_1}(\tilde{\sigma}_{\text{eq}})^{\alpha}\dot{\tilde{\sigma}}_{\text{eq}} \tag{13}$$

in which $\alpha = 2s_1 + 1/n - 1$. If $\tilde{\sigma}_{\text{eqM}}$ and $\tilde{\sigma}_{\text{eqm}}$ represent the maximum and minimum equivalent effective stress in the cycle number N, respectively. The increment of damage per cycle can be obtained by integrating (13) in one cycle, we have

$$\begin{aligned}\frac{\partial D}{\partial N} &= \frac{1}{nK^{1/n}}\left(\frac{R_{\text{v}}}{2Es_0}\right)^{s_1}\int_{\tilde{\sigma}_{\text{eqm}}}^{\tilde{\sigma}_{\text{eqM}}}(\tilde{\sigma}_{\text{eq}})^{\alpha}\,\mathrm{d}\tilde{\sigma}_{\text{eq}} + \frac{1}{nK^{1/n}}\left(\frac{R_{\text{v}}}{2Es_0}\right)^{s_1}\int_{\tilde{\sigma}_{\text{eqM}}}^{\tilde{\sigma}_{\text{eqm}}}(\tilde{\sigma}_{\text{eq}})^{\alpha}(-\mathrm{d}\tilde{\sigma}_{\text{eq}}) \\ &= \frac{2[(\sigma_{\text{eqM}})^{\alpha+1} - (\sigma_{\text{eqm}})^{\alpha+1}]}{(\alpha+1)nK^{1/n}(1-D)^{\alpha+1}}\left(\frac{R_{\text{v}}}{2Es_0}\right)^{s_1}\end{aligned} \tag{14}$$

At the constant temperature, the cyclic strength coefficient K, the cyclic strain hardening component n, and the parameters s_0 and s_1 are the material constants. The unknown parameters in (14) can be summarized as an undetermined parameter

$$C = \frac{2}{(\alpha+1)nK^{1/n}(2Es_0)^{s_1}} \tag{15}$$

and then the increment of the damage per cycle is rewritten as

$$\frac{\partial D}{\partial N} = C[(\sigma_{\text{eqM}})^{\alpha+1} - (\sigma_{\text{eqm}})^{\alpha+1}]\frac{R_{v}^{s_1}}{(1-D)^{\alpha+1}} \tag{16}$$

3. FE Simulation

In this section, the cell damage evolution in the module of 1580mm×808mm is simulated via the finite element method. The laminated configurations of the module from the top to the bottom are 3.2mm glass panel, 0.5mm Ethylene-Vinyl Acetate (EVA) layer, 0.19mm cell layer, 0.5mm EVA layer and 0.35mm Tedlar/Pet/Tedlar (TPT) layer, respectively. Since the cells are separated by small gaps in the module, the calculation unit is cut down at the middle of the module with the size of 129mm×129mm which contains one 125mm×125mm cell only as shown in Figure 1(a).

The loads acting on the calculation unit are shown in Figure 1(b). The corresponding material parameters and the geometry of the layers are shown in table 1 where E is young's modulus, t is the thickness of each layer, μ is Poisson's ratio, and a is the side length. According to the experimental results (Eitner et al, 2010), the shear storage modulus of the EVA layer used in calculation is about 6MPa at the temperature of $T = 306$K.

Since the size of the cell is about one order smaller compared to that of the glass panel of the module, it is reasonable to assume that the bending moment of the module within the calculation unit range is uniform. The x-axis of the cell is along the short side direction of the module while the y-axis is along the long side direction of the module in calculation. Under the wind pressure of 0.65kN/m^2, the corresponding bending moments at the middle of the module are $M_x = 55.88Nm$ and $M_y = 37.45Nm$, respectively (Li and Yang, 2016).

Table 1. The parameters for the layers.

	E(GPa)	t(mm)	μ	$a(mm)$
Glass panel	72	3.2	0.20	129
Cell	180	0.19	0.27	125
TPT layer	1.5	0.35	0.30	129
EVA layer	G_{EVA} = 6MPa	0.5	0.2	129

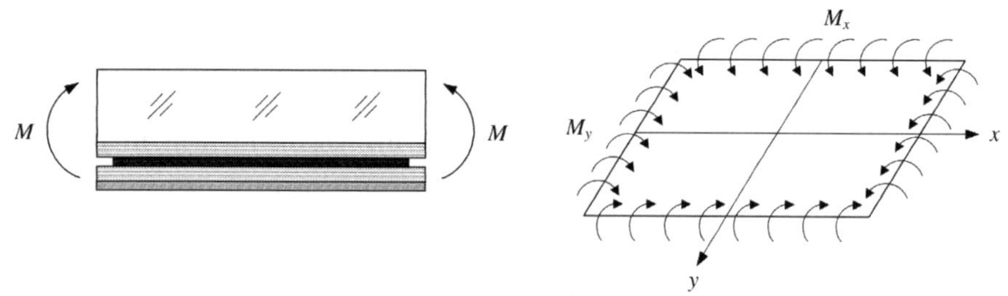

Figure 1. (a) The calculation unit of the cell; (b) The coordinates and the loads of the unit.

It should be noted that the uniform bending moment M_x and M_y in Figure 1(b) are applied to the glass panel, the upper and lower EVA layers, and the TPT layer, leave the cell free from the bending moment. In this manner, the loads of the cell are from the shear deformation of the upper and lower EVA layers, which is closed to the real situation for a separated cell layer.

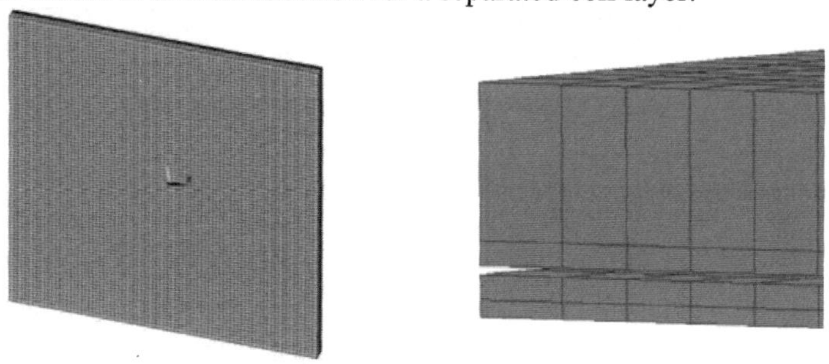

Figure 2. (a) The meshed finite element model; (b) the details.

The upper and lower surfaces of the calculation unit are divided into 80×80 meshes, and in the thickness direction it is divided according to the material of each layer. The established and meshed finite element model is shown in Figure 2. In order to avoid rigid body movement and rotation during the analysis, the nodes located on the x-axis of the unit are constrained to move in y-direction while those located on y-axis of the unit are restrained from moving in x-direction.

The commercial software ANSYS is used to establish the fatigue damage model. The cycle-by-cycle method to calculate the damage accumulation is used in analysis. For each cycle, the material properties are considered as the constants to simplify the problem. The effect of the damage for each element in the cell is introduced by the reduction of the Young's modulus. The module is subjected to pulsation load, i.e., $\sigma_{eqm} = 0$, in this case, the increment of the damage per cycle (16) reduces to

$$\Delta D = C(\sigma_{\mathrm{eqM}})^{\alpha+1} \frac{R_{\mathrm{v}}^{s_1}}{(1-D)^{\alpha+1}} \tag{17}$$

For *j*th element at the *i*th cycle, the steps to evaluate the damage accumulation for each element in the cell as follow:

(1) For *j*th element, the initial damage of the *i*th cycle is given by

$$D_j(i) = D_j(i-1) + \Delta D_j(i-1) \tag{18}$$

(2) The effective Young's modulus of the element is replaced by $E_i^j = [1-D_j(i)]E$ in calculation of the stress σ_{eqM}

(3) The damage increment $\Delta D_j(i)$ produced by this cycle is then evaluated by

$$\Delta D_j(i) = C(\sigma_{\mathrm{eqM}})^{\alpha+1} \frac{R_{\mathrm{v}}^{s_1}}{[1-D_j(i)]^{\alpha+1}} \tag{19}$$

In this manner, the damage evolution for each element in the cell can be calculated cycle by cycle. After finished the *i*th cycle, the damage of the *j*th element in the cell is expressed as

$$D_j = D_j(i) + C(\sigma_{\mathrm{eqM}})^{\alpha+1} \frac{R_{\mathrm{v}}^{s_1}}{[1-D_j(i)]^{\alpha+1}} \tag{20}$$

in which $\alpha = 2s_1 + 1/n - 1$. The average damage D of the cell is calculated by

$$D = \frac{1}{V_0} \sum_{j=1}^{m} D_j V_j \tag{21}$$

where V_j and D_j are the volume and damage variable of the *j*th element, respectively, and V_0 is the total volume of the cell.

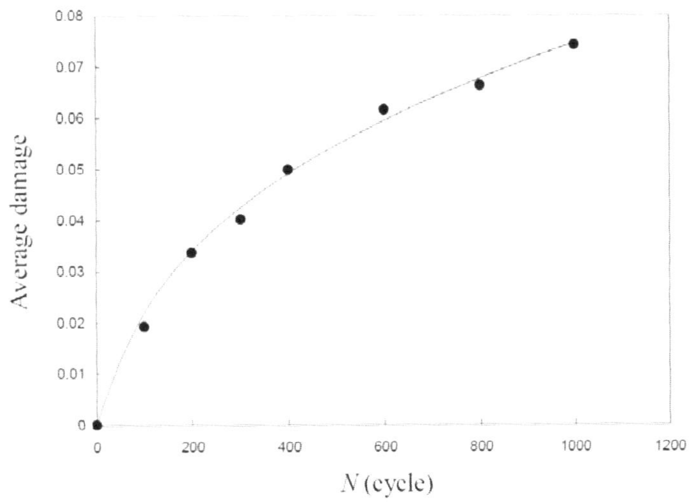

Figure 3. The average damage from the FE simulation and compared to experiment.

The parameters C, α, n and s_1 in (21) are obtained by fitting the experimental data in Li (2013) and the results are shown in table 2.

Table 2. The fitted parameters of the cell.

C	α	n	s_1
3.7×10^{-10}	−0.5	0.034	−14.456

The comparison of the average damage evolution of the cell evaluated from FE and the experimental results in Li (2014) are plotted in Figure 3. It can be seen that the results from the FE simulation are in good agreement with those of the experiment.

4. Conclusion

The damage evolution equation based on the Lemaitre's model for solar cell was derived. The damage accumulation formula for cycle-by-cycle calculation was established and the ANSYS was employed to simulate the damage evaluation in the cell. The numerical results are in good agreement to those of the experiment.

Acknowledgements

This work was supported by the Key Laboratory for Damage Diagnosis of Engineering Structures of Hunan Province.

References

[1] Chaboche J L and Lesne P M 1988 A nonlinear continuous fatigue damage model *Fatigue and Fracture of Engineering Materials and Structures* Vol 11 No 1 pp1-7.
[2] Eitner U, Kajari-Schroder S, Kontges M and Brendel R 2010 Non-linear mechanical properties of Ethylene-Vinyl Acetate (EVA) and its relevance to thermomechanics of photovoltaic modules, *Proceeding of 25th European Photovoltaic Solar Energy Conference* Valencia Spain pp 1-3.
[3] Lemaitre J A 1996 Course on Damage Mechanics *Springer-Verlag* Berlin.
[4] Lemaitre J and Desmorat R 2005 Engineering Damage Mechanics *Springer-Verlag* Berlin.
[5] Lemaitre J, Desmorat R and Sauzay M 2000 Anisotropic damage law of evolution *European Journal of Mechanics A/Solids* Vol 19 pp 187-208.
[6] Li Degang, Liang Yingchun and Bai Qingshun 2008 Nano–indentation experiments on anisotropic mechanical properties of single crystal silicon *Aviation Precision Manufacturing Technology* Vol 44 No 2 pp 9-12 (In Chinese).
[7] Li Z Q 2014 Investigation of fatigue damage of crystalline silicon solar cell in PV module for BIPV *Acta Energiae Solaris Sinica* Vol 35 No 6 pp 1004-1009 (In Chinese).
[8] Li Z Q and Yang F C 2016 Stress Analysis of Crystalline Cell in PV Modules Using ANSYS *Journal of Hunan University* (Natural Sciences) Vol 10 No 11 pp 149-153 (In Chinese).
[9] Richard B, Ragueneau F, Cremona C and Adelaide L 2010 Isotropic continuum damage mechanics for concrete under cyclic loading: stiffness recovery, inelastic strains and frictional sliding *Engineering Fracture Mechanics* Vol 77 pp 1203-1223.
[10] Richard B and Ragueneau F 2013 Continuum damage mechanics based model for quasi brittle materials subjected to cyclic loadings: Formulation, numerical implementation and applications *Engineering Fracture Mechanics* Vol 98 pp 383-406.
[11] Rozgonyi G, Youssef K and Kulshreshtha P 2011 Silicon PV Wafers: Mechanical strength and correlations with defects and stress *Solid State Phenomena* Vol 178-179 pp 79-87.
[12] Xiao Y C, Li S and Gao Z 1998 A continuum damage mechanics model for high cycle fatigue *International Journal of Fatigue* Vol 20 No 7 pp 503-508.
[13] Yadav I N and Thapa K B 2020 Fatigue damage model of concrete materials *Theoretical and Applied Fracture Mechanics* Vol 108 p 102578.
[14] Yang X H, Li N, Jin Z H and Wang T J 1997 A continuous low cycle fatigue damage model and its application in engineering materials *International Journal of Fatigue* Vol 19 No 10 pp 687-692.
[15] Yang Z and Liu H 2020 A continuum fatigue damage model for the cyclic thermal shocked ceramic-matrix composites *International Journal of Fatigue* doi: https://doi.org/10.1016/j.ijfatigue. 2020.105507.

IOP Conf. Series: Materials Science and Engineering **964** (2020) 012023 doi:10.1088/1757-899X/964/1/012023

Application of Morgenstern-Price Method in Reversion Analysis of the Loess Strength Parameters

Xiaoyan Lin[1,2,*], Linfan Zhang[3], Ze Yang[1,2], Ping Li[3] and Tonglu Li[3]

[1]School of Earth Sciences and Engineering, Xi'an Shiyou University, Xi'an, China
[2]Shaanxi Key Laboratory of Petroleum Accumulation Geology, Xi'an Shiyou University, Xi'an, China
[3]School of Geological Engineering and Surveying, Chang'an University, Xi'an, China

*Corresponding author email: xylin@xsyu.edu.cn

Abstract. It is difficult to obtain reliable shear strength parameters for the stability analysis and evaluation of high loess slopes. Hence, this paper determines 17 slopes with a height of 40-120 m based on the measured loess slopes in Jingbian-Ansai subregion of northern Shaanxi, China. The model is established using the finite element method, and the comprehensive shear strength parameters of different slope heights of the high slope in the study area are obtained by inversion using the Morgenstern-Price method. The results show that when the height of the slope increases, the cohesion c in the soil increases, and the internal friction angle φ decreases. According to the calculation results, a comprehensive inversion curve of strength parameters under the slope height of 40 to 120 m is drawn, which can directly evaluate the stability of the loess high slope, which improves the work efficiency and also provides reliable data support in the high side Slope geometry design.

1. Introduction

With the country's construction of the loess plateau has become increasingly active, these high slopes pose a great threat to the safety of people's production and life, and the stability of high loess slopes has gradually been given more attention. Much research shows that the relationship among the position of the sliding surface of the slope, the shear strength parameters and the stability coefficient are key to the evaluation analysis. Methods for obtaining strength parameters through inversion include strength reduction method that gradually reduces the strength parameters of sliding surfaces[1], numerical simulation methods (FLAC3D), genetic algorithms[2], etc. The above studies have conducted much research on medium and low loess slopes with slopes below 30 m, but the research on the stability of high loess slopes above 30 m has not reached a unified understanding [3] and lacks an accurate method for analysis.

In addition, most studies take a single slope or landslide as the object, and there is little systematic analysis and research on the variation law of the strength parameters of regional high loess slopes [4]. Hence, this paper analyzes the limit state slopes with obvious signs of damage in the natural high loess slopes in Jingbian-Ansai subregion of northern Shaanxi and the correlation between the slope height and the width of the slopes. We model slopes of different heights and apply the Morgenstern-Price method to inverse their corresponding comprehensive strength parameters. According to the regression equation curve of the slope height and slope width and the comprehensive strength parameters that correspond to the slope height, the relationship curves between the cohesion of different slope heights and the internal friction angle are drawn to calculate the stability factor of the known slope shape and the design of the proposed slope, also can analyze the safety factor and stability of the known high slope and design the slope shape of the slope to be constructed.

2. High-Slope Measurement and Slope Height-Slope Width Fitting

2.1. Measurement and Statistical Analysis of Limit-state Slopes

In this paper, the regionally distributed loess slopes are mostly distributed along the rivers and ditches of the Xing Zi River, Luo River, and Wuding River (see figure 1). These slopes are located in the Loess Plateau and its surrounding areas; the underlying bedrock has a large burial depth. All slopes in the statistical analysis adopt Wu Weijiang's classification criteria for pure loess landslides[5]. According to incomplete statistics, landslides in the loess layer account for more than one-third of the total number of landslides in the Loess Plateau. Therefore, the types of landslides in the loess layer selected in this paper are typical.

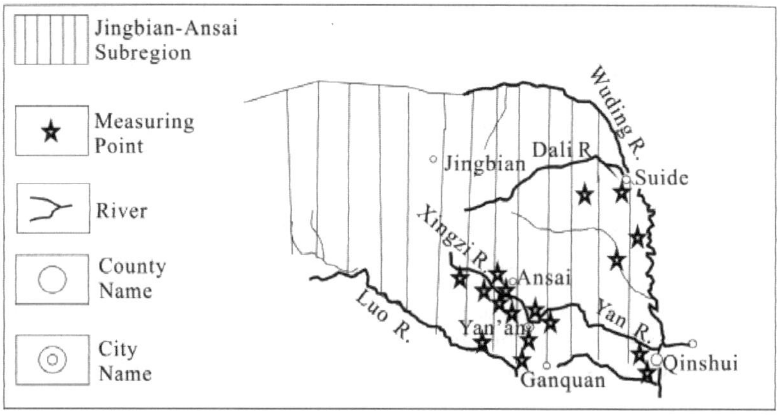

Figure 1. Slope measurement and sampling points of the loess limit state in the study area.

The premise of the strength parameter inversion is that the selected slopes is in the limit equilibrium state. For a given slope, one can determine whether it is in the limit equilibrium state through field surveys. In response to this problem, this article uses 4 criteria by [6] proposed that a clear concept of the limit-state slope and combined with the natural history analysis method, considering the regional geological background of the study area, this study investigated the signs, scale and morphology of high-slope deformation, sampled different types of slopes to determine their severity, and analyzed and controlled their deformation and damage. In the end, 17 slopes in the limit state were determined in study area, and the slope height was greater than 40 m and less than 130 m (as shown in ☆of figure 1).

2.2. Research on the Slope Height and Slope Width Fitting of High Loess Slopes

A correlation analysis was performed on the slope height and slope width of 17 limit-state high slopes in the study area. The exponential model of equation (1) was used to fit them through repeated trial calculations, and all degree-of-correlation coefficients are above 0.89. This fitting result provides a basis for arbitrarily setting the slope height when the slope model is established in the later period. Knowing the slope height, the corresponding slope width can be calculated using equation (1) to determine the slope shape.

$$L = 0.79291H^{1.10242} \tag{1}$$

where H is slope height, which is measured from the foot of the slope to the top of the slope; L is slope width, which is the distance between the horizontal projection of the slope shoulder and the slope foot.

3. Inversion Model and Result Analysis

According to the basic principle of [7] inverse strength parameter algorithm, two or more cross-section equations can be simultaneously solved under similar conditions. In the same area, when the slope heights are not greatly different, the water content, slip state, and mechanical parameters are similar. Statistical analysis methods are more commonly used in the study of landslide kinematics models like [8] and [9]. Therefore, the cross-section height difference of the joint is controlled at 5 m, and a set of inversion values of shear strength parameters are taken at a height interval of 10 m to represent the comprehensive strength parameters in this range of slope height. With the set slope height, the corresponding slope width is obtained by the slope width-slope height fitting equation (1). The internal friction angle φ and gravity γ are set. The value of φ is 5-35°, and the value gradient is 5°. The values of gravity γ within the range of 20m from the top of the slope is 16.1 kN/m^3, the value from 20 to 70m from the top of the slope is 18.0 kN/m^3, and the distance from the top of the slope is more than 70m to take the value of 21.0 KN/m^3.

The slope model is constructed using the Slope/w module of the finite element software GeoStudio, and the Morgenstern-Price method is used to invert the comprehensive cohesion c. Figure 2 shows a slope model with a height of 80 m in the Jianbian-Ansai subregion, just list this step, the slope models of each area are successively established, the three c-φ relationship curves in each 10-m-slope-height section of each area are intersected at one point, and the coordinates of this intersection point are the comprehensive strength parameters of the slope in the limit state within the range of 10 m, as shown in figure 3. Thus, slope heights of 40 m, 45 m, 50 m, 55 m, 60 m, 65 m, 70 m, 75 m, 80 m, 85 m, 90 m, 95 m, 100 m, 105 m, 110 m, 115 m and 120 m are obtained within the 40 ~ 120 m slope heights of the Jianbian-Ansai subregion inversion of 17 shear strength parameters. Then, the comprehensive strength parameters in the 10-m range of similar slope heights in each subregion are obtained, as shown in table 1.

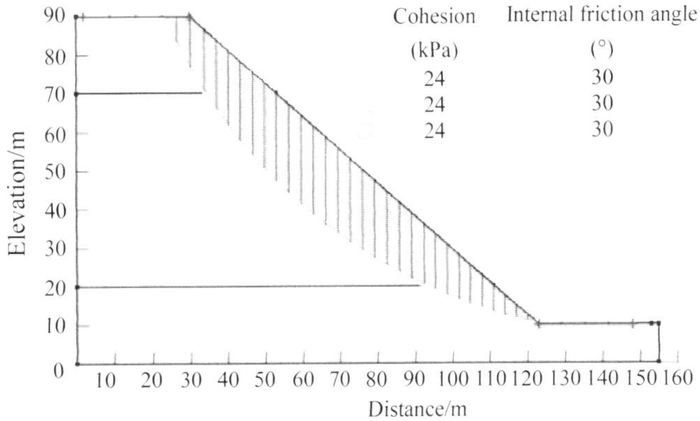

Figure 2. Using slope/w to establish the strength parameters of 80 m slope model in the Jingbian-Ansai subregion.

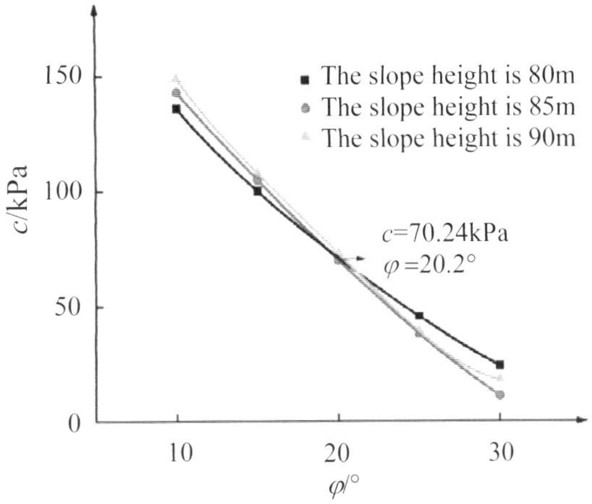

Figure 3. Parameters inversion results of slope height of 80~90 m in the Jingbian-Ansai subregion.

4. Inversion Results

According to the above inversion process, the shear strength parameter values of 8 slope height ranges of the Jingbian-Ansai subregion are obtained (see table 1). The inversion values of shear strength parameters of different slope heights are shown in figure 4.

Table 1. Inversion of comprehensive strength parameters for different slope heights in Jingbian-Ansai subregion.

Strength parameters / Slope height/m	Jingbian-Ansai subregion	
	Internal friction angle φ/°	Cohesion c/kPa
40~50	34.92	4.53
50~60	35.08	8.25
60~70	24.63	37.25
70~80	21.12	61.6
80~90	20.2	70.24
90~100	17.48	91.54
100~110	15.01	114.8
110~120	14.98	118.6

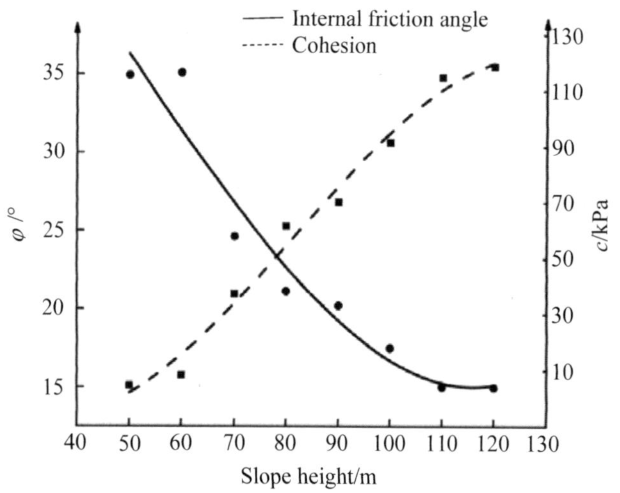

Figure 4. Inversion of strength parameters of different slope heights in the Jingbian-Ansai subregion.

The curve in figure 4 show that when the slope height increases, the increase in cohesive force becomes larger, and the decrease in internal friction angle becomes smaller. However, the older the stratum, the lower the φ value, which is different from the current understanding. There are two reasons for the author's analysis: (1) The stress state of the conventional test is inconsistent with the actual potential sliding surface stress. In conventional tests, 400 kPa is used as the maximum confining pressure, and the stress on the potential sliding surface of high slopes may be much higher than this value. (2) From low stress to high stress, the Mohr-Coulomb envelope does not change nonlinearly, and its slope gradually slows down. Therefore, the results of routine tests overestimate the internal friction angle under high stress (>400 kPa).

When the slope height is lower, the fluctuation of the internal friction angle slightly affects the friction force (anti-sliding force) and hardly affects the stability change. When the slope height is higher, the fluctuation of internal friction angle has a greater impact on friction (anti-sliding force), i.e., the effect of the internal friction angle on the slope stability becomes more obvious when the slope height increases. From the analysis of the principle of limit equilibrium calculation, the potential slip surface of a slope with a small slope height is shallow and steep, the normal stress of the slip surface is small, the anti-sliding effect of the internal friction angle is small, and the stability is mainly a function of cohesion. The potential slip surface of a slope with a large slope height is deep and gentle, the positive stress of the slip surface is large, and the effect of cohesion on the entire potential slip surface is no longer as obvious as that of a small slope height. At this time, the internal friction angle mainly controls the stability of the slope.

5. Conclusions

1. The high loess slopes in the Jingbian-Ansai subregion increase with the increase in slope height, the cohesion of the loess increases, and the internal friction angle decreases. In contrast, when the slope height increases, the contribution of the internal friction angle to the shear strength becomes more prominent, and the increase in cohesion also increases. When the slope height is in the range of 40 ~ 80 m, the cohesion will increase by 2 ~ 24 kPa for every 10-m increase in slope height; when the slope is in the range of 10 ~ 25m, the cohesion will increase by 25 ~ 45 kPa for every 10-m increase in slope height.
2. Based on the slope height and strength parameter curve obtained by the inversion method, the strength parameters of the high slope can be directly obtained according to the slope height. This method can be used for the initial judgment of the stability coefficient of known slopes and to design the slope shape and strength parameters.

Acknowledgement
The research is supported by National Natural Science Foundation of China (Grant No. 41402274, 41877242).

References

[1] Deng J H, Wei J B and Min H 2003 Three-dimensional analysis method of landslide stability based on strength reduction concept (I): inverse analysis of shear strength parameters of sliding zone soil vol 24 issue 6 (Rock and Soil Mechanics) pp 896-900
[2] Tang L S, Yin K L, Zhou L and Miu H B 2012 Research on back analysis of shear strength parameters of landslide based on numerical simulation and displacement monitoring vol 39 issue 4 (Hydrogeology and Engineering Geology) pp 32-37
[3] Jiao L and Li K Y 2010 Study on structural characteristics and design method of high slope of loess cutting in huangyan highway vol 8 issue 2 (Journal of Water Resources and Architectural Engineering) pp 136-138
[4] Liu C 2014 Inversion study on shear strength parameters of loess high slope (Dissertation, Chang'an University) chapter3 pp 17-32
[5] Wu W J and Wang N Q 2002 Basic types and activity characteristics of loess landslides vol 13 issue 2 (Chinese Journal of Geological Hazards and Prevention) pp 38-42
[6] Li P, Wang B G and Li T L 2009 Application of natural analogy method in loess cutting slope design vol 26 issue 2 (Highway Transportation Technology) pp 1-5
[7] Zheng X M 2003 On the inverse algorithm of the strength characteristics and strength parameters of slide zone soil vol 24 issue 4 (Rock and Soil Mechanics) pp 528-532
[8] Kenneth J. HSU 1975 Catastrophic debris streams (sturzstroms) generated by rockfalls vol 86 issue 1 (Geological Society of America Bulletin) pp 129-140
[9] Okuda S 1984 Features of debris deposits of large slope failures investigated from historical records (Kyoto: Kyoto University) pp 353-368

Structural Design of Prefabricated Sidewalk in Sponge City

Xuejun Wen[1] and Ming Huang[2,3]*

[1]Vice-Chief Engineer, Shanghai Municipal Engineering Design Institute (Group) Co., Ltd., Shanghai 200092, China
[2]Senior Engineer(R&D), PhD, Postdoctoral, Greenland Infrastructure Group Co., Ltd., Shanghai 200010, China
[3]Senior Engineer(R&D), PhD, Postdoctoral, GIG Technology Research Institute, Shanghai 200010, China

*Corresponding author email: huangming@tongji.edu.cn

Abstract. In the traditional pavement design, rainwater is mainly discharged by surface runoff, and in the structural design, the key point is the pavement waterproof. Therefore, the traditional sidewalk design is inconsistent with the requirements of sponge city. The research basis of this paper comes from a field engineering, the structural parameters of prefabricated sidewalk and the corresponding design methods are studied and summarized. The structural function of sidewalk is realized by superposition and combination of each structural layer. For the realization of prefabricated sponge sidewalk technology, the function of each structural layer must be fully considered, and the function of each structural layer should be designed and combined according to the design purpose. The results show that, by discussing the function of each layer of prefabricated sidewalk, the structural layer of prefabricated sidewalk is preliminarily determined(see Figure 2 in main text); according to the strength requirements and water permeability requirements of the structure, the thickness range of the structural layer is 13.11cm-21.3cm; finally, the relationship between plate thickness and critical position load, temperature fatigue stress, and finally determines the thickness of precast cement concrete slab is 15cm.

1. Introduction

Sidewalks are an important part of urban infrastructure and urban environmental carriers, and the construction and maintenance of sidewalks not only meet the traffic requirements of municipal roads, but also reflect the modernization and civilization level of a city [1,2]. With the implementation of the sponge city concept one by one, sidewalks, as an important part of slow traffic, have received widespread attention [3-5]. In order to improve the quality of the construction, the water permeability and speed up the paving speed, a comprehensive upgrade of the original pavement structure of the sidewalk is to meet the urgent need. Factory-based prefabricated assembly technology can make the sidewalk surface layer have good durability, flatness and sponge characteristics [6,7]; the corresponding base layer is changed from the original cast-in-place concrete to a permeable and non-cast-in-place base layer to keep the construction fast and Less cast-in-place guarantees the advantages of environmental friendliness while minimizing traffic interference. As the prefabricated technology has been popularized and applied in construction, bridges, tunnels and other projects [8-10], this research and development technology is the first research and application of prefabricated technology in urban slow traffic, which expands the application range of superior technologies. Due to structural changes, the structural design of the sponge prefabricated sidewalk is inconsistent with the structure of conventional paving tiles. Therefore, the research in this paper is to study the structural parameters of the prefabricated sidewalks required by sponge.

2. Comparative Analysis of the Two Structures

2.1. Existing Structure

Most of the current sidewalks are composed of "soil foundation + concrete foundation + sand leveling layer + block brick paving layer" [11,12]. Conventional sections have adopted this structure, and the specific structure is shown in Figure 1.

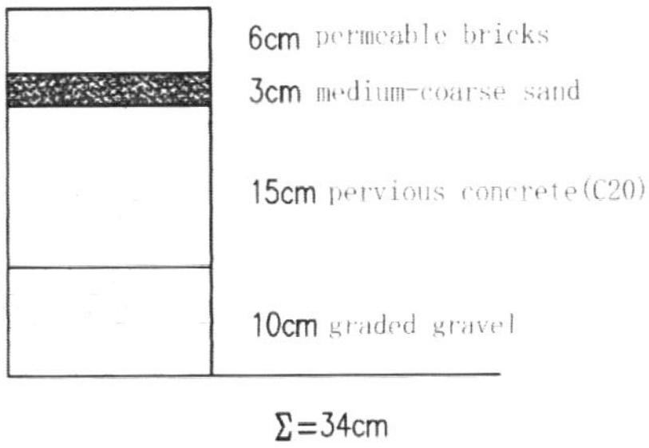

Figure 1. Representative drawing of existing sidewalk structure drawing.

According to Figure 1, the base layer of the existing structure adopts C20 concrete base layer. This kind of structure layer needs to reserve a large area on site, add cement concrete mixing or storage equipment and pour it, and it will take a long time to cure. The surrounding environment will have an impact and the construction period will be longer.

2.2. Innovative Structure Design

Analysis of the structure and function of the sidewalk, different structures have different materials to choose from, which will produce many structural combinations. The structural combination is not a simple arrangement and combination but gives full play to the characteristics of the materials used. And each material should be compatible [13], the structure layer should be coordinated, and the final design combination should be economical and practical [14].

1) Surface layer: For the surface layer, it should be ensured that the layer has sufficient mechanical strength [15], and it needs to be suitable for prefabricated construction. Therefore, a composite structure of reinforced cement concrete frame + permeable cement concrete is designed;
2) In view of the further increase in the thickness of the prefabricated sidewalk as the surface layer and the improvement of the structural strength, the strength of the base layer can be appropriately reduced. Therefore, the convenient technology replacement of the base layer is a problem that can be solved in the design of the fabricated sidewalk structure layer.
3) Base course and cushion: For drainage sidewalks, the base course and cushion should not only bear the load transmitted from the surface layer, but also take into account the functions of water permeability and water storage. Considering that the surface layer already has higher mechanical properties, the original. The concrete is poured into the base, and a gravel base is designed to have the function of a cushion. At the same time, a permeable pipe is set in the gravel base to collect rainwater and drain. The designed structure is shown in Figure 2.

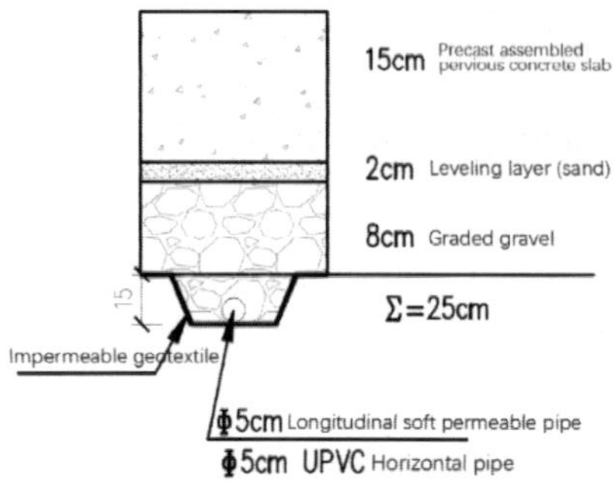

Figure 2. Structure design drawing of prefabricated sponge sidewalk.

3. Parameters of the Prefabricated Permeable Cement Concrete Structure Scheme

3.1. Rainfall Intensity

The prefabricated sponge sidewalk belongs to the category of permeable sidewalks, and its structural design should be combined with local hydrological conditions. According to the current design requirements for permeable sidewalks, the first consideration is rainfall conditions [16]. Therefore, local weather data should be collected and sorted out during design. The main content includes parameters such as rainfall intensity and return period. This paper takes Shanghai as an example, combined with local precipitation data, to determine the relevant parameters of permeable sidewalks.

Shanghai is located between 120°52′ to 122°12′ east longitude and 30°40′ to 31°53′ north latitude. The land area is 6340.5 square kilometers. It is located on the west coast of the Pacific Ocean, the east coast of the Asian continent, and the center of the north and south coasts of China. Shanghai has a subtropical monsoon climate with four distinct seasons, abundant sunshine and abundant rainfall. The climate in Shanghai is mild and humid, with short spring and autumn and long winter and summer. In 2013, the city's average temperature was 17.6°C, sunshine was 1885.9 hours, and precipitation was 1173.4 mm. More than 60% of the annual rainfall is concentrated in the flood season from May to September [17]. This paper uses the rainstorm intensity formula in the Shanghai local standard "Heavy Rain Intensity Formula and Design Rain Pattern Standard" for calculation. The calculation formula is shown in Formula 1.

$$q = \frac{1600\ (1+0.846 lgp)}{(t+7)^{0.856}} \tag{1}$$

Where:

q——Design rainfall intensity, L/(s·hm^2);

p-design return period, years;

t-rainfall duration, min.

According to Formula 1, we can determine the rainfall intensity through the rainfall return period and rainfall duration. As refer to the guidelines [12,18] that "the water storage capacity of permeable surface facilities should not be lower than the rainfall with a return period of two years". Therefore, the return period P=2 is chosen, and set the condition that “one hour of rainfall will not cause surface runoff on the sidewalk”. Incorporating these data into Formula 1, the design rainfall intensity q=54.9mm/h. The reason why the rainfall time is 1h is: because this article only considers the impact of short-term heavy rainfall on drainage sidewalks, because slow traffic only controls surface runoff, short-term heavy rainfall is the main cause of runoff, and the duration of heavy rainfall is usually Not too long.

3.2. Thickness Range of Structural Layer Based on Water Permeability Requirements

In common permeable sidewalks, the base layer is often used as a temporary water storage structure, and rainwater seeps through the soil foundation in the base layer and is finally drained away. The water storage structure of the base layer can be divided into saturated and unsaturated layers at any time, which is a dynamic division of the base structure. The mechanical properties of the pavement can be ensured by measuring the thickness of the base layer. At the same time, the calculation of water storage thickness is also a reference index for the design of structural layer thickness. The drainage sidewalk can regard the side water outlet as the bottom water outlet of the permeable sidewalk, so as to derive the thickness calculation formula with reference to the permeable sidewalk [19]. Set rainwater infiltration strength Wt, cm/s and rainwater infiltration strength Vt, cm/s, Figure 3 is a simplified schematic diagram of the structure.

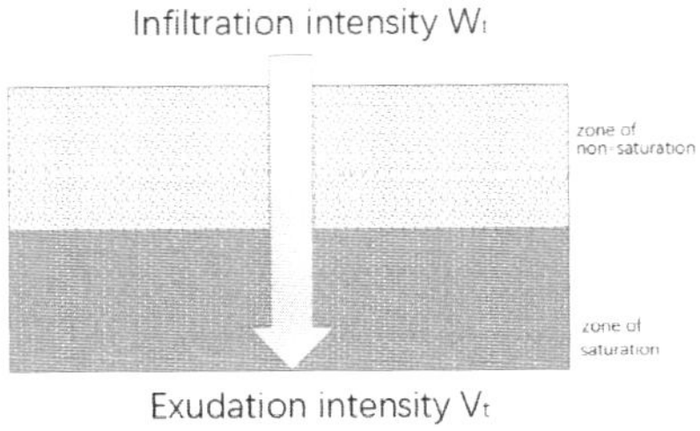

Figure 3. Expansion rate of foamed asphalt.

According to the water balance theory:

$$\mathrm{Akh(t)} = \mathrm{A}\int_0^t (Wt - Vt)dt \tag{2}$$

$$\mathrm{h(t)} = \frac{1}{t}\int_0^t (Wt - Vt)dt \tag{3}$$

Where:
k = effective void ratio of the base layer;
t=precipitation duration;
h(t)=storage height at time t.

Simplify the above integration and take the average rainfall intensity i as the rainfall intensity. Because there is no runoff on the road surface, it means that the infiltration intensity is greater than the rainfall intensity. Take w(t)=i, and the infiltration intensity is taken as the intensity of rainwater discharged into the soil foundation. It can be regarded as the average velocity $\overline{\mathrm{V}}$, and we get:

$$\mathrm{H} = (0.1\mathrm{i} - 3600\mathrm{q})\mathrm{t}/(60\mathrm{v}) \tag{4}$$

Where:
H——Thickness of permeable sidewalk structure, cm;
i——Intensity of regional rainfall, mm/h;
q——The average permeability coefficient of soil foundation;
t——The duration of rainfall, min;
v——The average air void ratio of permeable sidewalk structure layer.

The permeable sidewalk materials will be studied later, and the calculations will be performed after determining the value of each parameter. The designed air void ratio of the permeable surface layer is about 25%, the air void ratio of the base layer is between 35%-40%, and the average permeability coefficient of the soil foundation is 1*10-4cm/s, which can be obtained after substituting the above formula. The thickness range is 13.11cm-21.3cm.

3.3. Accurate Calculation of Plate Thickness Design

The thickness of the cement concrete slab relates to the carrying capacity of the precast slab. To calculate the structural strength, the following assumptions are made:

(1) Since the applied situation are the urban roads, the calculation is carried out according to the medium traffic load level, and the structural reliability is 1.2;

(2) According to the sidewalk surface structure, it is slab surface + gravel base (10cm) + soil foundation (modulus ≥ 20MPa).The temperature warping stress of the concrete road slab at the maximum temperature gradient is calculated according to Formula 6,

$$\sigma_t = \frac{E_c \alpha_c \Delta T}{2\left(1^- \mu_c\right)} \tag{5}$$

Where:

σ_t —Concrete temperature expansion coefficient

α_c =1×10-5/°C

ΔT —Temperature difference between concrete pavement slab and lower layer,
ΔT=Tg×h, his the thickness of the road slab, the recommended value and correction coefficient of the maximum temperature gradient value Tg are shown in Table 1;
μc—Poisson's ratio of concrete, μc=0.15

Table 1. The maximum temperature gradient of each highway's natural zoning.

Natural division	II, V	III	IV, VI	VIII
Maximum temperature gradient (°C /cm)	0.83-0.88	0.90-0.95	0.86-0.92	0.93-0.98

For the pavement of concrete slabs for sidewalks, the fatigue stress of the critical position of the pavement structure under different precast slab thicknesses need to be calculated, in the state of the action of driving load and temperature load (which are multiplied by the corresponding reliability) [20]. The relationship between panel thickness and critical position load+temperature fatigue stress result is shown in Figure 4.

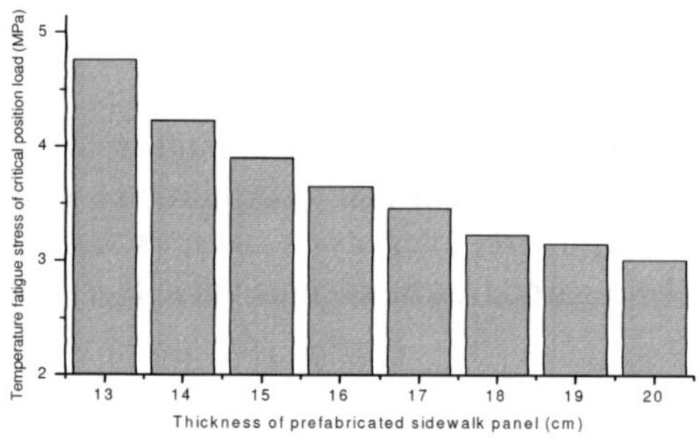

Figure 4. The relationship between plate thickness and critical position load and temperature fatigue stress.

When the thickness of the precast slab is greater than 15cm, the fatigue load stress under the critical position load and temperature of the precast slab is less than 4MPa of the flexural tensile strength of the permeable concrete. When the thickness of the precast slab is greater than 15cm, the fatigue load stress under the critical position load and temperature of the precast slab is less than 4MPa of the flexural tensile strength of the permeable concrete. Therefore, considering factors such as cost and weight, the thickness of the precast cement concrete slab is selected as 15cm.

4. Conclusions

1) By discussing the role of each layer of the prefabricated sidewalk, the structure layer combination of the prefabricated sidewalk is initially determined, see Figure 2.
2) According to the rainfall intensity in Shanghai, the calculation method for the thickness of the structure layer of the sidewalk was determined, and the thickness of the structure layer was finally determined to be between 13.11cm-21.3cm.
3) According to the strength requirements of the structure and the requirements of water permeability, the thickness of the precast cement concrete slab is finally determined to be 15cm.

Acknowledgments

The authors would like to thank 2017, 2020 Technology Program of Shanghai Municipal Engineering Design Institute (Group) Co., Ltd. No. K2017J018, K2020K070, Technology Program of Shanghai Construction Group Co., No. 19YJKF-26 for funding of this project, Tian Jianjun(SMEDI), and Prof. XU Jian(SMEDI) for helpful and constructive prophase studies.

References

[1] Wei Baoban. The humanized space of urban roads. *South China University of Technology*, 2013.
[2] Yachtani Yoshitaka, Matsuzaki Kazuhiro, Ito Akahiko, etc.. Application of high strength rc precast slab pavements in airports. *Papers on Paving Engineering*, 2003, 8: 183-193.
[3] Du Zhonghua. Application of Sponge City Concept in Urban Road Engineering. *Construction & Design for Engineering*,2016(03):69-71+73.
[4] Erhan Güneyisi, Mehmet Gesoğlu, Kareem Q , et al. Effect of different substitution of natural aggregate by recycled aggregate on performance characteristics of pervious concrete. *Materials and Structures*, 2014, 49(1-2):521-536.
[5] Chandrappa A K , Biligiri K P . Pervious concrete as a sustainable pavement material – Research findings and future prospects: A state-of-the-art review. *Construction and Building Materials*, 2016, 111:262-274.
[6] Zhou Zemin. Construction of prefabricated cement concrete pavement. *Central South Highway Engineering*, 1991(3):20-23.
[7] Zhang Yi. Research on prefabricated high-strength permeable concrete pavement. *Changsha University of Science and Technology*. 2017.
[8] Liu Yongjian, Gao Yumin, Zhou Xuhong, Zhang Zejun, Zhang Yuanfang. Technical and economic analysis of small and medium span steel-concrete composite girder bridges. *China Journal of Highway and Transport*,2017,30(03):1-13.
[9] Yao Yiwen, Jiang Lihua, Fan Yiqun. Overview of prefabricated assembly technology of underground space structure. *Urban Roads Bridges & Flood Control*,2012(09):286-292+344.
[10] Wang Jun, Zhao Jida, Hu Zongyu. Development status and thinking of my country's construction industrialization. *China Civil Engineering Journal*,2016,49(05):1-8.
[11] Ma Hailong. Analysis on the construction of road prefabricated sidewalks. *Science and Technology Innovation and Application*, 2013(15):223.
[12] DGTJ08-2241-2017. Technical specification for permeable sidewalk. Shanghai: *Press of Tongji University*
[13] Ito Akahiko, Yamawaki Hiroshi, Tanaka Hideki, etc. Basic examination about the design of the high strength rc precast pavement slab in airport. *Collection of Pavement Engineering*,2003, 8: 173-181.
[14] Novak J, Kohoutková A, Křístek V, et al. Precast concrete pavement – systems and performance review[J]. 2017, 236(1):012030.
[15] Wang Xiangying, Chen Pengfei, Yang Chao, et al. Analysis of factors affecting the performance of prefabricated road slabs. *Low Temperature Construction Technology*, 2019(5).
[16] Liu Wen,Feng Qi,Deo Ravinesh C,Yao Lei,Wei Wei. Experimental Study on the Rainfall-Runoff Responses of Typical Urban Surfaces and Two Green Infrastructures Using Scale-Based Models. *Environmental management*,2020.
[17] http://www.china.com.cn/aboutchina/zhuanti/09dfgl/2009-09/08/content_18488374.htm
[18] GB 50400-2016. Technical code for rainwater management and utilization of building and sub-district. Beijing: China Construction Engneering Press
[19] Wu Shitao. Structure and structure of prefabricated cement concrete pavement. *Tongji University*, 2017.
[20] Huang Ming, Wen Xuejun, Lu Hewei. Analysis on Mechanical Performance of Precast Assembled Pervious Concrete Sidewalk Slab. *Urban Roads Bridge & Flood Control*, 2019(5):237-240

Study on Control Standard of Frost Heave Deformation and Freeze-thaw Strength of Cohesive Soil of Expressway Widening Subgrade in Seasonal Frozen Region

Shengchuan Liu[1,*], Liming Yu[2] and Chun Xu[3]

[1]Research Institute of Highway Ministry of Transport, Beijing 100088, China
[2]Xinjiang Beijing Sifang Engineering Testing & Consulting Co,Ltd, Wurumuqi 830000, China
[3]Beijing Zhongshi Hengtong Construction Engineering Co., Ltd, Beijing 102200, China

*Corresponding author email: 3398194433@qq.com

Abstract. This paper is aiming at control standard of frost heave deformation and freeze-thaw strength of cohesive soil of expressway widening subgrade in seasonal frozen region. Research was proposed to study the deformation law of high water content clay subgrade in freeze-thaw condition through laboratory test, field monitoring and theoretical analysis. According to the results, frost heave deformation control standard of cohesive soil subgrade has been proposed. Considering the overload load, the subgrade's differential deformation rate should be less than 3.07‰ and the differential deformation rate must be less than 2.27‰. Considering the allowable bending tensile stress and the pavement structure service life, the differential deformation rate of the top surface of the subgrade under the effect of frost heaving of the high water content cohesive soil subgrade in the seasonal frost zone should be controlled within 2‰. After that, control standard for freeze-thaw strength of cohesive soil subgrade is put forward that, the subgrade-soil strength ratio of the high water content cohesive soil subgrade in the seasonal frost zone should be controlled within 0.75 under the standard load and within 1 under the overload load considering the allowable flexural stress and service life of the pavement structure.

1. Introduction

In recent years, China has carried out a number of highway expansion project, but a more general problem is that the new and old roadbed settlement of differences, and thus lead to more pavement and subgrade diseases, affecting driving comfort and quick, reducing road capacity, seriously affecting the economic and social benefits of the highway [1-3]. The key issue is the deformation of widening subgrade of high-grade way. It affects the life-span of the widened highway and driving comfort. As the area of seasonal frozen soil is more than half of the total land area in China, research on influence of widening in high-grade highway of seasonal frozen area on the deformation of subgrade makes a particular theoretical and practical significance on fully considering the influence of seasonal freezing-thawing in widening engineering, controlling the whole stability and asymmetrical settlement of new and old subgrade effectively, reducing the resulting cracking and damage of the layer on upper pavement [4-6]. Therefore, it is necessary to change the old and new highway extension difference subgrade settlement control technology [7-8].

Based on the study of the changes and influences of the strength, deformation and other road characteristics of the high-water content cohesive soil roadbed under freezing and thawing, this paper deeply studies the long-term performance of the high-water content cohesive soil roadbed in the seasonal freezing area, and clearly The performance change law of high water content cohesive soil roadbed in seasonal frost area and its influence on the pavement structure layer are discussed, and the frost heave deformation control standard of high water content cohesive soil roadbed is proposed.
The project is the widening section of the old road, and the design standards are two-way four lane (Figure 1). The design speed of the new section is 100km / h, the subgrade width is 26m, the design speed of the widening old road is 80km/h, and the width of roadbed is 24.5m.

Figure 1. Layout of settlement observation instruments of subgrade widening section.

2. Engineering Characteristics of Cohesive Soil with High Water Content Subgrade Widening Section

2.1. Physical Characteristics of Cohesive Soil with High Water Content

The soil sample of K132+850 in the Expressway, and the test results are shown in Figure 2.

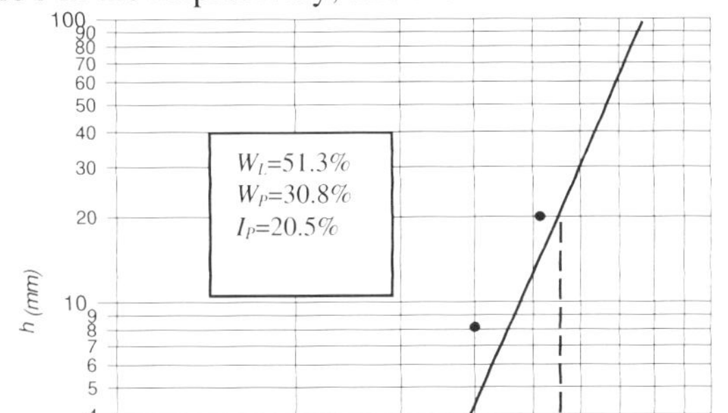

Figure 2. The relationship between penetration depth and moisture content.

According to the engineering classification of the soil, the soil sample is named high liquid limit clay.

2.2. The Road Properties of Cohesive Soil with High Water Content of Subgrade Widening Section

The relationship between soil dry density, CBR strength and consistency is shown in Figure 3., which shows: 1) The soil density-moisture content curve and strength-moisture content curve are separated from each other for high water content cohesive soil, which has different peak points respectively. This is one of the important characteristics of high-moisture clay soils different from general fine-grained soils. 2) The maximum strength moisture content is generally 0.2 to 0.3 less than the best moisture content in consistency according to the test statistics,, that is, the moisture content is 6% to 10% greater. On-site control of subgrade rolling according to the maximum strength moisture content is beneficial to the long-term stability of the roadbed .

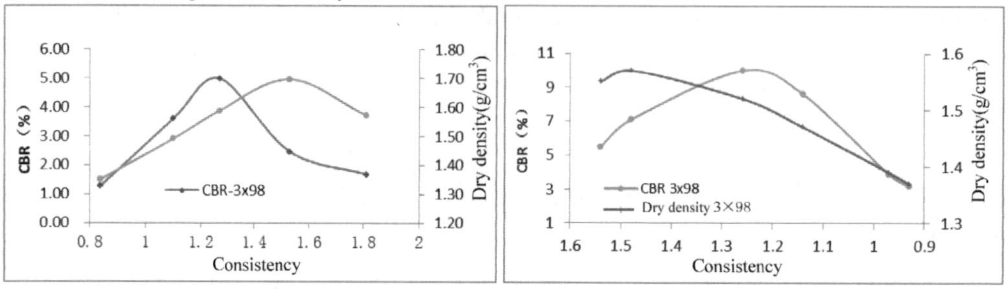

Figure 3. The relationship between strength, dry density and moisture content.

2.3. Freeze-thaw Characteristics of Cohesive Soil with High Water Content

For the resilient modulus of soil under different compaction degrees, as shown in Figure 4.

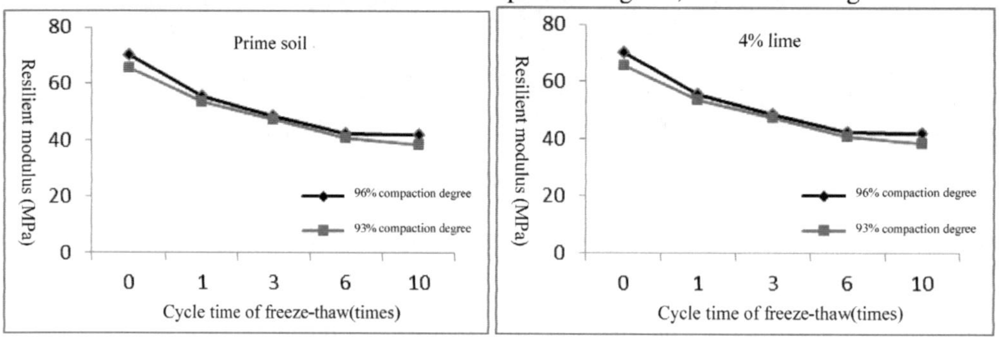

Figure 4. The relationship between the number of freeze-thaw cycles and the modulus of resilience at different degrees of compaction.

It can be seen that ash dosing follows that the resilience modulus at each degree of compaction decays with the increase in the number of freeze-thaw cycles.

3. The Moisture Content Characteristics of High Water Content Cohesive Soil Subgrade

After a complete year of subgrade water content monitoring, the results are summarized in figure 5.

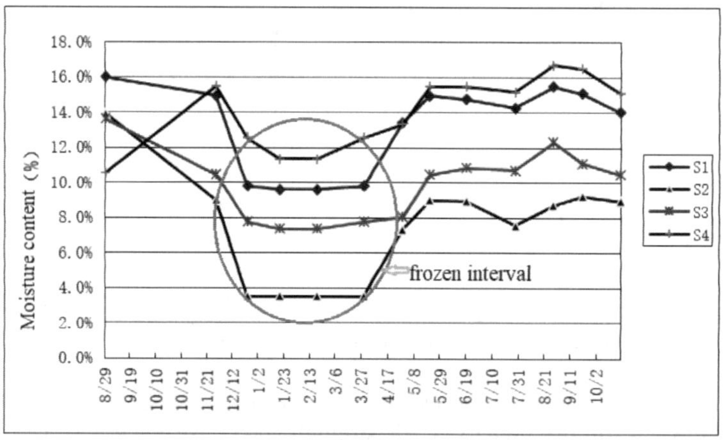

Figure 5. Monitoring of subgrade moisture content in different depth from slope.

Analysis of monitoring results shows that:

(1) The water content of each point in the subgrade basically keeps the same change cycle. From the

distribution of water content, the closer the distance to the outside of the slope, the greater the water content, indicating the greater the degree of influence by the external climate.
(2) According to the statistics of the annual water content fluctuation of monitoring points, it can be seen that the range of water content fluctuation increases with the decrease of the distance from the outside of the slope.
(3) S1 and S2 points are more than 3m away from the outside of the slope, and less than 1.5m away from the top of the road, but the change range of moisture content is not large, which shows that the impact of atmospheric precipitation on the humidity of the road bed through the pavement structure layer is very small.

4. Control Standard of Frost Heave Deformation and Freeze-thaw Strength of Cohesive Soil Subgrade

4.1. Frost Heave Deformation Control Standard of Cohesive Soil Subgrade

Based on design service life (15 years), the calculation results of differential deformation rate of subgrade are shown in table 1. From the calculation results, in order to ensure the service life of 15 years, the differential deformation rate of subgrade should be less than 2.61 ‰; and in the reverse bending deformation model of subgrade uplift, the maximum bending tensile stress is located in the subbase. Therefore, considering the standard driving load, the differential deformation rate of subgrade should be less than 3.07 ‰, so as to ensure the service safety within the design life. Considering the overload load, the subgrade's differential deformation rate should be less than 3.07 ‰ and the differential deformation rate must be less than 2.27 ‰.

Table 1. Calculation results of differential deformation rate of subgrade.

Computational model	Position of maximum bending tensile stress	Allowable bending tensile stress /MPa	Bending tensile stress under standard axle load /Mpa	Bending tensile stress under overload axle load /Mpa	Differential deformation rate /%	
					Standard axle load	Overload axle load
Positive deflection of middle uplift	Upper layer	0.424	No consideration	No consideration	0.261	
Shoulder uplift reverse deflection	Subbase	0.284	0.096	0.144	0.307	0.227

According to the fatigue equation of semi-rigid material and asphalt surface layer obtained from the split fatigue experiment of domestic asphalt and semi-rigid material, the service life of cohesive soil subgrade with high moisture content in the seasonal frozen area can be calculated from the different deformation rate of subgrade. The relationship between differential deformation rate of subgrade top and service life is shown in Table 2.

Table 2. Prediction of highway service life under different deformation rates.

Differential deformation rate	Maximum bending tensile stress /MPa	Splitting strength/MPa	Number of fatigue actions / Ten thousand times		Service life /years	
			Standard axle load	Overload axle load	Standard axle load	Overload axle load
1‰	0.064	0.6	1.43E+06	3.32E+04	86.1	46.7
2‰	0.121	0.6	1.81E+04	1.03E+03	40.5	13.5
3‰	0.223	0.6	71.85	9.62	1.8	0.3
4‰	0.278	0.6	7.34	1.30	0.2	0.0

From the calculation results, the subgrade differential deformation rate is 1 ‰ only considering the frost heave factor. Under the standard driving load and the overload effect, the service life of highway can reach 86 years and 46 years respectively, while when the differential deformation rate of subgrade is 4 ‰, the service life of both loads is less than 1 year. From the relationship between service life and differential deformation rate, it can be seen that when the deformation rate is more than 3 ‰, the relationship between the two is linear decreasing; when the deformation rate is more than 3 ‰, the service life changes little to zero. Considering the allowable bending tensile stress and the pavement structure service life, the differential deformation rate of the top surface of the subgrade under the effect of frost heaving of the high water content cohesive soil subgrade in the seasonal frost zone should be controlled within 2‰.

4.2. Control Standard for Freeze-thaw Strength of Cohesive Soil Subgrade

The prediction results of highway service life under different subgrade strength ratios are shown in Table 3.

Table 3. Prediction of highway service life under different subgrade strength ratios.

Strength ratio	Load condition	Maximum bending tensile stress /MPa	Splitting strength /MPa	Number of fatigue actions / Ten thousand times	Service life /years
0.5	Standard axle load	0.30	0.6	1.82E+02	4.0
	Overload axle load	0.34	0.6	2.98E+01	0.8
0.75	Standard axle load	0.25	0.6	2.24E+03	20.0
	Overload axle load	0.29	0.6	3.55E+02	6.8
1	Standard axle load	0.22	0.6	1.48E+04	38.5
	Overload axle load	0.25	0.6	2.37E+03	20.5
1.5	Standard axle load	0.18	0.6	2.64E+05	68.4
	Overload axle load	0.20	0.6	5.82E+04	52.6

From the calculation results, only considering the influence of freeze-thaw, the service life of the road can reach 38.5 years under the standard driving load and 20.5 years under the overload effect when the strength is not attenuated, and the service life of the road under the overload effect is less than 1 year under the strength ratio of 0.5. From the relationship between subgrade strength ratio and service life, we can see that under standard axle load, the relationship between the two is basically linear decreasing, while under overload axle load, the relationship between the two is quadratic polynomial. Considering the allowable flexural stress and service life of the pavement structure, the subgrade-soil strength ratio of the high water content cohesive soil subgrade in the seasonal frost zone should be controlled within 0.75 under the standard load and within 1 under the overload load.

5. Conclusions

Through analysis of control standard of frost heave deformation of cohesive soil of expressway widening subgrade in seasonal frozen region, the following conclusions can be drawn:

(1) The water content of each point in the subgrade basically keeps the same change cycle. From the distribution of water content, the closer the distance to the outside of the slope, the greater the water content, indicating the greater the degree of influence by the external climate. According to the statistics of the annual water content fluctuation of monitoring points, it can be seen that the range of water content fluctuation increases with the decrease of the distance from the outside of the slope.

(2) Considering the overload load, the subgrade's differential deformation rate should be less than 3.07 ‰ and the differential deformation rate must be less than 2.27 ‰.

(3) Considering the allowable bending tensile stress and the pavement structure service life, the differential deformation rate of the top surface of the subgrade under the effect of frost heaving of the high water content cohesive soil subgrade in the seasonal frost zone should be controlled within 2‰.

(4) Considering the allowable flexural stress and service life of the pavement structure, the subgrade-soil strength ratio of the high water content cohesive soil subgrade in the seasonal frost zone should be controlled within 0.75 under the standard load and within 1 under the overload load.

Acknowledgments
Authors wishing to acknowledge assistance or encouragement from colleagues, special work by technical staff or financial support from organizations should do so in an unnumbered Acknowledgments section immediately following the last numbered section of the paper.

References

[1] Pranshoo Solnki. Characterization of cementitiously stabilized subgrades for mechanistic -empirical pavement design [D]. Norman: University of Oklahoma, 2010.
[2] Basma A A, Tuncer E R. Effect of lime on volume change and compressibility of expansive clays[J].Transportation Research Board, Washington D C, TRR No, 1296, 1991, 54-61.
[3] Bell F G. Lime stabilization of clay minerals and soil[J]. Engineering Geology, 1996, 42, 223-237.
[4] Arabi M, Wild S. Property changes induced in clay soils when using lime stabilization [J]. Municipal Engineering, 1989, 6, 85-99.
[5] Konrad, J. M. Physical processes during freeze-thaw cycles in clayey silts [J]. Cold Regions Science and Technology. 1989.16 (3), 291-303.
[6] [11] Nishimura, T., Ogawa, S. Influence on freezing and thawing on volume change of unsaturated soils[C]. Proceedings of the International Symposium on Prefigure Deformation Characteristics of Geomaterials, 1994: 335-340.
[7] Wang R L, Zhu D P, Liu X Y, et al. Monitoring the Freeze-thaw Process of Soil with Different Moisture Contents Using Piezoceramic Transducers[J]. Smart Materials and Structures, 2015, 24(5): 57003.
[8] Romanoschi S A. Empirical Models for Permanent Deformation of Subgrade Soils from the Data Collected at the Pavement Subgrade Performance Study[J]. Journal of Materials in Civil Engineering, 2017, 29(3): 4016236.

Bridge Vibration Virtual Sensor Based on Eulerian Video Magnification and Gray Mean Difference

Jiayan Zheng[1, 2,*], Qiumo Li[2], Ruoyu Mao[2], Yuzhou Wang[2] and Zhixiang Zhou[1, 2]

[1]State Key Laboratory of Mountain Bridge and Tunnel Engineering, Chongqing Jiaotong University, Chongqing, China
[2]School of Civil Engineering, Chongqing Jiaotong University, Chongqing, China

*Corresponding author email: jiayanzheng@cqjtu.edu.cn

Abstract. A non-contact measurement method is proposed to overcome the shortcomings of traditional contact sensors in this paper. The camera is used as a non-contact virtual sensor of the whole field vibration. The motion in raw video was amplified by Eulerian Video Magnification technology and the gray mean of the selected region from processed video is obtained. The gray value difference between frames represents the dynamic response signal of the local area of the bridge structure under the excitation. The actual displacement amplitude in the time domain of the target structure can be obtained by conversion. A model bridge experiment was carried on, the results show that good coincidence in time domain and good accuracy in frequency domain, the error of the proposed method is 2.2% compared with the traditional measurement, indicating that the method without manual mark has the advantages of accurate, easy to obtain, low cost and high efficiency.
Keywords: Vibration; Bridge Structure; Eulerian Video Magnification; Gray mean difference; Virtual sensor.

1. Introduction

Vibration information is the key dynamic data of bridge structure and bridge health monitoring[1]. At present, the most conventional method is to obtain the relevant parameters (such as structural displacement, strain, curvature, vibration frequency, etc.) through the sensor network arranged on the bridge for analysis. The sensor has the advantages of mature technology, reasonable price and accurate data. However, the traditional sensor life is generally shorter than the service life of civil structure, and it is inconvenient to replace, and it will lead to the interruption of monitoring data. The traditional sensor is usually arranged in the key position of the structure, which can only realize single point measurement, and the sensor points in the whole field measurement are wide and complex in distribution. During the measurement process, it produces load effect due to contacting with the tested object directly[2]. For this, some new non-contact measurement methods have emerged, such as laser scanner, radar, vision measurement and so on. They can measure the position, size, shape and other parameters of the target without touching the measured object.

With the rapid development of photographic equipment and computer vision technology, the measurement based on machine vision becomes one of the most popular technologies because it does not need laser light source, magnetic field interference, complex interference light path and other auxiliary devices. There are a lot of researches on non-contact measurement of engineering structures. Wahbeh[3]measured the vibration displacement of large-scale bridges by arranging two laser points with fixed relative positions on the civil structure for tracking. Lee[4]introduced an advanced dynamic real-time displacement measurement system based on vision and the displacement of multi-layer frame structure during vibration was analyzed in real time. Feng[5][6] studied the vibration signal of the bridge structure when the train passes through the railway bridge by finite element analysis and visual measurement, showing that the visual method has good robustness. Yang[7]used the high-speed video to extract the vibration of the cantilever structure under the impulse excitation, to obtain the displacement signals and estimated the modal parameter. Rao[8]analyzed the impact of underwater environment on the structure vibration and calculated the natural frequency and damping of the structure by extracting the amplitude information of underwater structure under the earthquake excitation. Zhou[9] proposed method based on SIFT point matching without artificial target to obtain the displacement time history of the target. Zheng[10] used edge operator to extract the contour of the vibration bridge in the magnified video, which proves the accuracy of the method to obtain the structural holographic shape data.

Most of these vibration measurements based on computer vision need to be manually marked on the tested structure in advance, such as arranging and pasting target points, LED point light source or laser point light source. Besides, it is hard to obtain structural deformation information directly from videos or images due to the tiny deformation of bridge structure, and the accuracy and reliability of identifying displacement only by pixel points on the edge contour line are low. For problems above, this paper proposed a method to extract structural displacement, frequency and other information without manual marking for a micro motion. Eulerian video magnification and image processing technology (marginal gray value method) were applied to quantify the vibration information of structure, and greatly reduced the error caused by noise and other reasons through pixel gray mean difference method. The validity and accuracy of the method are verified by comparing with the data from dynamic displacement sensor and vibration sensor in a self-anchored suspension bridge modal. This method provides a new non-contact measurement technology for NDT (Non-Destructive Testing) and SHM (Structural Health Monitoring).

2. Related Work

2.1. Eulerian Video Magnification (EVM)

Researchers from the computer science and Artificial Intelligence Laboratory (CSAIL) of Massachusetts Institute of Technology put forward a series of algorithms to amplify the weak change signals in the video, which can exaggerate the weak brightness changes and small object motion information in the original video.

In 2005, Liu[11] proposed a method to enlarge the motion changes in video objects. This method tracks the motion track of feature points and enlarges their motion amplitude. It is a Lagrangian algorithm, and needs to fill the background, so the algorithm is very complex and time-consuming. Wu[12] proposed a technique called Eulerian video magnification, different from the Lagrangian perspective, the Euler perspective focus on the signal change of fixed pixel coordinates. In 2013, on the basis of Eulerian magnification algorithm, Wadhwa[13] proposed a phase-based image motion processing technology, which can translate the noise and achieve better amplification effect.

The first step of EVM is to use image pyramid to filter the video sequence image in order to get the baseband of different spatial frequencies. This process helps to reduce noise and facilitate the approximation of image signals. The phase-based motion amplification algorithm uses the pyramid to decompose the video signal into the amplitude and phase of the local space group, the local space phase signal is decomposed into a series of spatial phase signals by Fourier transform, and the sine wave represents the harmonic motion. Then, the band-pass filter in the time domain is applied to each baseband to extract the part of change signal that we are interested in. Then amplify and recombine the changed signal, and finally synthesize the amplified video. The result is that the motion in the video is amplified within the specified band-pass range, while the signals in the rest band-pass range remain unchanged and are combined with the amplified signals into the amplified video. The result is that the motion in the video is amplified within the specified band-pass range, while the signals in the rest band-pass range remain unchanged and are combined with the amplified signals into the amplified video. The flow is shown in Figure 1.

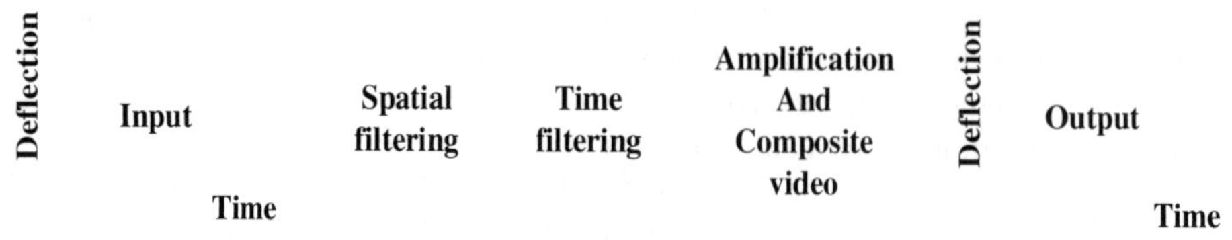

Figure 1. EVM processing flow.

2.2. Gray-scale Theory of Image

Digital images are often stored in RGB (red-green-blue) color mode measured by camera image sensor. Each of the three channels has 256 levels of brightness, each pixel can combine 16.77 million colors through 256 levels of brightness of each of the three channels.

In the amplification methods mentioned in the previous section, they are all completed in YIQ color space. Firstly, RGB is transformed into YIQ, then amplified, and then presented as RGB. Y component represents brightness (intensity) information of image, I component represents In-phase, color from orange to cyan, Q component represents Quadrature-phase, color from purple to yellow green (as shown in Figure 2). YIQ color space has the following linear relationship with RGB color space

$$\begin{bmatrix} Y \\ I \\ Q \end{bmatrix} = \begin{bmatrix} 0.299 & 0.587 & 0.114 \\ 0.596 & -0.275 & -0.321 \\ 0.212 & -0.528 & 0.311 \end{bmatrix} \begin{bmatrix} R \\ G \\ B \end{bmatrix} \tag{1}$$

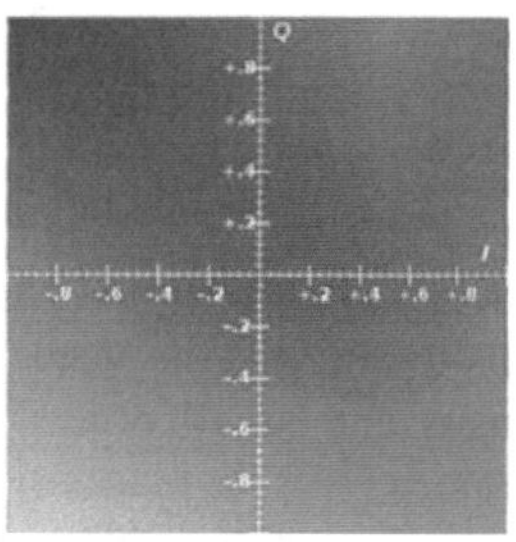

Figure 2. YIQ colour space.

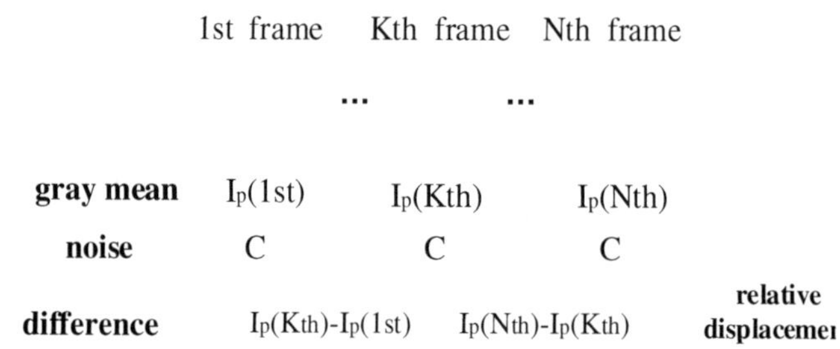

Figure 3. Gray value difference.

The pixel information contained in the micro motion is mainly concentrated in the Y channel in most cases, namely brightness information, and the colour has little impact on the results. Li[14] put forward the method of only amplifying the signal of Y channel, not amplifying the I and Q channels, and then combining the amplified Y channel signal with the original I and Q channel signal, which greatly improves the operation speed and hardly affects the amplification effect. He[15] converts the original video to grayscale and then enlarges the original video, the speed is significantly improved and the effect is good. In the gray value, 0 and 255 represent black and white respectively. This process can eliminate the hue and saturation information while maintaining brightness. For the amplification of motion, the color has almost no effect, so gray image signal can improve the efficiency and not affect the correctness of the results.

3. Method of Box Selection and Subtraction

Under the background of EVM and gray-scale image, a method based on the subtraction of gray mean of edge regions was proposed in this paper to obtain the required engineering physical quantities such as displacement and frequency from the video.

3.1. Gray Mean Obtaining in Selection Area

Firstly, the RGB image of video is converted to gray image. The part of first frame of video image was selected to calculate the gray mean within the selection range, and then the vibration parameters are obtained by the change of gray mean within the same frame selection region, as shown in Figure 3. The assumptions of this method are as follows:

- there are obvious differences between the tested object and the background color in the video.
- the vibration range of the tested object in the video should be within the selection range.
- assuming that the noise value generated by the image is the same in the short time, that is, the noise value in each frame after video decomposition is approximately equal, then the noise value can be offset by subtracting the gray mean of the front and back frames, and only the relative change of the gray value generated by the image due to the upper and lower vibration is retained.

Taking the model experiment in section 4 as an example. Assuming that the background is dark while the interested part is light, the observed structure moves only in the vertical direction. The gray mean $I_p(t)$ of selected areas change with the up and down vibration of the target structure. The gray mean of edge region is to average the intensity of a selection region from image edge, which can not only solve the problems of low recognition accuracy and large error caused by the blurry image edge contour, but also improve the signal-to-noise ratio of images.

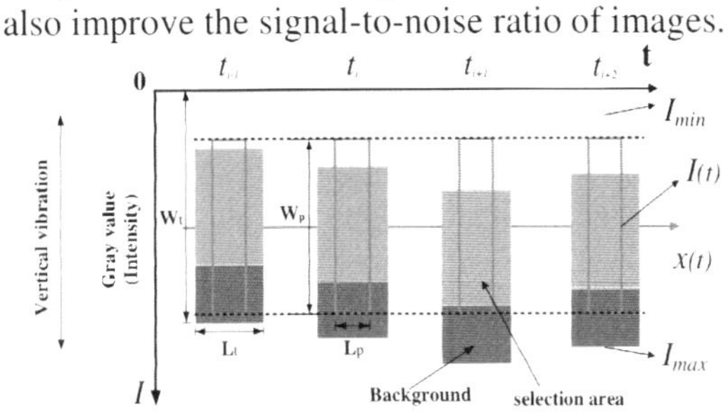

Figure 4. Gray mean in time domain.

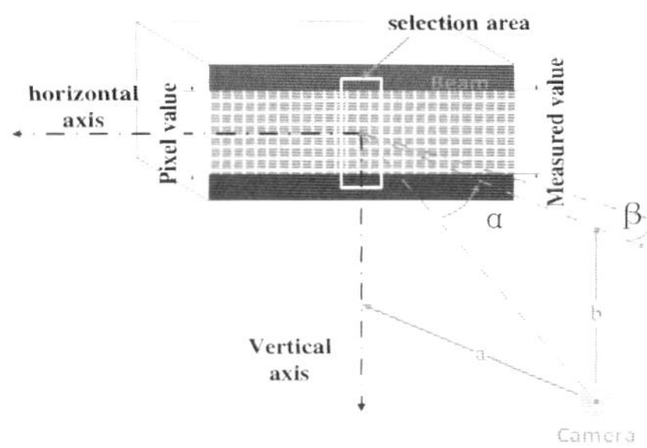

Figure 5. Position between beam and camera.

Figure 4 further clarified the function and significance of selection, the size of $L_t \times W_t$ (pixel × pixel) is regarded as the overall video area of the target structure, a smaller area with size of $L_p \times W_p$ (pixel × pixel) at a strong contrast edge from whole image was selected. The location and size of the selected area ($L_p \times W_p$) are unchanged, while the captured target area ($L_t \times W_t$) vibrates after excitation. The gray mean in the previous frame is taken as the reference, then the differences of the gray mean of two adjacent frames and the relative change of the gray value is calculated. The gray function is:

$$I(t) = \frac{I_{max} - I_{min}}{L_p} x(t) + n(t) \tag{2}$$

where $I(t)$ is the gray value of selected area, L_p is the length of the selected box, $n(t)$ is the noise of the selected area. $x(t)$ represents the position of I on the x-axis over time. I_{max} and I_{min} are maximum and minimum pixel intensity of the box selection area. The length W_p should be large enough to ensure that the maximum displacement amplitude be included in selected area, namely, the movement range of the main beam edge shall not exceed the selection area to avoid nonlinear.

3.2. Displacement Transformation

Due to the actual vibration has been amplified and there is an inclination angle between the camera (in figure 5) and the measured object in the measurement, there are errors in the test results, so it is necessary to consider the influence of amplification and geometric factors.

In order to establish the relationship between the gray value and the deflection of structures, the constant B (mm / pixel) is introduced, which is realized by dividing the actual physical length h (mm) of the frame selected part in the image by the counterpart of length L_i (pixel) in video frame.

$$B = h / L_i \quad (3)$$

When the shooting direction is not perpendicular to the displacement direction, i.e. $\alpha \neq 0$, the geometric correction factor *C* (dimensionless) is applied:

$$C = \frac{1}{\cos\left[\tan^{-1}(\frac{b}{a})\right]} \quad (4)$$

where a and b are the horizontal and vertical distance between the camera and the target center respectively.

The vertical axis of the camera should be aligned with the vertical coordinate axis, and the angle β (the horizontal angle between the camera and the structure center) can be ignored due to minimal effects on the vertical vibration test.

Considering that the Eulerian amplification technology has an amplification effect on the original video displacement, a reduction factor *K* is adopted, whose value is the reciprocal of the Eulerian amplification factor. The target dynamic displacement $u_{act}(t)$ (mm) can be obtained:

$$u_{act}(t) = B \cdot C \cdot K \cdot I_p(t) \quad (5)$$

3.3. Noise Processing

The existence of image noise increases the error of the test results and reduces the test accuracy. In this paper, subtraction difference theory and pixel gray mean are used to reduce the influence of noise. We introduced $n_p(t)$ to represent the noise value of selected area in this paper:

$$n_p(t) = \frac{1}{N}\sum_{i=1}^{N} n_i(t) \quad (6)$$

where *N* is the total number of pixels in the selection area, and n_i is the noise value in the pixel *i*, *t* represents the video image at time *t*.

The gray value in equation (2) is also averaged, and its expression is

$$I_p(t) = \frac{1}{N}\sum_{i=1}^{N} I_i(t) \quad (7)$$

where I_i is the gray value of pixel *i*.

Assuming that the noise is independent and the signal can be represented by a static process, the expression of the standard deviation of the pixel gray value can be obtained:

$$\sigma^2(I_p) = \frac{(I_{\max} - I_{\min})^2}{L_p^2}\sigma^2(x) + \frac{\sigma^2(n_p)}{N} \quad (8)$$

from the second item on the right side, it can be seen that with the increase of the number of pixels *N*, the noise decreases; that is to say, the smaller the standard deviation is, the more concentrated the numerical results are, the closer they are to the average value, indicating that the smaller the error caused by noise is.

We can replace *N* with $W_P \times L_P$ (width and length of image selection area), the expression of SNR(signal to noise ratio) is as follows:

$$SNR = \frac{\frac{(I_{\max} - I_{\min})^2}{L_p^2}\sigma^2(x)}{\frac{\sigma^2(n_p)}{W_p L_p}} = \frac{W_p}{L_p}(I_{\max} - I_{\min})^2 \frac{\sigma^2(x)}{\sigma^2(n_p)} \quad (9)$$

Equation 9 shows that in order to achieve the maximum SNR, it is necessary to maximize the W_p to L_p ratio, that is to say to keep the length as short as possible and the width as wide as possible, meanwhile that L_p must cover the maximum displacement amplitude. For a fixed camera and a short-time stable illumination condition, it can be assumed that the pixel noise power is constant, and an appropriate selection size can be specified. Another factor of SNR is difference between I_{max} and I_{min}, which has a more significant effect on SNR. We can get a higher the SNR with greater the brightness difference between the target and the background.

4. Test on Suspension Model Bridge

In order to further explore and demonstrate the feasibility and applicability of non-contact detective method proposed in the last few section, a modal of self-anchored suspension bridge was chosen to test.

4.1. Test Overview

The test bridge is a scaled model bridge of Taohuayu Yellow River Bridge on Wuxi expressway, with a proportion of 1:30. The main span of the model bridge is 13.534m and the side span is 5.3m. The main girder is steel box girder, and the deck is consisted of prestressed concrete. Acceleration dial indicator are arranged in 1#-5# measuring points, as shown in Figure 6. The instrument of this test includes Sony FDR-AX700 camera, dial indicator, DH3818Y static acquisition instrument, DH5902 dynamic acquisition instrument. The layout of each instrument is shown in Figure 7.

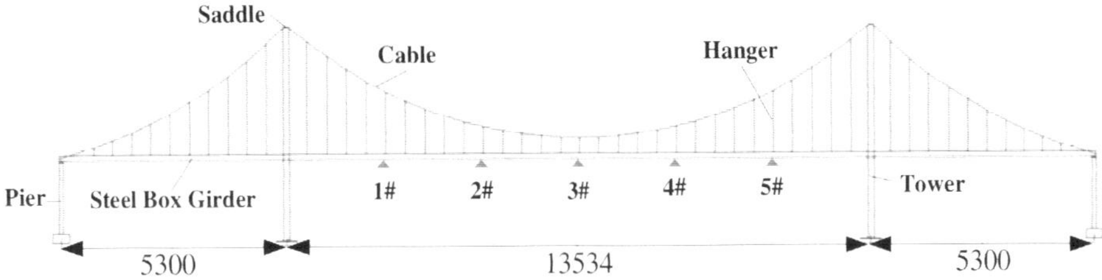

Figure 6. Laboratory suspension bridge. (mm).

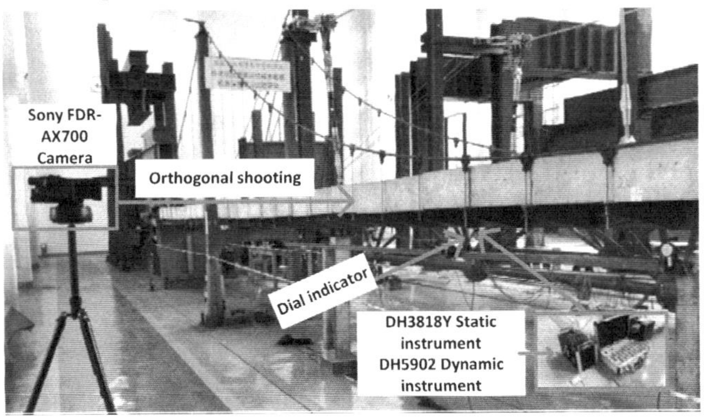

Figure 7. Experimental instrument layout.

In order to make the experiment more effective and less difficult, the good color contrast between the girder structure and the background, i.e. the (I_{max}-I_{min}) value is large, is benefit to capture the vibration of the structure. Meanwhile, most of the experiments are carried out under ideal conditions, which reduces the difficulty. Considering the characteristics of small stiffness and large flexibility of the suspension bridge, only a vertical impact is applied to the side span as the excitation to produce free vibration until the structure tends to be stable under the damping effect.

In this test, dial indicator and DH3818Y catch the deflection of main beam, and DH5902 collected the fundamental frequency information. Sony FDR-AX700 camera was used to obtain the image data, the frame rate is 100 fps (frames per second), which can meet the requirements of the vibration signal acquisition of the structure. The shooting direction is orthogonal to the side of the structure, mainly shooting the vibration of the main beam in the mid span area (around measuring point 3#), so as to reduce the error caused by the angle effect, and an appropriate shooting distance was selected to avoid major lens distortion.

4.2. Data Pre-processing

4.2.1. Gray processing and amplification processing. The video is processed by Eulerian amplification and the clear holographic deformation is obtained. Then the video of suspension bridge is decomposed into frames, after that, the RGB images of these frames were converted into gray image. One of the gray frames is shown in Figure 8. (a) is raw image before amplification, and (b) is after amplification. The obvious effect of magnification can be seen from the dotted line in (b).

(a) raw image (b) magnified image

Figure 8. Before and after magnifying in the same frame.

4.2.2. Box selection and mean value acquisition. The experimental image data is processed (selection-decomposition-mean processing) automatically by the program written in MATLAB. During the vibration of the main beam, the variation range of beam is within the range of selection(figure 9), so as to avoid introducing non-linear data and signal interception. Different selection region can be set to extract displacement of different parts of bridge. Each selection area can be considered as a "virtual vibration sensor" mounted on the main beam to provide dynamic signals for the structure. The function is equivalent to the traditional sensor, which can obtain the dynamic response of the structure, and its acquired signals are applied to time-frequency analysis to get the basic frequency. Frame 1, frame 5, frame 10, frame 15 and frame 20 are selected for display.

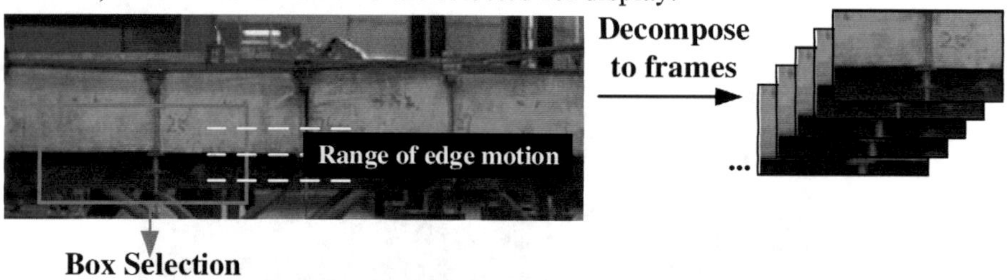

Figure 9. Box selection and decomposition.

4.2.3. Acquisition of actual displacement. For connecting the gray value with the actual displacement, two steps are taken: (a) signal synchronization in the time domain, (b) the parameter *B, K, C* was taken into account to calculate the true displacement. The vertical displacement is a sensitive index under the excitation, firstly, the variation can be obtained by subtracting the gray mean of two adjacent frames. Because it is a preliminary exploration, the error caused by the shooting angle should be minimized as much as possible. At this time, the position of the experimental camera was orthogonal to the shooting surface, the lens was aimed at the center of the target, and the height of the lens and the shooting target were the same, the geometric factor $C=1$. Then the displacement signal is transformed into displacement by equation 5. In the last, the deflection data from the dynamic dial indicator was acquired as standard. Taking the peak value of two sets of data as a unified index (figure 10), the time delay between two sets of data is calculated, and the time axis coordinate is unified accordingly. The correlation between the two groups of data was analyzed.

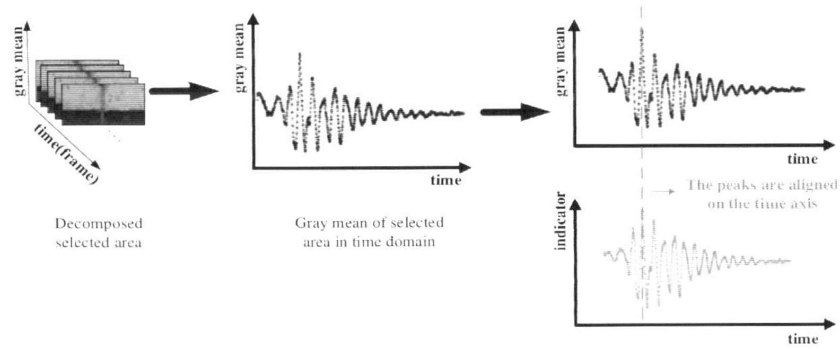

Figure 10. Gray time history calibration.

4.3. Analysis and Result

4.3.1. Displacement analysis. The displacement measured and gray value after unified coordinates are shown in Figure 11. It can be seen that the displacement curves from camera and dial indicator(3#) match well in the larger amplitude stage. With the attenuation of the amplitude, the error increases gradually, which may be caused by the image amplification. The signals details of the camera are less than the displacement data of the dial indicator, and had more noise. The result of correlation analysis of the two groups is shown in the right side of Figure 11. The figure illustrates that the correlation between the camera data and the dial indicator data is 0.9797(close to 1), indicating the high correlation between the two groups of data.

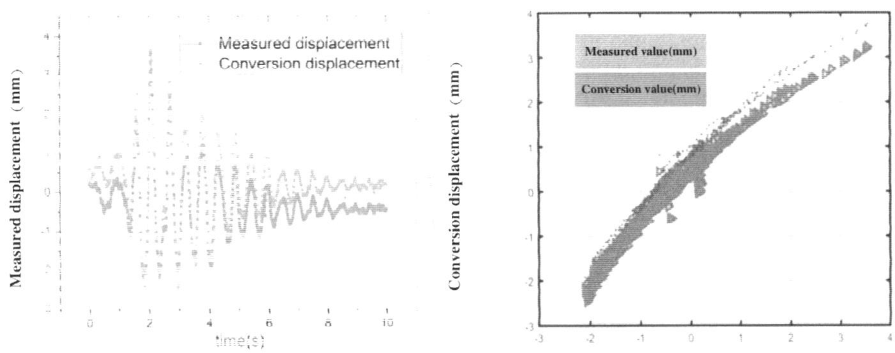

Figure 11. Result contrast diagram.

4.3.2. Frequency acquisition. The dynamic parameters (fundamental frequency) can be obtained by the time history data of acceleration and displacement through Fourier transform. The following content (figure 12) shows the structure frequency value after FFT of the displacement or acceleration time history data obtained in three ways: gray mean from video (a), dial indicator (b) and acceleration sensor (c). The fundamental frequency obtained by the three methods are summarized in Table 1. The fundamental frequency of the structure obtained by the camera method is 1.80Hz, the value of the dial indicator is 1.76Hz after Fourier transform, and the measured value of the vibration sensor is 1.76Hz. The difference is 0.04Hz, the relative error is only 2.2%. The results demonstrate that the edge area gray mean processing method proposed in this paper is able to obtain the displacement and fundamental frequency with good accuracy and feasibility. The reason of the error may be the image accuracy, and noise introduced by amplification processing. Later, the higher resolution camera and more appropriate magnification factor will be adopted to increase accuracy.

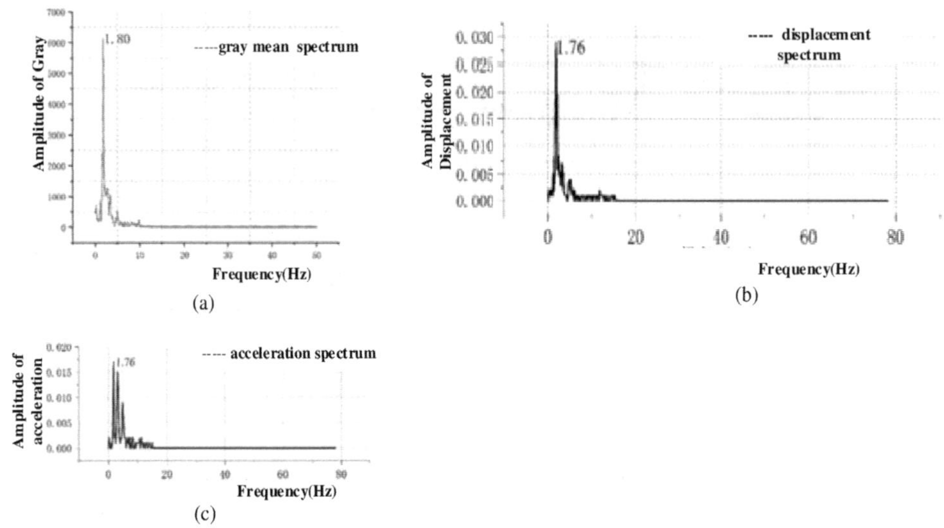

Figure 12. Frequency spectrum from three methods.

Table 1. Frequency data comparison.

Method of Acquisition	Fundamental frequency (Hz)
Image	1.80
Dial indicator	1.76
Vibration sensor	1.76

5. Discussion and Limitations

According to the low-frequency vibration characteristics of bridge structure, the specified band-pass range signal is amplified by EVM algorithm, and the method of subtracting of pixel gray value in edge area is proposed. The displacement of structure is obtained by the way of difference of gray value in adjacent frames, which can effectively obtain the change information of structure and avoid random error from single pixel, and also reduce the impact of noise in video image information. Compared with the traditional measurement, the high-speed camera is cheaper and more convenient. At the same time, the measurement results also have good accuracy, which meets the requirements of engineering measurement.

The experiment is carried out in the ideal indoor environment. The indoor light is stable, the camera almost has no jitter, the shooting direction is orthogonal to the beam elevation and the horizontal height is the same. Based on the three assumptions proposed in the third section, the frame selection subtraction method is adopted. In the actual outdoor measurement, due to the change of sun position or cloud illumination and background conditions, the wrong signal may be introduced. Secondly, in this experiment, only a part of the beam is photographed, and an area is selected as a single sensor for analysis. To obtain the overall information of the structure, the shooting distance needs to be larger, and the area selected by the frame needs to be more. In addition, sampling rate, resolution, noise, image sensor quality, size, lens type and other factors will affect the accuracy of measurement. In this case, we must pay attention to avoid too many changes of video sequence. In the actual outdoor measurement, the influence of shooting angle and light is great, it is necessary to carry out further research for applying this method to the long-term monitoring of structures.

6. Conclusion

In this paper, a new method is proposed to obtain bridge vibration information by Eulerian micro motion amplification technology and gray mean difference method. A high-speed camera is used as a non-contact virtual sensor of full field vibration. The displacement time history curve of the structure is obtained by the virtual sensor, and the fundamental frequency of any point of the structure is calculated by FFT. The results of image data are compared with the measured results from traditional sensor, including the transformed displacement curve and the foundation frequency of the structure. The results show that the displacement curves are in good agreement and the accuracy of fundamental frequency is high. Therefore, this work provides a new non-contact measurement method for the use of camera as a remote virtual sensor to measure the relevant parameters, and makes a preliminary exploration and preparation for further structural state evaluation.

Acknowledgments
The authors acknowledge the support provided by NSFC (51778094; 51708068). We also acknowledge State Key Laboratory of Mountain Bridge and Tunnel Engineering for providing experimental equipment.

References

[1] Ji Y T, Lei Z, Ran Z H and Jia Y 2018 Research on dynamic analysis theory for bridge structure based on experimental modal *Highway* **no.03** p78-83
[2] Zhang D S 2017 Research and applications on vision-based structural motion extraction algorithms University of Science and Technology of China
[3] Wahbeh A M, Caffrey J P and Masri S F 2003 A vision-based approach for the direct measurement of displacements in vibrating systems *Smart Materials and Structures* **vol 12** p785-794
[4] Lee J H, Ho H N, Shinozuka M and Lee J J 2012 An advanced vision-based system for real-time displacement measurement of high-rise buildings *Smart Materials Structures* **vol 21** 125019.
[5] Feng D M, Sun H and Feng M Q 2015 Simultaneous identification of bridge structural parameters and vehicle loads *Computers and Structures* **vol 157** p76-88
[6] Feng D M and Feng M Q 2015 Vision-based multipoint displacement measurement for structural health monitoring *Structural Control and Health Monitoring* **vol 23** p876-890
[7] Yang H, Takaki T and Ishii I 2011 Simultaneous dynamics-based visual inspection using modal parameter estimation *Journal of Robotics and Mechatronics* **vol 23** p180-195
[8] Rao R G V, Sreekala R, Kumar S K, Gopalakrishnan N, Muthumani K, Iyer N R, Lakshmanan N and Reddy G R 2016 Seismic response measurement of an under-water model through high speed camera and feature tracking *Experimental Techniques* **vol 40** p83-90
[9] Zhou Y, Zhang L X, Liu T and Gong S M 2018 Structural system identification based on computer vision *China Civil Engineering Journal* **vol 51** p17-23
[10] Zheng J Y, Mao R Y, Wu T, Zhou Z X and Tang L 2019 Eulerian-based micro motion amplification method to measure holographic deformation form of bridge *Science Technology and Engineering* **vol 19** p347-354
[11] Liu C, Torralba A, Freeman W T, Durand F and Adelson E H 2005 Motion magnification (Los Angles: SIGGRAPH) p519-526
[12] Wu H Y, Rubinstein M, Shih E, Guttag J, Durand F and Freeman W 2012 Eulerian video magnification for revealing subtle changes in the world *ACM Transactions on Graphics* **vol 31(4)** p1-8
[13] Wadhwa N, Rubinstein M, Durand F and Freeman W T 2013 Phase-based video motion processing *ACM Transactions on Graphics* **vol 32** p1-10
[14] Li L P, Lin L, Sun S F, YIN H and Dong F M 2015 Improved video small motion magnification processing *Computer Engineering and Applications* **vol 51** p195-200
[15] He Y M and Chen X Y 2019 Gray image magnification method for amplifying small motions in the video *Computer and Digital Engineering* **vol 47** p2022-2026

Design and Application of Drilling System Digital Prototype for Deepwater Drilling Platform

Xiuju Yang[1,2,*], Juan Sun[1,2], Zhiwei Zhang[1,2], Langjun Xu[3], Chunchun Wang[1,2], Zhitong Liu[1,2] and Lixian Zhang[1]

[1]Baoji Oilfield Machinery Co., LTD, Baoji, China
[2]CNPC National Engineering Research Center for Oil&Gas Drilling Equipment Co., LTD, Baoji, China
[3]Yantai CIMC Raffles Offshore Limited, China

*Corresponding author email: yangxiuju999@cnpc.com.cn

Abstract. With the rapid development of offshore oil drilling equipment industry, PDMS (Plant Design Management System) software has been widely used in offshore platform Design. Firstly, this paper combines the application of PDMS in designing the drilling system for deepwater semi-submersible offshore platform, systematically introducing the digital prototype design process based on PDMS software for the design of supporting platform (ship) drilling system. Secondly, we get the application effect of Marine engineering design based on PDMS. Finally, it is concluded that the overall layout of the platform with tight design space and high design density can be better optimized by using PDMS software, which enable multiple professionals to jointly complete the complex design process and explore an efficient Marine production mode.
Keyword: Deep water drilling platform; Drilling system; Drilling module; PDMS software; Digital prototype design.

1. Introduction

With the acceleration of the globalization process for the petroleum equipment manufacturing industry, accelerating the informatization of the marine equipment manufacturing industry, and actively exploring the business collaboration model has become a trend, and the transformation of the three-dimensional plant design system PDMS will play a powerful role in the design of the offshore platform. Promote the role. The software integrates engineering planning, design, construction, and management, and mainly covers piping, equipment, structure, electrical, instrumentation and other specialties. It can realize collaborative design and perform collision check in real time. Using this design tool can shorten the design cycle and ensure the design quality while reducing the project cost and improving competitiveness. At the same time, it can liberate the designer from the heavy drawing work, laying a foundation for the development process of modern engineering project management from passive to active [1]~[8].

The advantages of PDMS software in the design of complex engineering pipelines such as offshore engineering are obvious. The specific advantages are summarized as follows:

- Introduce the idea of project management;
- Multi-disciplinary collaborative design;
- Simple pipeline path planning;
- Real-time collision check;

- Automatic and accurate material statistics;
- Easy and fast drawing.

2. Digital Prototype Design of Drilling System

2.1. Top Layer Design of Digital Prototype

Digital prototype design can be divided into equipment, structure, piping, outfitting, electrical and other majors according to major, and can be divided into library building, modeling, interference inspection, production information input inspection and drawing, etc. according to function. When multiple designers collaborate, the project design process is more complicated, and the designers need to be professionally divided and authorized. The staffing plan is: administrator, project leader, key designer, ordinary designer, temporary designer. The flow chart of digital prototype modeling design for the drilling system is shown in Figure 1.

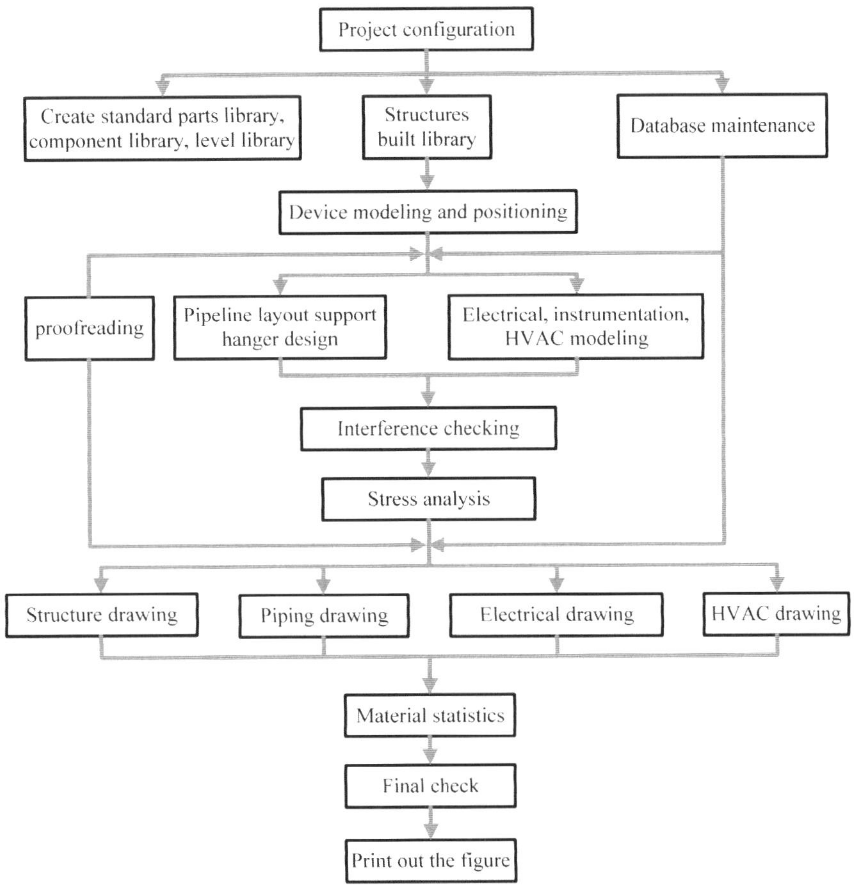

Figure 1. Flow chart of MEP digital prototype modeling design.

2.2. Structure, Outfitting and Mechanical Modeling Design

Structural modeling takes the database as the core. When building the database, the data and the graph are unified, and will not have the mistake; the design process collision inspection may also carry on at any time, carries on the design under the three-dimensional entity environment, it can control the display and hide of the model, measure or judge the distance between objects, find the collision at any time and check the rationality of the assembly. In the same project platform, collision analysis can be done between different professional models, producing both graphics and collision interference reports. The structure and outfitting modeling flow diagram is shown in Figure 2.

It is convenient and quick to model the equipment that the owner has provided drawings. It can define any complex graphics like other 3D software and can accurately define the interface information for each equipment. The flow chart of drilling bag mechanical modeling design is shown in Figure 3.

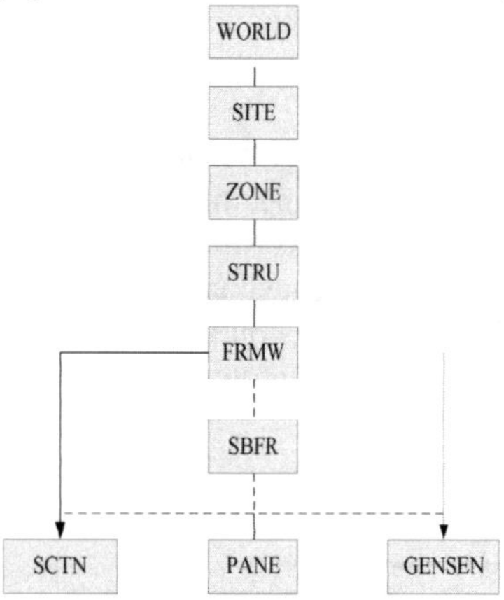

Figure 2. Flow chart of structure and outfitting modeling design.

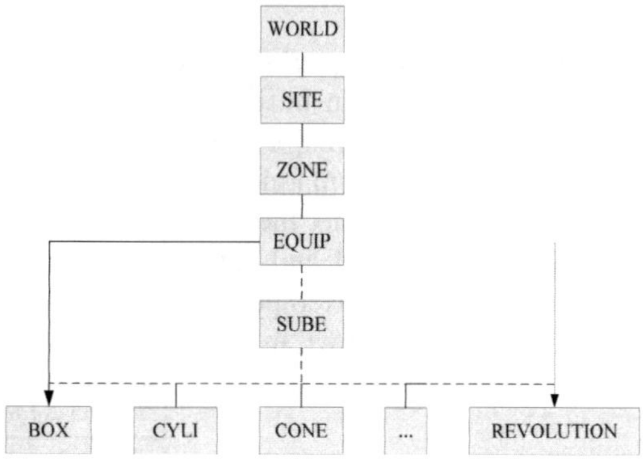

Figure 3. Flow chart of mechanical modeling design.

2.3. Piping Modeling Design

In the project design, pipeline layout is generally the most problematic link, and in PDMS, pipeline modeling is one of its most powerful functions. It can improve the design quality, strictly control the selection of materials, and avoid professional collision interference and other errors of pipelines with others as much as possible. Pipeline modeling usually adopts the method of "pipe guiding pipe"[8]. When modeling, you only need to define the beginning and end positions, and the software will automatically complete the pipe layout path of the pipe. The flow chart of piping modeling design is shown in Figure 4.

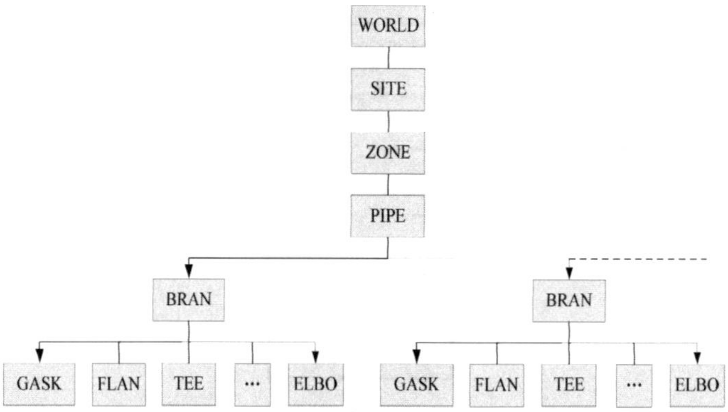

Figure 4. Flow chart of Piping modeling design.

The complex structure of the pipeline network, coupled with the large number of platform equipment and structural parts, interference will occur when the pipeline path is planned with little care. The software comes with a collision check function. When a collision occurs, the collision part will be highlighted in red. This function is very helpful for the design of the pipeline. The interference parts shown in the collision inspection of the pipeline and the equipment are shown in Figure 5. At the same time, the pipeline model can also show the engineering design parameters and additional parameters such as pressure, temperature, insulation and heat tracing required by the pipeline required in engineering design.

The pipe support plays the role of adjusting the pipe flexibility and supporting the pipe, which is essential in pipe layout design. PDMS sets standard pipe supports as templates and integrates them into MDS. From THE MDS into the project, it is convenient to choose the form, material, specification and adjusting position of pipe supports. The template-type pipe support modeling and drawing are the special functions of PDMS, which save a lot of time for the model drawing of pipe support. Figure 6 shows several typical tubular support models in the project.

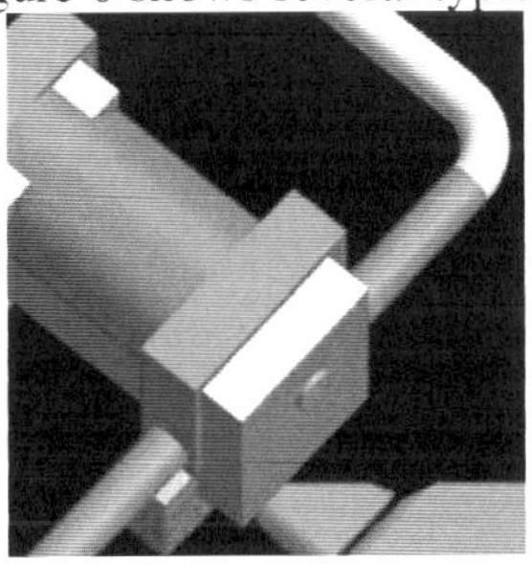

Figure 5. Pipeline interference check.

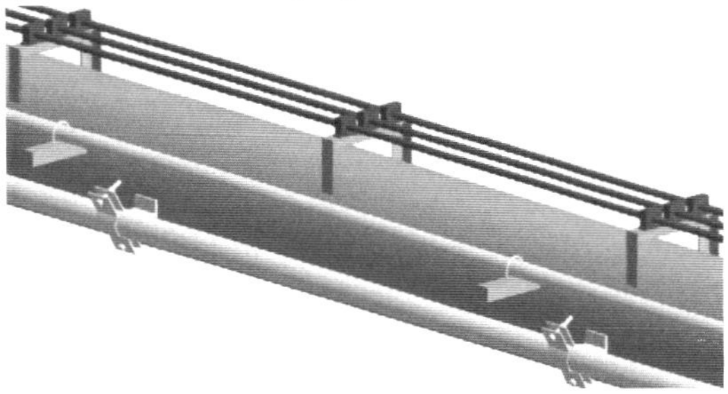

Figure 6. Pipe support modeling.

2.4. Electrical Modeling Design

In the project of electrical equipment, electrical fittings and path in the main cable, modeling, and according to the shipyard mode, to guide the construction. Such modeling and figure out ways to produce electrical construction work ahead of time to design, and drawing content elaboration to each size precision, improve the efficiency of field construction, improve the quality of the product.

The structural data is stored according to the electrical data hierarchy.The electrical data hierarchy cannot be saved by crossing as shown in Figure 7. The modeling of electrical equipment and cables is not possible with structures or piping and must be completed by starting the modeling of equipment.

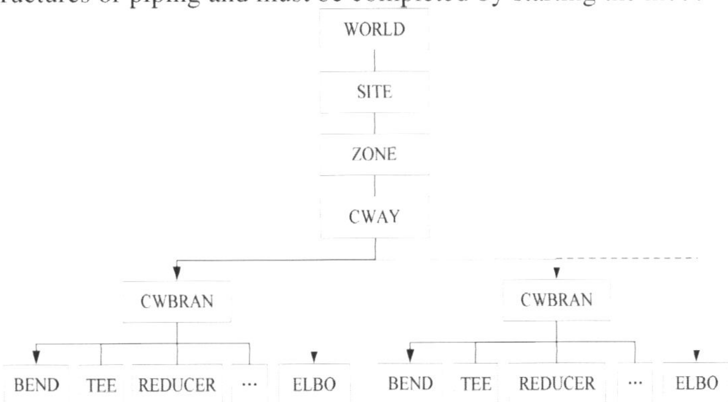

Figure 7. Flow chart of electrical modeling design.

2.5. Production Design

PDMS has good scalability and secondary development capability. According to users' needs, we can customize the drawing style and various forms of reporting according to the shipyard or users' requirements.

Automatically according to the requirements of the design development of PDMS parts named to a small program, CAD drawings automatically in order to merge small procedures, 2 d graphics automatic tagging each view coordinates small programs, from CAD to create any custom graphics import PDMS small programs, as well as the structure, iron snoop-proof, piping, electrical, report forms according to the demand of the drawing template automatically drawing program, etc.

With the continuous use and design of PDMS software, there will be some suitable for the design needs, convenient application of small programs have been developed and used. These can effectively improve production efficiency and reduce the labor intensity of designers.

The secondary development program of PDMS software can be used to input and check the production information of the parts. The input information can be reflected in the title bar and details of the drawing after the drawing. The program can also be used to export the production information of the parts to obtain the production information table and material list statistics for the subsequent production and processing. And PDMS has the function of quickly generating reports on has designed a 3 d model, can be in the form of EXCEL file output information that users need, especially the design of the structure of the bill of materials can be set according to the selected modules, assembly space and a good template output, direct forming production materials, statistics and rapid time efficiently.

The production drawing output preparation is summarized as follows:

- Establishment of structure, outfitting, equipment, piping and electrical models
- Each profession completes the interference inspection, modifies the interference model
- Add part production information
- Input and check production information, modify parts production information
- Enter the DRAFT module
- Process check, return the DESIGN module to modify the relevant production information
- Output production drawing

According to the needs of each profession, automatically generate all kinds of required drawings. The drawing style of this software is slightly different from the traditional CAD drawing style. This drawing style can reduce the labor intensity of the machining personnel and make the pipelining operation mode in the machining process possible.

To have designed the 3 d model, PDMS has the function of quickly generating reports can be output in the form of EXCEL file users the information they need to, especially the design of the structure of the bill of materials can be set according to the selected modules, assembly space and a good template output, direct forming production materials, statistics and rapid time efficiently.

3. Application and Effect

In the process of using PDMS software to design the digital prototype of the drilling package, the following problems are mainly solved and certain technical accumulation is obtained.

3.1. Realizing the "Collaborative Office" Project Management Mode

In terms of configuration optimization of projects, PDMS can break the conventional design habit. In the process of project design, it is not necessary for every designer to be familiar with every major, but to make his/her own 3D model and modify the model until the drawing is made. PDMS is a pipelined workflow pattern, which divides the roles of designers according to the design process. Let each professional designer focus on their own field of work, improve the work efficiency of the project. With PDMS, the number of 3D designers will be reduced, and the overall optimization of the personnel allocation throughout the project will result in fewer designers and efficient operation compared with conventional designs [9].

The REVIEW interface can provide realistic pictures with photo characteristics, anti-blur and projection, and can generate and record complex dynamic roaming diagrams. It can make various observations from the perspective of people, enabling the project staff to check engineering information at any time. This feature can improve design efficiency and reduce design errors.

PDMS has good scalability and secondary development capability. According to users' needs, we can customize the drawing style and various forms of reporting according to the shipyard or users' requirements. An important aspect of PDMS development is that users are constantly improving themselves in the process of using PDMS. Users can define a large number of common components and devices, which makes the project more reusable. Enterprises can also customize PDMS enterprise standards and PDMS operating guidelines according to their own project structure needs, to urge designers to organize the done generic data into the corresponding database, so that the PDMS in the process of use is constantly enriched.

3.2. Exploring Efficient Marine Production Model

According to the needs of each profession, automatically generate the drawings required by each profession. The drawing mode of this software is different from the traditional CAD drawing mode. Meanwhile, the processing means are somewhat heavy, such as heavy welding, light machine addition

and precision control, etc., which make the product design more in line with the habit of Marine construction.
The advantages of the production mode lie in that the drawing mode can reduce the production burden, reduce the labor intensity of workers, and make the pipelined operation mode in the processing process possible.

3.3. Presenting an Innovative 3 D Dynamic Design Method for Cable

In accordance with the SFI system coding rules, the code standard of drilling encapsulation system is formulated. In the establishment of piping system model, each pipe branch and pipe fitting is named and classified according to the code, so as to make the piping more orderly and standardized and improve the appearance quality of products [10].
Model the electrical equipment, outfitting and main cable path, guide the field construction. The electrical production and construction work was advanced to the design stage, and the drawing content was refined to each precise size, which improved the on-site construction efficiency. While improving product quality, it also improves the appearance of product quality and makes electrical design more standardized.

Figure 8. Dynamic design results of piping system.

Figure 9. Dynamic design results of cable.

4. Application

The problem of improper layout caused by incorrect lofting of piping and cables was solved, and the problems caused by improper lofting were solved at the design stage in advance, so as to better guide the production and processing, reduce the rectification rate and improve the appearance quality of products.
Before the new prototype is installed and debugged, the modules are installed and butted in the virtual interface, so as to optimize the whole scheme of interference or unreasonable structure, standardize the modeling and drawing mode, and carry out multi-professional cooperative design, improve the quality of interface design, shorten the product development cycle, reduce on-site debugging safety risk and correction rate.

5. Conclusion

PDMS in semi-submersible drilling system for Marine platform design, the application of professional designers can make each coordination office, the tacit understanding coordination, so as to reduce design errors, effectively prevent interference, reduced the modification and rework time, improve the efficiency of the figure, and ensure the quality of the work, makes the steady and efficient design of offshore drilling system. The application of PDMS software in the design of offshore drilling platform can enable the integration of design and management, so as to establish the offshore oil platform drilling system database in the design of offshore engineering. The powerful parameterization and exploitable performance of PDMS can lay a solid foundation for the research and development of offshore engineering projects.

Acknowledgments
Ministry of Industry and Information Technology" Seventh Generation Super Deepwater Drilling Platform (Ship) Innovation Special"-"Application Research of Drilling Package Integration and Some Key Equipment" Project (Ministry of Industry and Information Technology Joint installed[2016]24) ..

References

[1] Dingya Wang, Ping Li. Current Status and Development Trend of offshore drilling Platform Technology. Petroleum Machinery[J]. 2010 (4): 71-73.
[2] Aveva. PDMS Training Materials[M]. 2009
[3] Xuan Liu,Yanfeng Han. Application of PDMS 3d software in Marine engineering[J]. China shipbuilding supplement, 2011, 52 (1): 224-228.
[4] Meili Cui. Application of PDMS in the THREE-DIMENSIONAL Design and Management of General Drawings[D]. Master thesis of Xi 'an University of Architecture and Technology, 2011.
[5] Xiangyong Wu, Yi Li, Qinwu Lu, Yalin Luo, Jie Mao, Zhen Li. Research and development of module 3d design system based on PDMS platform[J]. Nuclear power engineering supplement, 2011, 32 (2): 127-132.
[6] Jia Jin, Xuan Liu. Assumption and Research on Pipeline Automation Based on PDMS[J]. Petroleum Engineering Construction, 2011, 37: 52-54.
[7] Xiuju Yang, Yongsheng Liu,Yahui Li, Bo Zhao, Lan Wang. Offshore Platform Drilling System Design based on PDMS Software[J]. Mechanical Engineer, 2016 (2): 168-171.
[8] Huan Xu, Fei Xu, Xiang Shen. Some Thoughts on Optimization of Ship Pipeline Layout[J]. Ship Materials and Market, 2020 (05): 1-2.
[9] Liang Pang. Discussion on the collaborative design of PDMS software platform[J]. Petrochemical design, 2020, 37 (01): 56-58+66+8.
[10] Yongsheng Liu, Chenghua Mai, Peng Zhang, Nan Hu, Xiuju Yang, Jingxiang Wang. Application of PDMS software in pipeline design supporting platform[J]. Chemical equipment and pipelines, 2017, 54 (3): 80-82.

Features of the Temperature Cracks in a Tunnel Final Lining

Jianqin Ma[1,*], Chen Xue[1], Le Wang[2], Pan Fang[1] and Ke Ren[2]

[1]School of Highway, Chang'an University, Xi'an, Shaanxi, 710064, China
[2]Xi'an Railway Engineering Co., Ltd, CRSG, Xi'an, Shaanxi, 710032, China

*Corresponding author email: majq@gl.chd.edu.cn

Abstract. Cast-in-place concrete lining, which is the major structure of a tunnel with composite or monolithic lining, requires high safety, durability and waterproof performance. The prevention and control of lining concrete cracks are of great concern. However, the temperature and shrinkage cracks in the arch and wall of a cast-in-place concrete lining are common construction quality defects in practical cases. This contribution presents the changing features of temperature crack widths in responding to the temperature variation with seasons in several cast-in-place final ling panels, where there are no waterproof and drainage sheets in their composite linings. The crack width changing features imply that the constraints to the temperature drop and shrinkage deformation of the concrete in its arch and wall in construction stage still work in the period of post-construction. Considering the cracks beyond the planned allowable values, the information on the width changing features presents clue to determine the rule on the occurring and developing of the temperature and shrinkage cracks in the cast-in-place lining concrete, as well as crack resistance design and construction quality control.

1. Introduction

The numbers and dimensions of the tunnels and underground structures are increasing, especially in the fields of transportation in China. On the other hand, safety and durability are of the concern of the involved groups in a tunnel project. For example, the cracks in the cast-in-place concrete final lining should be under control to meet the specified quality control requirement, such as in terms of water leakage preventing.

However, the cases with defects in lining concrete seem not rare due to various reasons[1-2]. For example, although the reasons of the concrete cracking during construction are of project unique, temperature and shrinkage cracking is one of the common reasons in practical cases[3-5]. In simplicity, crack appears as the imposed loads exceeding the bearing capacity of a concrete member. For example, lining crack appears as the tensile stress is over concrete tensile strength. Of the temperature and shrinkage cracks, which occur and develop in a cast-in-place concrete lining during construction, is closely related to the features of the constraints to the temperature drop and shrinkage deformation of the concrete. Mainly due to the project unique of the constraints in temperature and shrinkage cracking, the rule on the occurring and developing of the temperature and shrinkage cracks in the cast-in-place lining concrete needs further study.

This contribution presents the changing features of temperature crack widths in responding to the temperature variation with seasons in several cast-in-place final ling panels, where there are no waterproof and drainage sheets in their composite linings. The crack width changing features imply that the constraints to the temperature drop and shrinkage deformation of the concrete in its arch and wall in construction stage still work in the period of post-construction. Considering the cracks beyond the planned allowable values in terms of types and size, the information on the width changing features of the cracks in these lining panels is beneficial to determine the rule on the occurring and developing of the temperature and shrinkage cracks in the cast-in-place lining concrete. The results would serve for the cast-in-place lining concrete crack resistance design and construction quality control in tunnel and underground engineering.

2. The Tunnel and the Final Lining Building

The 1022-m-long tunnel is one of the loess tunnels in the Menghua Railway line passing through the west of the Henan province. The landform of the tunnel site is of tableland feature and the maximum overburden of the tunnel is about 100 m. The surrounding rocks of the tunnel consists of Pleistocene sandy clay, silty clay or silt, which is characteristics of the old loess in China, with water content less than their plastic limits. The ground around the excavations is mainly silty clays. Top-heading and bench method is applied in tunnelling, with special reference to the principles of New Austrian tunnelling Method.

2.1. Lining System

A composite lining system with invert is applied in the mined section of the tunnel. The primary lining is 30-cm-thick shotcrete, with steel sets and steel meshes embedded, and the final lining is 55cm-thick cast-in-place reinforced concrete. The strength grades of the shotcrete and cast-in-place concrete are C30 and C45, respectively. There is a waterproof and drainage sheet between the linings in the original design.

2.2. Building of the Final Lining

2.2.1. Construction procedure in general. The tunnel lining system is built with conventional method, in a two-pass mode. The final lining is built in two stages (Figure 1). A section of invert is first casted in a length of two times of the formwork jumbo length and then followed by the filling concrete in a short time period. And then the low wall of the final lining is built in a few days for each section of the corresponding invert. The sidewall and arch of the final lining are late, with a time lag varying from around ten days to more than a month. In general, the sidewall and the arch of a final lining is 30 m to more than 100 m behind the shotcrete lining in a section.

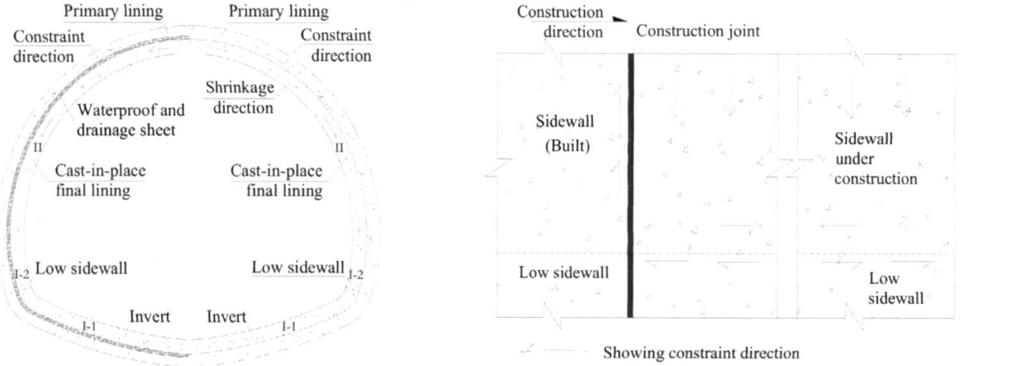

(a) Cross section of the composite lining with or without water-proof and drainage sheets, respectively.
(b) Showing the constraint of sidewall from its foot, side-end and outer surface

Figure 1. Sketch showing the building procedure of the final lining and the constraints to the temperature drop and shrinkage deformation of the concrete in its arch and sidewall.

2.2.2. Cancelling the interlayered waterproof and drainage sheets. During tunneling, no water dripping appears in the excavations. It seems that the functions of the waterproof and drainage sheets in the composite lining system have no working chance. On the other hand, it is favorable to the project in terms of time schedule and cost saving, as well as avoiding voids behind the final lining, the waterproof membrane and geotextile sheets was canceled and the composite lining structures were tuned accordingly in the following lining panels.

2.2.3. Applying the interlayered waterproof and drainage sheets. It is a pity that cracks was identified in the final lining panels, in which the waterproof and drainage sheets had been canceled in the composite lining system. Because the monitoring results of the crown settlement and the convergence of the shotcrete linings in some sections of the tunnel indicate that the excavations under the primary lining system are stable, final lining concrete cracking is not due to the ground pressures and the shotcrete lining deformation[6]. Considering the features of the potential thermal stresses in the final panels, with interlayered sheets cancelled, the constraint conditions of the final lining (Figure 1) would have a strong influence on the cracking of the built panels. So, the structures of the composite lining system were set back to the original design and the cracking in the new built lining panels, with waterproof membrane and geotextile sheets applied, is under control.

3. The Occurrence Features of the Concrete Temperatures Cracks in the Final Lining

Cracks occurred in the lining panels with interlayered sheets cancelled. The cracks are observed in the sidewall and arch of the final lining concrete shortly after formwork jumbo removing and are of temperature drop and shrinkage cracking features[1-2].

3.1. Type and Occurrence

Based on the relationship between the crack extension directions and the tunnel axis line, three types of cracks are recognized in the tunnel lining arch and sidewall concrete, i.e., longitudinal, circumferential, vertical and diagonal cracks (Figure 2a), respectively.

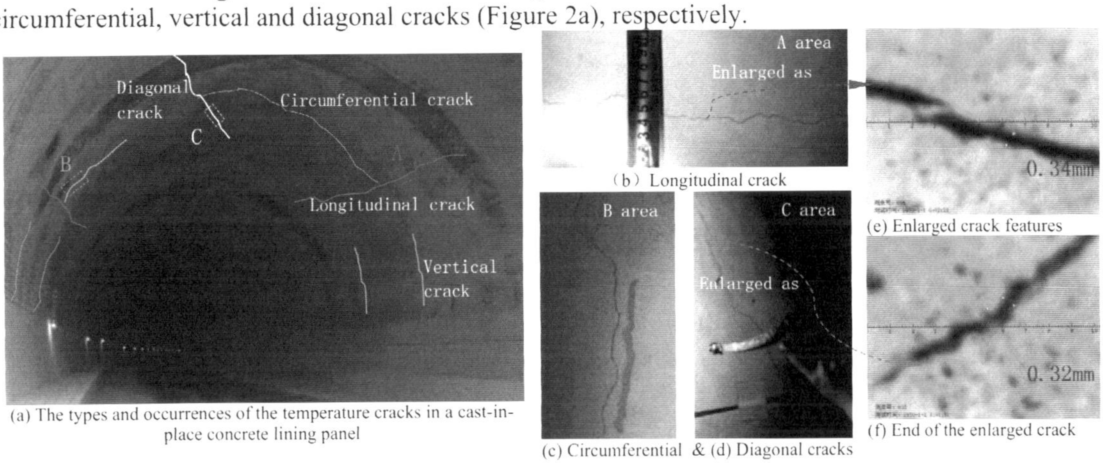

(a) The types and occurrences of the temperature cracks in a cast-in-place concrete lining panel
(b) Longitudinal crack
(c) Circumferential & (d) Diagonal cracks
(e) Enlarged crack features
(f) End of the enlarged crack

Figure 2. The types, occurrences and the features of the temperature cracks in a final lining panel.

3.1.1. Longitudinal cracks. The longitudinal cracks are nearly horizontal and parallel to the tunnel axis, mainly in the transitional position between the lining crown arch and sidewall (Figure 2a). The longest longitudinal crack run through a concrete panel, but most of the cracks are several meters in length or two cracks along a line. A longitudinal crack does not extend into neighbor panels.

3.1.2. Circumferential cracks. The circumferential cracks extend as a vertical crack in upper part of the tunnel sidewalls (Figure 2c) while as circumferential cracks in the lining arch. In general, the cracks are above a longitudinal crack and extend upward, with a length of 1.5 m to 4.5m. In most of the lining panels, circumferential cracks appear nearly around the middle of the panel length. There generally is one circumferential crack at one sidewall in a panel.

3.1.3. Vertical cracks. The vertical cracks appear in the lower parts of the sidewalls (Figure 2a). There is one or two cracks in one sidewall, with a feature of nearly halving or trisecting the length of the concrete panel, respectively. In general, the vertical cracks starts from about one meter above the top of the final lining low wall and extend upward about 2 m.

3.1.4. Diagonal cracks. Diagonal cracks mainly appear in the lining crown (Figure 2d). A diagonal crack extends along a line with an oblique angle to the lining panel edge line and the features of intermittent extension. There often is a branch in the extension of a diagonal crack.

3.2. Features of the Cracks

The cracks are of the following general features: (1) the cracks appear in specified positions, respectively; (2) two longitudinal cracks generally appear in each panel; (3) the cracks may extend in an intermittent pattern, with snake shape and width varying. These features are of the characteristics of cracks due to tensile stresses. Under a crack width meter, the sidewalls of a crack are much rough and in a wavy pattern (Figure 2e, f). The bored cores indicate that the longitudinal cracks cut through the thickness of the final lining and the crack width is almost same along the whole thickness (Figure 3).

4. Width Changing of the Cracks in Responding to the Temperature Variation with Seasons

4.1. Crack Width Monitoring in General

The cracks in the lining panels are a type of defect in terms of quality control, since the widths of the cracks (Figure 4) are beyond the target value, no more than 0.2 mm. To observe the changing features of the cracks, vibrating strain gauges, with temperature meter, are applied to monitoring the variation of the crack widths (Figure 5). The longitudinal, circumferential, vertical and diagonal cracks in three lining panels, numbered as panel I, II and III, were monitored with 11 strain gauges, respectively. The gauges were installed (stuck to) perpendicularly crossing the monitored crack at a site.

The monitoring results include the strain gauge recordings, which are read as changes of the widths of the cracks in strains in responding to the temperature variation with seasons. The temperature variation with seasons is presented in Figure 6, which is changing with the air temperature at the tunnel site. The features of the strain gauge recordings are discussed in the following chapter.

Figure 3. Crack features in a bored core.

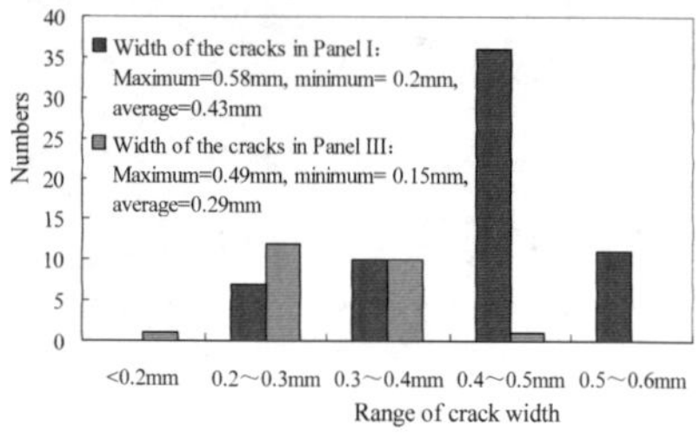

Figure 4. The width of the cracks.

4.2. Features of the Crack Width Changing with Season Temperatures

4.2.1. General features. In general, the widths of the monitored cracks increase with the air temperature decreasing with season variation, while the crack widths decrease with the air temperature rising, as shown in Figure 7. In a cycle of the seasons, the width changes of a monitored crack are minor. Therefore, the changes of the crack widths can generally be attributed to the concrete properties of concrete expansion and contraction due to the temperature rising and falling in responding to the season variation at the tunnel site and the contribution of the other reasons is negligible. On the other hand, the curves presenting the width changes of a monitored crack with time are not smooth and there are variations with a jump somewhere in the curves (Figure 7). The reasons for this feature will be discussed in details in the following chapter.

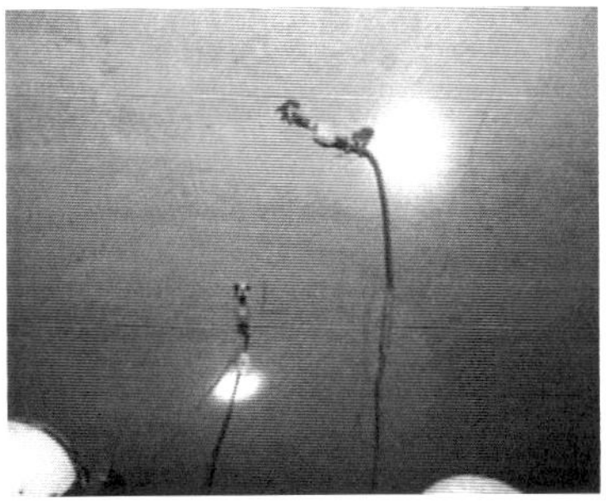

Figure 5. Crack width monitoring with vibrating strain gauges.

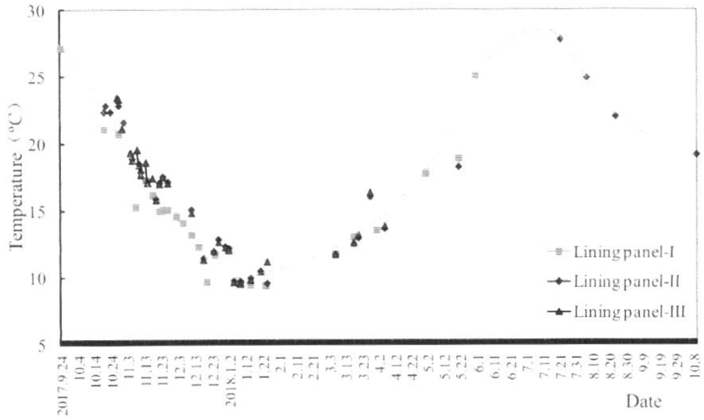

Figure 6. Temperature changing with the seasons.

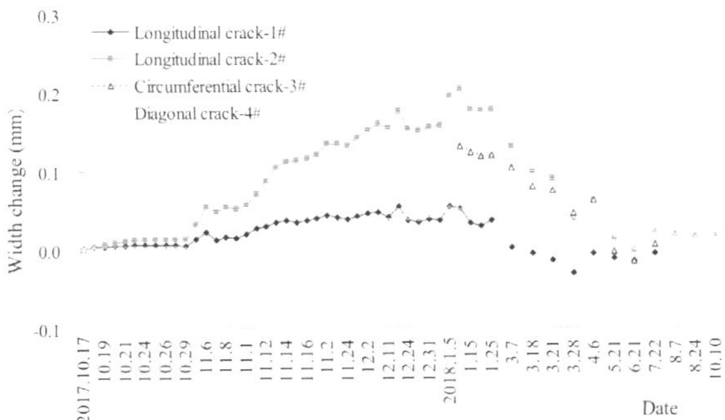

Figure 7. General features of the crack widths in responding to the temperature variation with seasons.

4.2.2. Discussion on the changing features of the crack width. As the above-mentioned, the monitored temperature and shrinkage cracks are in the cast-in-place lining panels, where there are no waterproof and drainage sheets interlayered in their composite linings. Of the mechanism of temperature and shrinkage cracks in construction, two parameters are of dominate features, i.e., the magnitude of temperature decreasing and constraints to the lining concrete contraction during temperature falling, which determine the features of thermal stress in a lining panel. In this case history, the magnitude of temperature decreasing can generally be considered same in different lining panels, since the structure features and building procedure are same for all of the panels, such as in terms of design and construction plan. So, the difference of the constraint conditions between the lining panels, with or without waterproof and drainage sheets, is the key factor. In simplicity, the constraint conditions of the lining system without waterproof and drainage sheets interlayered are favorable to the development of temperature and shrinkage cracks and the magnitudes of constraints are large enough to meet the requirement of cracking tensile stress.

To analyze the influence of the constraint conditions of the final lining panels on the changing features of the crack widths in the season variation, the width changes of a crack is presented in terms of temperature rising and falling, respectively. In Figures 8 to 10, the changing features of the crack widths in the season variation are of variations with a jump somewhere, when there is a temperature decrement about 5 to 10 °C or an increment about 15 °C. On the other hand, the width changing at the different parts of a crack is not synchronized in a temperature rising or falling period (Figure 8c). These features imply that the constraints to the lining concrete expansion and contraction due to the temperature rising and falling may vary with positions and that the influence of the constraints from the outer side and side-end the lining panels still work in the period of post-construction. There seems mutual confirmation between the multiple reasons of the constraints (Figure 1) and the types and occurrences of the cracks, as well as the difference in width changing under same temperature variation. The jumping features, i.e., the crack width changing on the conditions of certain temperature decrement or increment, imply that a large thermal stress is required to overcome the constraints to the expansion and contraction due to the temperature rising and falling.

Considering the cracks beyond the planned allowable values in this case, the influence of the constraint conditions of typical structure types on the mechanism of temperature and shrinkage cracking in the lining concrete is of importance, especially for a composite lining system, in terms of the cast-in-place lining concrete crack resistance design and construction quality control.

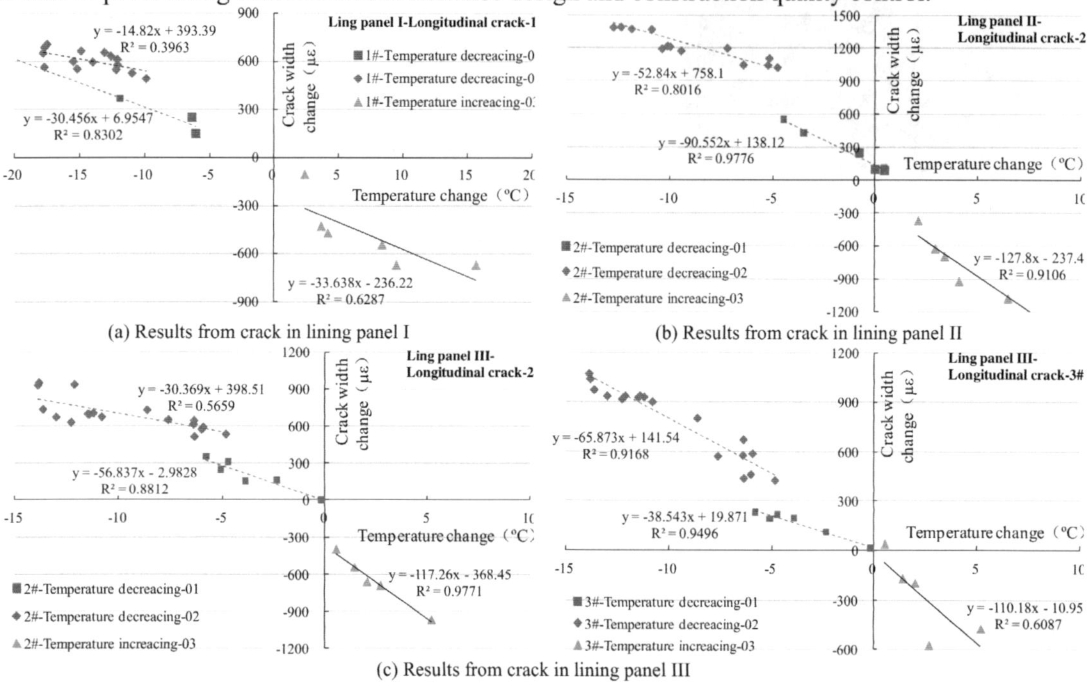

Figure 8. Features of the longitudinal crack width changes with season temperatures.

5. Conclusions

Based on the width changing features of the temperature cracks in the cast-in-place lining panels, where there are no waterproof and drainage sheets interlayered in their composite linings, in responding to the temperature variation with seasons, the following conclusions can be drawn.

(1) The widths of the monitored cracks increase with the air temperature decreasing with season variation while the crack widths decrease with the air temperature rising, and the changes of the crack widths can be attributed to the lining concrete properties of concrete expansion and contraction due to the temperature rising and falling in the season variation, respectively.

(2) The changing features of the crack widths in the season variation are of variations with a jump somewhere, when there is a temperature decrement about 5 to 10 °C or an increment about 15 °C. On the other hand, the width changing at the different parts of a crack is not synchronized in a temperature rising or falling period. These features imply that the constraints to the lining concrete expansion and contraction due to the temperature rising and falling may vary with positions and that the influence of the constraints from the outer side and side-end the lining panels still work in the period of post-construction.

(3) In terms of the cast-in-place lining concrete crack resistance design and construction quality control, the influence of the constraint conditions of a composite lining, with or without interlayered sheets, on the mechanism of temperature and shrinkage cracking in the lining concrete is of key factor.

Acknowledgments

The China Railway Seventh Bureau Group Co., Ltd. is acknowledged for financial support of this research project, under the No. 17A13.

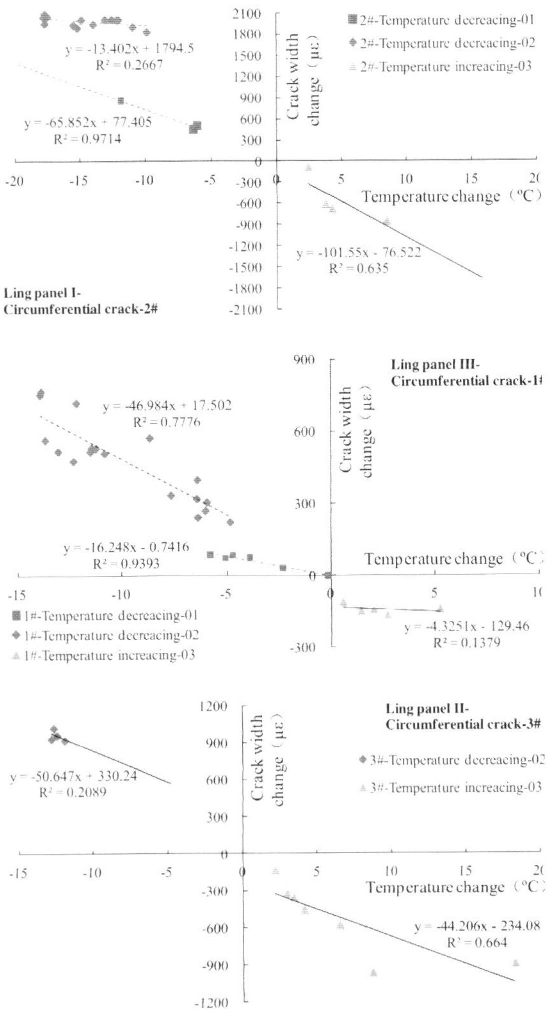

Figure 9. Features of the circumferential crack width changes with season temperatures.

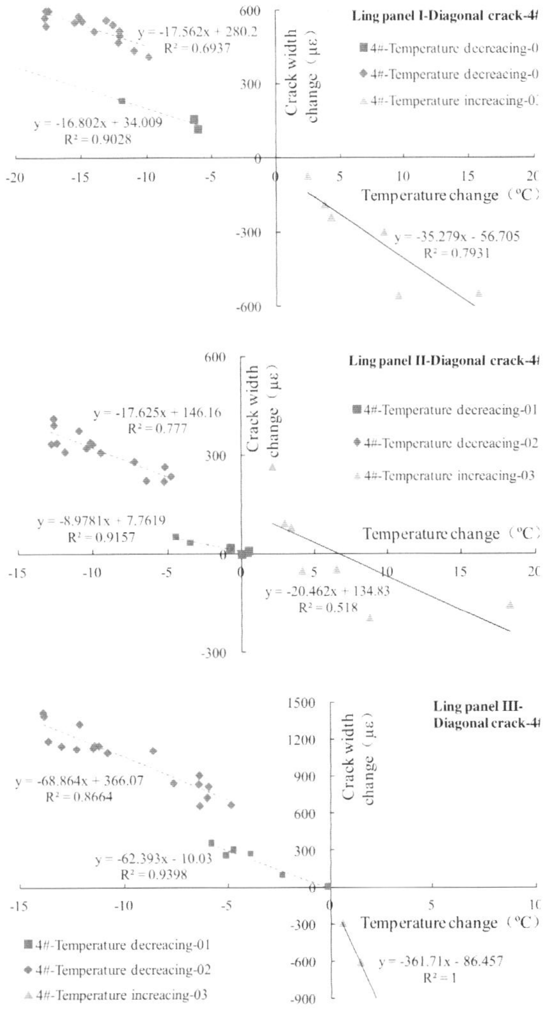

Figure 10. Features of the diagonal crack width changes with season temperatures.

References

[1] Fan, Q., Duan, Y. (2015) Hydraulic Tunnel Lining Concrete Temperature Control Crack Prevention Innovation and Practice. China Water Resources and Hydropower Press, Beijing.

[2] Wang, T. M. (2017) Control of Cracking in Engineering Structure, 2nd edition. China Architecture Publishing & Media Co. Ltd., Beijing.

[3] Duan, Y., Peng, Y., Luo, G., et al. (2018) Analysis of mechanism and development of cracks in portal lining concrete. Engineering Journal of Wuhan University, 51: 847-853.

[4] Zhu, H., Hu, Y., Li, Q., et al. (2020) Restrained cracking failure behavior of concrete due to temperature and shrinkage. Construction and Building Materials, 244, 118318.

[5] Li, M., Xu, W., Wang, Y.J., et al. (2020) Shrinkage crack inhibiting of cast in situ tunnel concrete by double regulation on temperature and deformation of concrete at early age. Construction and Building Materials, 240, 117834.

[6] Ma, J., Ren, K., Xue, C., et al (2019). Final lining concrete quality control of a loess tunnel with an integrated approach. IOP Conference Series: Earth and Environmental Science, 2019, 340, 052012.

Research and Application of the Composite Reservoir Stimulation Technology for Ultra-High-Temperature Ultra-Deep Naturally-Fractured Carbonate Reservoirs

Guifu Duan[1,2,*], Chunming He[1], Chong Liang[1], Bo Cai[1], Jun Yan[1] and Chao Wang[1]

[1]Research Institute of Petroleum Exploration & Development, PetroChina, Beijing, 100083, China
[2]China University of Petroleum, Beijing, 102249, China

*Corresponding author email: duanguifu@petrochina.com.cn

Abstract. The Ordovician buried-hill carbonate reservoir in the northern Jizhong Sag of China is seen with abundance in hydrocarbon resources. Nonetheless, attempts of multiple reservoir stimulation technologies have all failed to make breakthrough, due to its lithological complexity (mixture of dolomite and limestone), ultra-high temperatures (180°C) and high in-situ stress gradients (0.023 MPa/m), Given the aforementioned reservoir geological challenges, reservoir evaluation, technology optimization and plan design starting from the reservoir geological characteristics have been carried out, and the "composite" reservoir stimulation technology for ultra-high-temperature ultra-deep naturally-fractured carbonate reservoirs has been proposed. In terms of the short effective affected distance of acid under high temperatures (180°C), low conductivity sustainability owing to high closure stresses (induced by the burial depth of 5,000 m), extreme difficulties in expanding stimulated reservoir volumes (large lateral in-situ stress differences and long vertical spans), the practice of "multi-stage injection, temporary plugging and diversion, and propped acid fracturing" highlighting high pump rates and combination of low- and high- viscosity fluids (acid and fracturing fluids) has been developed and evolves into the network fracturing technology capable of inter-layer temporary isolation and within-layer diversion. The above technology has been applied in Well AT1x, with high volumes of fluid consumption and the maximum pump rate of 11 m^3/min. Safe and efficient operation over six hours featuring challenges such as 3,000 m^3 of composite fluids, ultra-high temperatures and ultra-high pressures is achieved. On-site monitoring such as the micro-seismic monitoring and simulation results show that compared with the previous reservoir stimulation, the developed technology presents SRV 3.5 times larger, and results in three-dimensional reservoir reconstruction. The post-fracturing daily gas production amounts to 409,000 m^3 and daily oil production, 71 m^3. In production testing, the wellhead pressure maintained over 25 MPa, and the daily gas production exceeds 100,000 m^3, which indicate stable pressures and production, and a safe wellbore. The application accomplishes the design goal of integrated exploration and exploitation, and geology and engineering, and provides references for reservoir simulation of analogous reservoirs.
Keywords: Ultra-high temperature; Ultra-deep; Composite reservoir stimulation; Carboxymethyl fracturing fluid; Temporary plugging and diversion.

1. Introduction

Over recent years, deep high-temperature oil and gas reservoirs have become an important frontier for energy source replacement, as targets of oil and gas exploration are increasingly complex [1]. The buried hill structures in the northern Jizhong Sag, such as the Yangshuiwu buried hill, possess hydrocarbon in place over one hundred million tons, and are considered as an emerging field for oil and gas resource replacement in the Bohai Bay area [2, 3]. Exploration in this sag starts in 1980s, during which Wells WG1, WG2 and WG4 are drilled successively and all present certain hydrocarbon show [4]. Given the reservoir characteristics and requirements of reservoir stimulation techniques, the small-scale propped fracturing, acid fracturing using gelled acid and composite acid fracturing practice of propped fracturing and clean acid fracturing are applied one after another. However, the effective stimulated time is limited, no stable production occurs, the stimulation result is unsatisfactory and no exploration breakthrough has been achieved, although production is to some extent stimulated. Over the past few years, constant progress is seen in the reservoir stimulation technique for deep reservoirs[5,6]. Ultra-high-temperature, ultra-deep reservoir stimulation technologies are found with breakthrough in Bakken, Eagle Ford, etc., of North America and also in the Tarim Basin and Sichuan Basin of China, which provides new inspiration for reservoir stimulation in this area [7]. Nonetheless, compared with reservoirs of marine facies in the Tarim Basin and Sichuan Basin in China, the carbonate reservoir of the study area is facing more challenges, such as the more complex lithology (interbedding of dolomite and limestone), higher temperatures (180°C), higher in-situ stress gradients (0.023 MPa/m), fast acid-rock reaction and limited stimulated distance. Reservoir stimulation techniques specific to the characteristics of the study area are in demand.

This paper, on the basis of experience of reservoir stimulation in analogous reservoirs both in China and other countries and also inspiration from reservoir reconstruction practice in the study area, adopts the fracturing fluid materials and acid system rated for 180°C, and develops the composite temporary-plugging technique using fibres + diverting acid + proppant that is able to improve the affected distance of acid fracturing and achieve within-layer diversion and inter-layer temporary plugging and "soft" layer separation. Moreover, the support technique for safe implementation of high-pump-rate, high-fluid-volume, multi-fluid alternate injection has taken shape. In field applications of the aforementioned techniques, Well AT1x presents high-rate oil and gas streams with post-fracturing daily oil production of 71 m3 and daily gas production of 400,000 m3, which shall provide advanced technical guidance upon high-efficiency stimulation of ultra-high-temperature ultra-deep complex reservoirs in China.

2. Reservoir Geological Features and Challenges for Stimulation

2.1. Overview of Geological Features

The previous drilling and exploration of the condensate gas reservoir in northern Jizhong area demonstrate that the reservoir features low permeability (0.1×10-3–0.3×10-3 μm2), low porosity (5%–8%), natural fractures dominated by micro and fine cracks that are highly filled and thus have inferior connectivity, high in-situ stress gradients (0.023 MPa) and large differences between the maximum and minimum horizontal stresses (7–10 MPa). With respect to such reservoir features, attempts of multiple reservoir stimulation techniques have been made, including propped fracturing, large-scale acid fracturing using gelled acid, repeated fracturing, large-scale acid fracturing using clean acid, etc. The resultant fracture is mainly single, with half length of only 80–120 m and limited connected distance, leading to failure to fulfill the purpose of deep-penetration reservoir stimulation by complex fractures. The reservoir simulation treatment turns out to provide short effective time and no stable production, and such results are unsatisfactory, although production stimulation is to some extent realized. A set of new reservoir simulation technique and practice applicable to the intrinsic reservoir geological features and technical difficulties of this block are demanded, for breakthrough in the new frontier of condensate gas reservoirs in the northern Jizhong Sag. Considering the reservoir geological features presented above, the reservoir stimulation difficulties are summarized as follows:

(1) The reservoir lithology is characterized by interbedding of limestone and dolomite. Acid-rock reaction in dolomite is simpler in terms of the process, and the reaction rate is higher than that of dolomite reservoirs under the reservoir condition. However, the effective stimulated distance of acid shortens, as acid-rock reaction accelerates.

(2) The reservoir, which is tight and deep buried, is seen with high breakdown pressures. Hence, it is

difficult to create and maintain effective fractures in the formation. The operation pressures during fracturing have reached over 80 MPa in multiple previous wells, and the fracture closure pressure exceeds 100 MPa. The pump rate in operation is limited, and it is found barely possible to fulfill the requirements of SRV-oriented reservoir simulation (SRV = Stimulated Reservoir Volume).
(3) The reservoir has a large vertical span (> 100 m), and thus effective selective fracturing and sufficient utilization of layers are extremely challenged.
(4) The burial depth of the reservoir is 5,000 m, accompanied by the formation temperature up to 180°C. This requires higher performances of fracturing fluids in terms of thermal and shearing resistance and of acid solutions regarding corrosion inhibition and reaction retardation.
(5) The closure pressure of the reservoir exceeds 80 MPa, and thus fracture conductivity is hard to maintain.
In reference to the aforementioned difficulties, integrated research on reservoir stimulation that incorporates the geological features needs to be carried out, in order to confirm geological factors affecting network fracturing, explore the networking fracturing technique capable of within-layer fluid diversion and inter-layer temporary plugging and its material system, and develop corresponding technical approaches for post-fracturing management and monitoring, which ultimately achieves systematic integration of geology and engineering, and exploration and exploitation.

3. Research on the Composite SRV-oriented Reservoir Simulation Technique

3.1. Enhancing Conductivity of Complex Fractures under High Closure Pressures

3.1.1. Acid Etching Characteristics under High Closure Pressures. Studies of both China and other countries have demonstrated that irregular etching of acid on fracture surfaces is mainly related to the reservoir lithology and heterogeneity and also injection process. The deep Ordovician buried-hill carbonate reservoir has strong heterogeneity in minerals, and thus fracture acid etching is seen with high irregularity, leading to high conductivity of etched fractures (Figure 1). In terms of the injection process, the "preferential flow channel" left after primary etching of acid can be further dissolved, by increasing the quantity of acid injection stages. As a result, the fracture width is expanded, fracture conductivity of multi-stage injection is higher than that of single-stage injection by 25%, and more importantly, the fracture conductivity maintained of multi-stage injection is increased by 40%. On the basis of the comprehensive analysis, it is believed that alternative injection involving 2–4 stages is suitable for acid fracturing in this block.

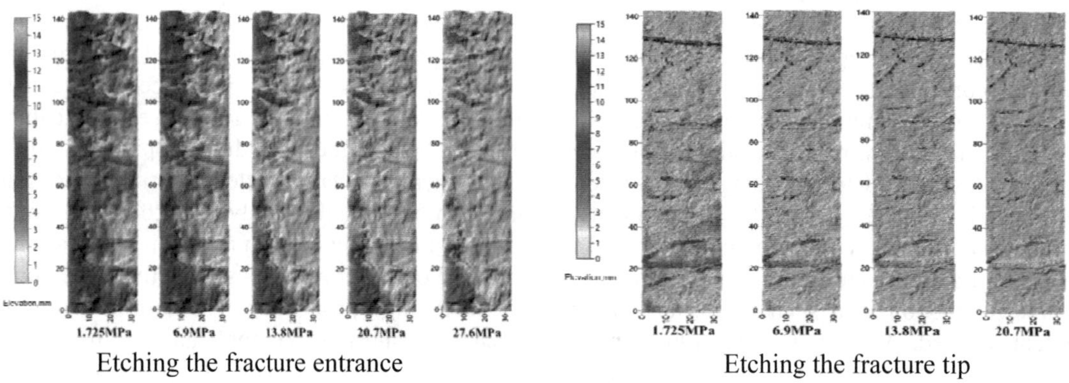

Figure 1. Acid etching characteristics at different acid injection stages.

3.1.2. Conductivity Evaluation of Complex Fractures. On the basis of the above acid etching experiment, conductivity of etched fractures has been measured. From Figure 2, it is seen that the maintaining capacity of fracture conductivity after large-scale acid injection is inferior for near-wellbore fractures. As the closure pressure grows, fracture conductivity declines significantly. Given this, laboratory tests of propped fracture conductivity after acid etching have been carried out, in which 50 g and 100 g of proppants (30/50 mesh) are placed across the rock sheet and conductivity variation features of acid etched rock sheets with varied proppant concentrations are experimentally captured. Results show that the rock sheets, with 50 g of proppants laid after acid etching, are able to present conductivity up to 190 μm²·cm even under the closure pressure of 60 MPa, which indicates great enhancement of the maintaining capacity of fracture conductivity. On the basis of these fracture conductivity experiments, a composite network fracturing approach, combining acid fracturing and propped fracturing and moreover suitable for deep Ordovician buried-hill reservoirs, has been proposed.

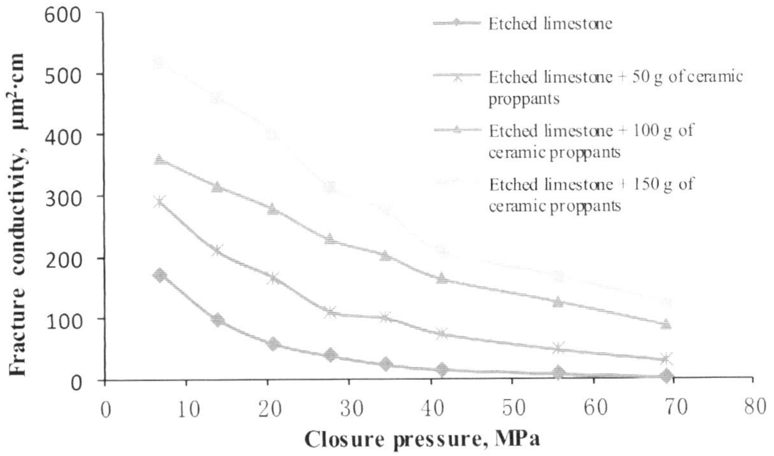

Figure 2. Fracture conductivity variation with different reservoir stimulation practices.

3.2. Research on the Prepad-cooling Multi-stage Temporary-plugging Technique

The reservoir temperature of Well AT1x is up to 177°C, and the reservoir matrix is seen with poor physical properties, with fractures serving as the main storage space and flow channel, which indicates an ultra-high-temperature, deep, naturally-fractured carbonate reservoir. In present China, the main reservoir stimulation techniques such as deep-penetration acid fracturing have come into being, with respect to fractured-vuggy carbonate oil and gas reservoirs represented by the Tahe oilfield and dissolution-pore carbonate oil and gas reservoirs represented by the Sichuan Longwangmiao gas reservoir, which facilitate effective exploitation of these types of reservoirs. Nonetheless, successful reservoir simulation treatments in naturally-fractured carbonate reservoirs are still less reported. On the basis of the concept and method of network fracturing in shale and also the features of carbonate reservoirs, this paper proposes the reservoir stimulation approach integrating "deep-penetration communication and volumetric stimulation". From a technical point of view, multi-scale natural fractures are developed in the reservoir, and it is difficult to etch and connect fractures at different scales under ultra-high temperatures, so as to expand the effectively stimulated volume; in the meantime, the target zone of stimulation has a large span, which results in great challenges for vertically uniform reservoir stimulation. Accordingly, the appropriate high-temperature-tolerable acid fracturing fluid system and fracturing practice are needed to ensure successful operation and high and stable production of Well AT1x.

3.2.1. The Prepad-cooling Technique. In order to slow down acid-rock reaction and extend the effective stimulated distance of acid, optimization is carried out in two aspects, namely the stimulation practice and acid system. First, the formation is cooled down by injecting a large volume of prepad fluids, and simulation suggests the best formation cooling performance occurs and within the effective stimulated distance of acid, the temperature is kept around 130°C, when the injection volume of fracturing fluids reaches 250 m³ (Figure 3).

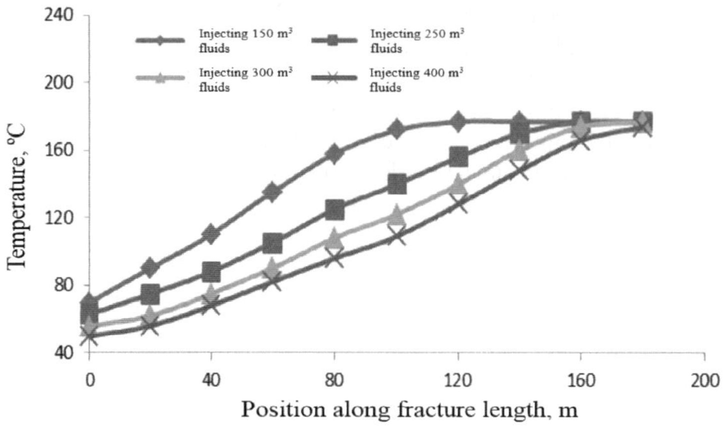

Figure 3. Formation temperature vs. injected fracturing fluid volume.

3.2.2. The Temporary Plugging and Fluid Diversion Technique. Natural fractures in the study area include both mostly-filled and partially-filled micro and fine cracks, and also large-scale near-wellbore fractures. It is required to achieve fluid (hydraulic fracture) diversion within the layer to improve natural fracture complexity and create fracture networks. Furthermore, the stimulation target section has a vertical span over 100 m, which demands inter-layer temporary plugging to enhance overall vertical recovery and fulfill the technical goal of three-dimensional reservoir stimulation. Within-layer diversion is to achieve dissolution and connection of natural fractures at different scales. For micro and fine cracks, low-viscosity slickwater is first injected at high pump rates to activate natural fractures and meanwhile cool down the formation, according to the results of the full three-dimensional hydraulic fracturing physical simulation experiment. Then, low-viscosity VES diversion acid is pumped into the formation also at high rates to dissolve and connect natural fractures, and build the natural fracture flow channels combing "hydraulic fracture networks and acid etching wormhole networks". In terms of the large-scale natural fracture, deep connection is achieved mainly by alternately injecting highly viscous fracturing fluids and acid solutions.
With respect to inter-layer temporary plugging, it is realized primarily by carrying a certain quantity of degradable temporary-plugging balls with acid solutions into the perforation tunnels of the layers that have received acid solutions and then sealing these perforation tunnels. By doing so, acid solutions are subsequently diverted to low-permeability layers, and the process is repeated for uniform acid placement. Regarding this technique practice, experimental studies of degradation performance and sealing capacity of fiber temporary plugging balls are conducted, which provide references for optimization of fibers and fiber temporary plugging balls.

4. Field Application and Performance Evaluation

Fracturing testing and main fracturing operation have been conducted in Well AT1x successively. Through the analysis of minifrac treatment, it is determined that the fracture extension friction resistance is about 1.3 MPa; the leakoff coefficient is 1.1; the fracture extension stress gradient is higher than those of the high-production reservoirs in previous cases of China; the pressure difference between two shut-in pressures is about 5 MPa, and the pressure decline rate drops from 14.6 MPa/h in the first shut-in to 9.4 MPa/h, while the pressure grows from 24.7 MPa to 30.4 MPa. The last observation implies that the scale of the good natural fracture development zone is limited, and dominated by tiny fractures. In the field, the reservoir (fracture) connection performance is improved by raising up the acid concentration and also pump rate, and the pressure drop is about 12 MPa. Fracturing testing suggests that continuous operation may lead to further elevation in operation pressures and it would be difficult to enable high-pump-rate operation and go against high production. The main fracturing operation adopts a large-scale plan involving 3,000 m^3 fluids, and one-time injection to improve the fracture connection distance and fracture network complexity. During the main fracturing operation, on the basis of the operation pressure variation and microseismic monitoring results, the originally designed five-stage injection is changed into three-stage; the original four stages of temporary plugging agents and proppants are changed into three stages; pump rates, proppant concentration and fluid properties are adjusted accordingly. By doing so, we manages to implement safe and highly efficient operation over six hours under various unfavorable conditions such as operation pressures over 90 MPa, a maximum pump rate up to 11 m^3/min and an injection fluid volume up to 2,500 m^3. 21.5 m^3 of proppants are injected, and the phenomenon, displayed as sudden pressure drops under stable-pump-rate conditions and suggesting fracture connection, occurs for five times during operation, with the maximum pressure drop of 23 MPa. The designed goals of deep connection have been fulfilled.

The microseismic monitoring result shows that the length of the artificial fracture is 280–350 m; the fracture height, 140–160 m; the width of the fracture network, nearly 200 m; the stimulated reservoir volume is expanded by 3.5 times, compared with those in previous cases. The three-dimensional reservoir stimulation is achieved, and the goal of design optimization is fulfilled. The operation achieves the design goal of integration between exploration and exploitation, and between geology and engineering, and is ground-breaking for exploration of Yangshuiwu buried hill structures. It overcomes the technical bottlenecks in achieving high and stable production in this block, and provides technical support to future reservoir stimulation in this buried hill reservoirs.

Acknowledgments
This research is supported by the CNPC Science and Technology Major Project (2017E-15) and the National Science and Technology Major Project (2017ZX05030-05) of China.

References

[1] Zhao Xianzheng, Wang Quan, Jin Fengming, et al.Re-exploration Program for Petroleum-rich Sags and its Significance in Bohai Bay Basin, East China[J]. Petroleum Exploration and Development，2015, 42(6): 723-733.

[2] Zhao Xianzheng, Zhu Jieqiong, Zhang Ruifeng, et al. Characteristics and Exploration Potential of Tight Calcilutite Rrudstone in Shulu sag, Jizhong Depression, North China reservoirs [J]. Acta Petrolei Sincia, 2014,35(4):613-622.

[3] Cai Bo,Zhao Xianzheng,Shen Hua.Hybrid Stimulated Reservoir Volume Technology for Tight oil in Shulu sag[J]. Acta Petrolei Sincia,2015，36(S1):76-82，90.

[4] Zhao Xianzheng, Jin Fengming. Precise Buried-hill Hydrocarbon Reservoir of Exploration Hydrocarbon of Subtle Enriched Depressions[M]. Beijing: Petroleum Industry Press, 2010

[5] Bale A, Smith M B, Klein H H.Stimulation of Carbonates Combining Acid Fracturing with Proppant (CAPF): a Revolutionary Approach for Enhancement of Final Fracture Conductivity and Effective Fracture Half-length[R].SPE 134307,2010

[6] Wang Yonghui,Lu Yongjun,Li Yongping,et al. Progress and Application of Hydraulic Fracturing Technology in Unconventional Reservoir[J]. Acta Petrolei Sinica, 2012 ,33(s1): 149-158.

[7] Ming, T. Design and Application of Expandable Casing Drilling Technology of Deepsidetrack Horizontal well in Tahe Oilfield[R].IPTC 16760,2013

Research of Shear Experiment of Interface between New Steel Fiber Self-compacting Concrete and Old Ordinary Concrete

Jinping Zhuang*, **Sizheng Shen and Zheng Zhang**

Department of Civil Engineering, Fujian University of Technology, Fujian, China

*Corresponding author email: 19781208@fjut.edu.cn

Abstract. In prefabricated concrete structure, the shear performance of interfaces between prefabricated parts and cast-in-place parts needs improvement. Steel fiber self-compacting concrete is a desirable kind of concrete to solve this problem. This paper introduced Z-shape specimens with interfaces between new steel fiber self-compacting concrete part and old ordinary concrete part as direct shear experiment objects. Factors including steel fibers volume ratio and inserting rebars were also studied. The results show that the shear strengths of interfaces increase with the increases of steel fibers volume ratio, and inserting rebars can improve shear strengths and change failure form from brittle failure to ductile failure.
Keywords: Steel fiber self-compacting concrete; Shear strength of interface; Direct shear experiment; New and old concrete.

1. Introduction

Prefabricated concrete structure (PCS) features energy saving, environmental protection and short construction period. The development of PCS is an inevitable trend for the sustainable development of construction industry.

However, some problems that inhibit the development of PCS still exist, such as poor seismic performance and weak connection between prefabricated parts and cast-in-place parts. Because the concrete of prefabricated parts sets earlier than that of cast-in-place parts, the interface between them becomes a vulnerable part of structures. So, it is of great significance to study the shear strength of the interfaces between cast-in-place and prefabricated concrete.

Nowadays, researches on the shear strength of interfaces between cast-in-place and prefabricated concrete have matured, but there are still some shortcomings. Steel fiber self-compacting concrete (SFSCC)is a suitable kind of concrete with prominent mechanical properties to cope with such problem. However, most of the previous researches focused on study of the shear strength of interfaces between ordinary cast-in-place concrete and old ordinary concrete (OOC), there are few researches on the shear strength of interfaces between cast-in-place SFSCC and OOC.

Therefore, this paper presents the research results of shear experiment on specimens with interface between SFSCC and OOC. Introducing 2 kinds of interface process methods including rebars-inserting and rough interface and grooves-setting interface and through the change steel fibers volume, its effect on shear strength is studied. Meanwhile, the effect of rebars-inserting on failure form is also studied.

2. Experimental Study

2.1. Specimen Design

According to existing reference [1] and considering the shear transmission and distribution on the interface, adopting Z-shape specimen to study the shear strength of interface between SFSCC and OOC.

The geometric dimensions and appearance of Z-shape specimen are shown in Figure 1.

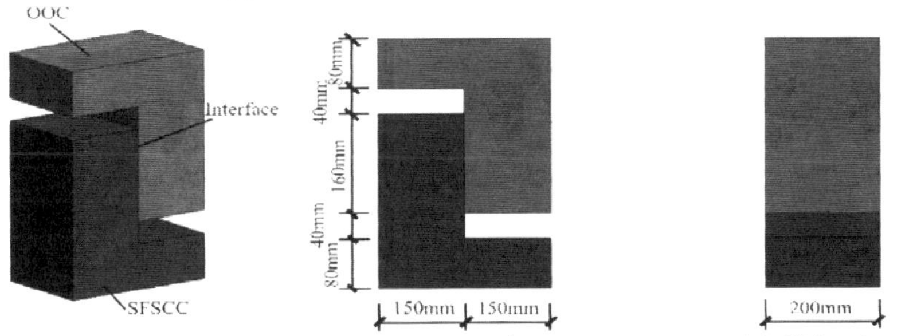

(a) 3D view (b) Front view (c) Side view

Figure 1. The geometric dimensions and appearance of Z-shape specimen.

2.2. Materials and Mix Proportion

The cement used for this experiment is grade 42.5 P·O. The coarse aggregate is crushed stone. Fine aggregate is river sand with good gradation. Superplasticizer is chosen as concrete admixture. Mineral admixture is grade II fly ash. The length of steel fiber is 30mm and its nominal diameter is 0.5mm. Mix proportion of ordinary concrete is designed according to *Code for Design of Mix Proportion of Ordinary Concrete* [2]. Mix proportion design of SFSCC refers to *Technical Specification for Application of Self-Compacting Concrete* [3]. Details of mix proportion is demonstrated in Table 1.

Table 1. Concrete mix proportion.

Strength grade	Amount of ingredients in concrete (kg/m^3)						
	Water	Cement	Sand	Stone	Fly ash	Superplasticizer	Steel fiber
C30	185	350	650	1190	—	—	—
	192	350	815	816	200	4.15	39
C45	192	350	815	816	200	4.15	78
	192	350	815	816	200	4.15	117

2.3. Process of Interface.

Ordinary concrete parts of Z-shape specimens were cast on January 12, 2017. After 3 around months, interfaces of ordinary concrete parts of Z-shape specimens were processed. Process methods contain 2 different types, including roughness with rebars-inserting and grooves-setting. Ordinary concrete parts of Z-shape specimens with rebars-inserting and rough interface (RIRI) are shown in Figure 2. Ordinary concrete parts of Z-shape specimen with grooves-setting interface (GSI) is shown in Figure 3.

Figure 2. Rebars-inserting and rough interface.

Figure 3. Grooves-setting interface.

2.4. Finish Specimens

SFSCC part of Z-shape specimens was poured in May 2017, 100 days after the ordinary concrete part was poured. As shown in Figure 4, ordinary concrete part whose interface was already processed was put in wood template. Simultaneously, put pearl cottons to make entire specimen form Z-shape after SFSCC set. Then 2 days later, Z-shape specimens were removed from templates and cured by water for 2 months. One of finished Z-shape specimen is shown in Figure 5.

Figure 4. Ordinary concrete part.

Figure 5. Finished Z-shape specimen.

Totally 6 different types of Z-shape specimens with different interfaces were made, and the number of each of them is 3, including 3 ordinary concrete specimens that were integrally cast and used for comparison. List of Z-shape specimens is shown as Table 2.

Table 2. Lise of Z-shape specimens.

Types	Codes	Groove depth (mm)	Groove width (mm)	Steel fibers volume ratio (%)	rebars inserting ratio (%)
Groove-setting interface	S1-G2	12	30	0.5	/
	S2-G2	12	30	1.0	/
	S3-G2	12	30	1.5	/
Rough and rebars-inserting interface	S1-CP2	/	/	0.5	0.32
	S2-CP2	/	/	1.0	0.32
Comparison specimens	ZT	/	/	/	/

2.5. Load Methods and Experimental Machines

According to *Specification for Seismic Test of Buildings* [4], Procedure of experimental load can be divided into two steps, which are pre-load and formal load.

Pre-load is important [5]. Because it can help to ensure that all parts of the specimens are in good contact with machines, the relationship between load and deformation is stable, the reliability of experiment machines and whether all devices work normally or not [6]. Generally, the value of pre-load should not be greater than the crack load, so the value of pre-load is 15KN.

Formal load contains force load and displacement load. Force load was divided into 3 stages. In first stage, load rate is 60N/s and maximum value is 5KN. In second stage, load rate is 100N/s and maximum value is 60KN. In lats stage, load rate is 40N/s and maximum value is 80KN. Finally, displacement loaded uniformly at the speed of 0.005mm/min until the specimen fail.

As shown in Figure 6, the shear load machine is 100 tons SANS computer controlled electro-hydraulic servo pressing machine. The SFSCC part of Z-shape specimens was set at the load side, OSC part was set at the fastened side. Strain gauges is set at the SFSCC part. Baffles were used to fasten specimens. Displacement can be measured by dial indicator.

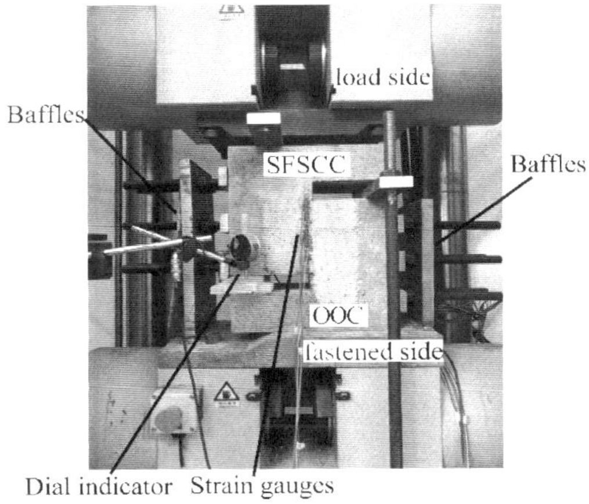

Figure 6. Experiment machines.

3. Experimental Result and Analysis

3.1. Failure Phenomena

3.1.1. Failure phenomena of specimens with GSI. At the beginning of load, small cracks began to appear from the load side. As the force load process went on, the cracks that appeared from load side gradually extended to the fastened side along the interface. Then, as displacement load went on, tiny vertical cracks began to appear at the bottom of bond surface. Finally, these cracks became wider. Meanwhile, the shear stress reached the peak value, the shear strength of specimens with GSI reduced sharply. Therefore, shear failure had happened. It also had the characteristics of brittle failure. The typical initial and failure states are shown in Figure 7.

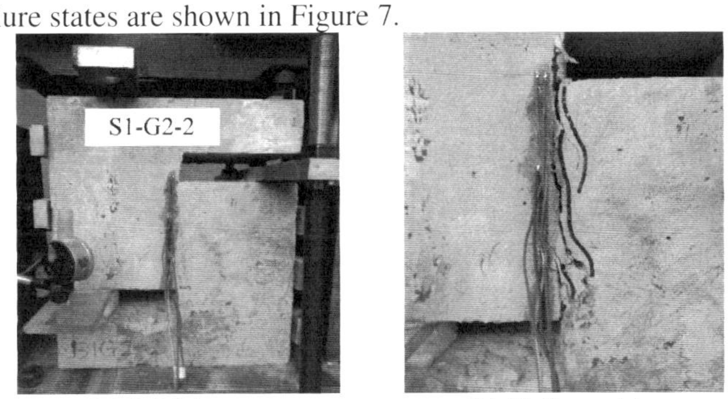

(a) Initial state (b) Failure state

Figure 7. Typical failure process of specimens with GSI.

3.1.2. Failure phenomena of specimens with RIRI. At the beginning of force load, small cracks began to appear from the load side. When the displacement load continued after the force load, cracks began to extend along the interface to the fastened side. As the displacement load process went on, the crack at the load side continued to widen to 0.5 mm. Meanwhile, the shear stress firstly reached the peak value, and cracks began to appear at the fasten side. When shear stress decreases sharply to about 85% of the first peak value, it increased again, which indicated rebars worked. Meanwhile, the surface along interface began to peel off, cracks became wider, and maximum width is up to 5mm. However, because of the effect rebars, the specimen remained as a whole, which represented ductile failure. The typical initial and failure states are shown in Figure 8.

(a) Initial state (b) Failure state

Figure 8. Typical initial and failure state of specimens RIRI.

3.2. Shear Strength

Shear stress is calculated by equation (1). Its values are shown in Table 3.

$$\tau = \frac{V}{A} \tag{1}$$

where τ is shear stress, V is shear force and A is area of interface.

Table 3. List of shear stress.

Specimens	Number	Interface type	Steel fibers volume ratio (%)	Shear force (KN)	shear stress (MPa)	Mean values (MPa)	Ratios to comparison specimens (%)
S1-G2	1	Grooves-setting	0.5	179	5.85	5.56	62.2
	2			170	5.32		
	3			185	5.78		
S2-G2	1	Grooves-setting	1.0	262	8.18	7.88	88.1
	2			242	7.57		
	3			-	-		
S3-G2	1	Grooves-setting	1.5	343	10.73	9.33	104
	2			245	7.67		
	3			307	9.60		
S1-CP2	1	Rebars-inserting and rough	0.5	250	7.82	7.74	86.6
	2			260	8.12		
	3			233	7.29		
S2-CP2	1	Rebars-inserting and rough	1.0	236	7.36	8.04	89.9
	2			276	8.64		
	3			260	8.11		
ZT	1	Comparison specimens	0	313	9.79	8.94	100
	2			239	9.17		
	3			252	7.86		

Note: Cracks appeared at the bottom of S2-G2-3 before load, which resulted in low shear strength, so it was not included in the list.

3.3. The Effect of Steel Fibers Volume Ratio of SFSCC on Shear Strength.

Shear strengths of specimens with RIRI are shown in Figure 9 (a), when specimens' interfaces are rough and inserted two rebars, as steel fibers volume ratio increases from 0.5% to 1.0%, shear strength increases 3.88% correspondingly. Shear strengths of specimens with GSI are Figure 9 (b), when grooves width and depth of interface are 30mm×2 and 12mm respectively, as steel fibers volume ratio increases from 0.5% to 1.0% to 1.5%, shear strength increases from 5.56MPa to 7.88MPa to 9.33MPa, the increase rate are 41.8% and 18.4% respectively.

It is obvious that steel fibers can effectively improve the shear strength of Z-shape specimens. As the steel fibers volume ratio increases, shear strength increases. Moreover, it reaches its peak value when steel fiber ratio is 1.5%, and there is still room for improvement [7]. Because, steel fiber can prevent extension of cracks by reducing the stress concentration at crack tips, remarkably reduce the shrinkage of SFSCC. Furthermore, part of steel fibers slowly extends into interface of OOC during water curing period. The bridge effect of steel fibers [8] across the interface improves specimens shear strength.

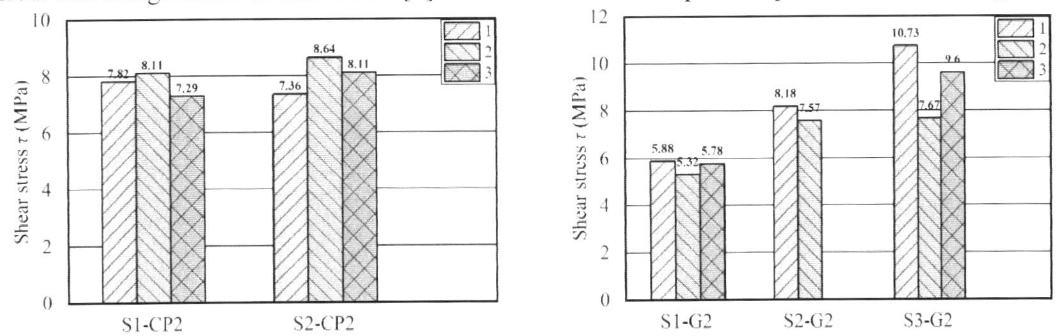

(a) Comparison between specimens with rough and rebars-inserting interface

(b) Comparison between specimens with groove-setting interface

Figure 9. Comparison of shear strength between specimens with different steel fibers volumes ratio. Note: 1, 2 and 3 denote the number of three identical specimens in the same group

4. Conclusion

Generally, cracks of different specimens appeared along the bond surface, then they extended from load side to fastened side. Finally, specimens failed because those cracks widened. The failure of specimens without inserting rebars at interface is typical brittle failure, nonetheless the failure of specimens with inserting rebars at interface is typical ductile failure.

Steel fibers can effectively improve the shear strength of interface of SFSCC and OOC. The shear strengths of interfaces of Z-shape specimens increase with the increase of steel fibers volume ratio. Shear strengths of specimens (S3-G2) with GSI and 1.5% steel fibers volume ratio exceed that of comparison specimens by 4%.

Acknowledgments

This work is supported by the National Natural Science Foudation of (51678152) and the Research Development Foundation of Fujian University of Technology (GY-Z13119)

References

[1] Li L, Wang Z L 2009 Study on Shear Properties for Adherence of New and Old Concrete and Its Practical Calculation *International Conference of Chinese Transportation Professionals* pp 1-7.

[2] 2011 *Code for Design of Mix Proportion of Ordinary Concrete* (Peking: China Construction Industry Press)

[3] 2012 *Technical Specification for Application of Self-Compacting Concrete* (Peking: China Construction Industry Press)

[4] 2015 *Specification for Seismic Test of Buildings* (Peking: China Construction Industry Press)

[5] Zhang W G, Chen T S and Chi W S 1995 Prefabricated frame steel fiber reinforced concrete cogging joint *Journal of architectural structure* vol 3 pp 52-58

[6] Resreepo, Jose L, Robert P and Andrew H B 1995 Design of connections of earthquake resisting precast reinforced concrete perimeter frames *PCI Journal* vol 40 pp 68-76

[7] Chen H Q 2007 *Experimental study on bond performance between steek fiber concrete and old concrete* (Zhengzhou: Zhengzhou University)

[8] Chen H Q, Liu G L and Gao L Y 2013 Experimental study on bond shear performance of steel fiber concrete and old concrete *Journal of hydropower* vol 4 pp 177-192

Application Research on Reliability Modeling of Multi-failure Modes Correlation System Based on Time Varying Copula Model

Ling Zhou* and Zhenxin He

Changchun Reliability Science and Technology Ltd., Changchun, China

*Corresponding author e-mail: ccruilaier@163.com

Abstract. The reliability modeling of dynamic correlation system with multi-failure modes is studied in this paper. Firstly, Patton model, one of the time-varying copula models, is introduced. Secondly, multi-dimensional time-varying copula function of multi-failure modes correlation system is transformed into two-dimensional time-varying copula functions based on vine Copula function, and then reliability modeling and solution procedure of multi-failure modes correlation system is presented based on time-varying copula model. Finally, the above reliability modeling and solution procedures are demonstrated through a vehicle suspension system engineering case, and their correctness are proved by the results obtained in engineering case.

1. Introduction

Now copula function has been more and more used in reliability modeling of correlation systems. In the initial research stage, the correlation parameters in copula function are constants, namely it is a static Copula function [1]. However, it is found that the correlation between failure modes in complex system changes with time in-depth study, i.e., correlation parameters in Copula function should not be constant, but change with time due to the changes of working environment, inherent properties and state of the system. Therefore, dynamic Copula function and model should be used for reliability modeling, analysis and evaluation of dynamic correlation systems.

Zhihua Wang et al. propose a general framework and corresponding methods to deal with the time-dependent reliability analysis of a mechanism effectively [2]. Time-variant copula function is used to estimate the cumulative probabilities of failure by Wang Zhonglai et al. [3]. Zhao-Xia Xu et al. apply the copula-based sampling method firstly to study 3D reliability of seismic slopes as well as the cross correlation between the geotechnical parameters [4]. Árpád Rózsás et al. qualitatively and quantitatively analyze the impact of Gauss (normal or Gaussian) copula assumption on failure probability [5]. Y. T. Sun et al. present a new reliability analysis method in which the mixed Copula is constructed to describe the correlations among multi-failure modes [6]. Pan Yue et al. develops a hybrid Copula-Bayesian-based approach to model the structural health of an operational metro tunnel in a dependent system [7]. Dan Xu et al. propose a method of multivariate failure behavior modeling and reliability assessment based on vine-copula and accelerated degradation data [8]. Hu Qiguo et al. propose the empirical distribution function-local maximum likelihood two-step method to estimate the time-varying parameters in the dynamic Copula function [9]. Yaping Wang et al. develop a dependent competing risk model for systems subject to multiple degradation processes and random shocks using time-varying copulas [10].

At present, the application research on dynamic copula function used in dynamic correlation system reliability modeling and analysis is rare, so it is urgent to be studied to meet the increasingly complex, huge and correlation system reliability modeling, prediction and analysis requirements.

2. Dynamic Copula Model and Time-varyingparameter Estimation

Patton, who studied dynamic copula model earlier, proposed that the current correlation between variables can be explained by the previous correlation and the historical average of the cumulative probability of the two variables according to Hansen's autoregressive conditional density model. In Patton model, Copula's correlation parameter is a function of lag data and a parameter transformed by autoregressive term. In Patton's article [11], based on Hansen autoregressive distribution model, dynamic correlation parameter in normal copula was presented as follows

$$\rho_t = \Lambda\left(\omega_\rho + \beta_\rho \rho_{t-1} + \alpha_\rho \frac{1}{m}\sum_{i=1}^{m} \Phi^{-1}\left(U_{r^1}(t-i)\right)\Phi^{-1}\left(U_{r^2}(t-i)\right)\right) \tag{1}$$

where $\Lambda = \left(1-e^{-x}\right)/\left(1+e^{-x}\right)$ is a transfer function, whcih is used to ensure correlation parameter ρ remains within the range of [-1,1]; $\Phi^{-1}(\cdot)$ is inverse function of cumulative distribution function of the standard normal function; The parameters ω_ρ, β_ρ and α_ρ are constant parameters estimated by the time series; m is the window length value, which represents the number of lag items, and generally it is set as 10.

Time-varying correlation parameter estimation process of bivariate copula function based on SP method is as follows:

1) Through the empirical distribution formula $u = \widetilde{F}(x) = \frac{1}{n}\sum_{k=1}^{n} I\left\{x^{(k)} \le x_t\right\}$, the original data series X, Y are transformed into [0,1] interval sequence u, v;

2) Time-varying correlation parameter is substituted into time-varying bivariate copula function;

3) The likelihood functions corresponding to time-varying bivariate copula function is obtained as follows

$$l(\delta) = \sum_{k=1}^{n} \operatorname{Inc}\left(\widetilde{F}\left(x^{(k)}\right), \widetilde{F}\left(y^{(k)}\right), \delta\right) \tag{2}$$

4) Time-varying correlation parameter is estimated by optimization method as follows

$$\hat{\delta} = \underset{\delta}{\operatorname{argmax}}\ l(\delta) = \min\left(-\ l(\delta)\right) \tag{3}$$

3. D-Vine Copula and Reliability Formula of Serial System

Vine Copula function as a convenient mathematical tool can be used for reliability modeling problem of multi-failure mode correlation system [12]. Vine Copula function is used to decompose joint distribution functions of multiple variables into marginal distribution functions and two-dimensional Copula functions. C-Vine and D-Vine functions are two common types of Vine Copula functions.

In this paper, D-Vine function is used to decompose joint distribution functions of multiple variables. $\boldsymbol{X} = \left(X_1, X_1, \cdots, X_n\right)$ is an N-dimensional random vector. Based on D-Vine function, The decomposition form of probability density function is as follows

$$f\left(x_1, x_2, \cdots, x_n\right) = \prod_{k=1}^{n} f_k\left(x_k\right) \prod_{j=1}^{n-1}\prod_{i=1}^{n-j} c_{i,j+i|i+1,\cdots,i+j-1} \left(F_{i|i+1,\cdots,i+j-1}\left(x_i \mid x_{i+1}, \cdots, x_{i+j-1}\right), F_{i+j|i+1,\cdots,i+j-1}\left(x_{i+j} \mid x_{i+1}, \cdots, x_{i+j-1}\right)\right) \tag{4}$$

And Copula function of multi-failure mode correlation system is decomposed as follows

$$c(F_1(x_1), F_2(x_2), \mathrm{L}, F_n(x_n)) = \prod_{j=1}^{n-1}\prod_{i=1}^{n-j} c_{i,i+j|i+1,\mathrm{L},i+j-1}$$

$$\left(F_{i|i+1,\mathrm{L},i+j-1}(x_i \mid x_{i+1},\mathrm{L},x_{i+j-1}), F_{i+j|i+1,\mathrm{L},i+j-1}(x_{i+j} \mid x_{i+1},\mathrm{L},x_{i+j-1})\right) \tag{5}$$

where $f_k(x_k)$ is marginal probability density function of each variable, k=1,2..., n. The subscript j is the tree T_j, and the subscript i is the edge in T_j. $c_{i,i+j|i+1,\cdots,i+j-1}$ is a two-dimensional Copula function acting on $F_{i|i+1,\cdots,i+j-1}(\cdot)$ and $F_{i+j|i+1,\cdots,i+j-1}(\cdot)$.

In this paper, the reliability of series system will be calculated, and its expression is introduced firstly as follows

$$\begin{aligned} R_S(t) &= P(T>t) = P\{\min(X_1,X_2,\cdots,X_n)>t\} \\ &= 1-\sum_{i=1}^{n}(1-R_i(t)) + \sum_{1\le i<j\le n} C_2(1-R_i(t),1-R_j(t)) - \end{aligned} \tag{6}$$

$$\sum_{1\le i<j<k\le n} C_3(1-R_i(t),1-R_j(t),1-R_k(t)) + \cdots + (-1)^n C(1-R_1(t),1-R_2(t),\cdots,1-R_n(t))$$

where $2\le k\le n$.

4. The Calculation Procedure of Multi-failure Mode Correlation System Reliability Based on Dynamic Copula

Based on Vine Copula function and bivariate time-varying Copula model, the reliability solution procedure of multi-failure mode correlation system is as follows:

1) According to the life data of each unit in the system, the marginal life distribution of each unit is fitted;

2) The reliability model of system is established according to the logical relationship between multiple failure modes, and then the calculation formula of system reliability is determined according to the reliability model of system . If it is a series system, the system reliability is expressed as formula (6);

3) The Vine Copula function is used to decompose the multiple Copula function into bivariate Copula function;

4) The correlation parameters in bivariate Copula functions are estimated by time-varying Copula model and static Copula model, and optimization Copula model is selected through the maximum values principle of logarithmic likelihood function, AIC principle and BIC principle.

5) The bivariate Copula model can be determined by substituting the time-varying correlation parameters ;

6) The reliability degree of multi-failure mode correlation system can be obtained by substituting bivariate Copula model into the calculation formula of system reliability.

5. Application Case

The suspension system of a vehicle is composed of balance elbow, torque shaft and hydraulic shock absorber. Referenc [1] provides 50 life data of each unit of the suspension system.

The reliability calculation process of multi-failure mode correlation system based on dynamic Copula model and Vine Copula function is as follows:

1) Weibull distribution, which is commonly used in the life distribution of mechanical products, is used to describe the life distribution of balance elbow, torsion shaft and hydraulic shock absorber. According to the life data of referenc [1], the Weibull distributions of balance elbow, torsion shaft and hydraulic shock absorber are fitted firstly;

2) According to the composition and working principle of the suspension system, the reliability model of suspension system is a series system. And reliability degree formula of the suspension system is as follows

$$R_S(t)=P(T>t)=1-\sum_{i=1}^{3}F_i(t)+\sum_{1\le i<j\le 3}C_2\left(F_i(t),F_j(t)\right)-\sum_{1\le i<j<k\le 3}C_3\left(F_i(t),F_j(t),F_k(t)\right) \quad (7)$$

where $R_S(t)$ is system reliability degree, $R_i(t)$ is reliability degree of each unit, $F_i(t)$ is failure probability function of each unit, C_2 is bivariate Copula function, C_3 is triple Copula function.

3) Based on Vine Copula function decomposition technique, triple Copula functions in formula (7) are decomposed into bivariate Copula function as follows

$$C_3\left(F_1(t),F_2(t),F_3(t)\right)=C_2\left(F_1(t),F_2(t)\right)C_2\left(F_2(t),F_3(t)\right)C_2\left(F_{1|2}(t),F_{3|2}(t)\right) \quad (8)$$

where the formula of $F_{1|2}(t)$ and $F_{3|2}(t)$ are expressed as follows

$$F_{1|2}(t)=\frac{\partial C\left(F_1(t),F_2(t)\right)}{\partial F_2(t)} \qquad F_{3|2}(t)=\frac{\partial C\left(F_3(t),F_2(t)\right)}{\partial F_2(t)} \quad (9)$$

4) In view of the correlation between the life of mechanical parts are usually positive correlation, also it is shown that Gumbel copula function can accurately describe the correlation of mechanical parts in reference [1]. Thus, Gumbel copula function is selected to describe the correlations between balance elbow, torque shaft and hydraulic shock absorber. At the same time, normal copula function is adopt to compare with Gumbel copula function. Correlation parameters are estimated by time varying Gumbel copula model and static Gumbel copula model, as shown in Fig.1. The logarithmic likelihood function values, AIC values and BIC values of the static model and time-varying model of Gumbel Copula function and Normal Copula function are shown in Table 1. The results show that time-varying model of Gumbel Copula function is more appropriate to describe the correlation between units of suspension system.

5) When the optimal model of Copula function is determined to be time-varying model of Gumbel Copula function, the value of bivariate time-varying Gumbel Copula function can be calculated at a specific time point. For example, when t=6000km, bivariate time-varying Gumbel Copula function can be determined as follows

$$\begin{aligned}
C_2\left(F_1(t),F_2(t)\right)&=\exp\left(-\left((-\log(1-F_1(t)))^{\delta}+(-\log(1-F_2(t)))^{\delta}\right)^{1/\delta}\right)=9.7233\times10^{-4}\\
C_2\left(F_2(t),F_3(t)\right)&=\exp\left(-\left((-\log(1-F_2(t)))^{\delta}+(-\log(1-F_3(t)))^{\delta}\right)^{1/\delta}\right)=9.7403\times10^{-4}\\
C_2\left(F_2(t),F_3(t)\right)&=\exp\left(-\left((-\log(1-F_1(t)))^{\delta}+(-\log(1-F_3(t)))^{\delta}\right)^{1/\delta}\right)=0.1998\\
C_2\left(F_{1|2}(t),F_{3|2}(t)\right)&=\exp\left(-\left((-\log(1-F_{1|2}(t)))^{\delta}+(-\log(1-F_{3|2}(t)))^{\delta}\right)^{1/\delta}\right)=0.9973
\end{aligned} \quad (10)$$

6) Substituting above bivariate time-varying Gumbel Copula function into (8) and (7), and reliability degree of suspension system at t=6000km can be obtained as follows

$$\begin{aligned}
R_S(t)&=1-\sum_{i=1}^{3}F_i(t)+\sum_{1\le i<j\le 3}C_2\left(F_i(t),F_j(t)\right)-\sum_{1\le i<j<k\le 3}C_3\left(F_i(t),F_j(t),F_k(t)\right)\\
&=1-F_1(t)-F_2(t)-F_3(t)+C_2\left(F_1(t),F_2(t)\right)+C_2\left(F_2(t),F_3(t)\right)+C_2\left(F_1(t),F_3(t)\right)-C_3\left(F_1(t),F_2(t),F_3(t)\right)\\
&=0.5208
\end{aligned} \quad (11)$$

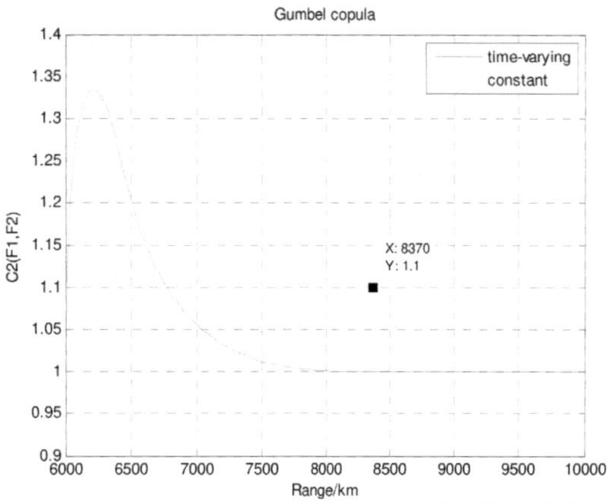

(a) correlation parameter in $C_2(F_1(t),F_2(t))$

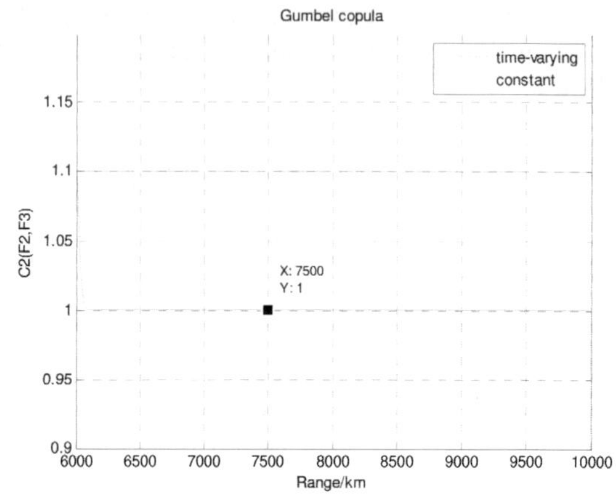

(b) correlation parameter in $C_2(F_2(t),F_3(t))$

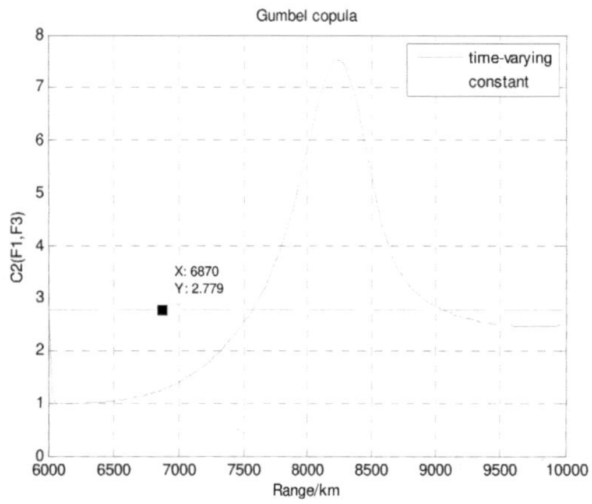

(c) correlation parameter in $C_2(F_1(t),F_3(t))$

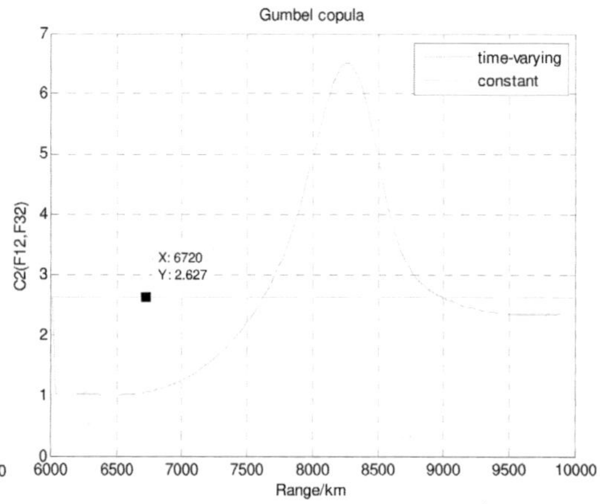

(d) correlation parameter in $C_2(F_{1|2}(t),F_{3|2}(t))$

Figure 1. Variation of correlation parameter in Gumbel Copula function.

Table 1. Logarithmic likelihood function values, AIC values and BIC values.

Copula function	Copula model	$l(\delta)$	AIC	BIC
$C_2(F_1(t),F_2(t))$	Dynamic Gumbel copula	5.6355	11.2411	11.1918
	Static Gumbel copula	-21.6195	-43.2489	-43.2653
	Dynamic Normal Copula	-76.2430	-152.5158	-152.5651
	Static Normal Copula	-2.4840e+04	-4.9681e+04	-4.9681e+04
$C_2(F_2(t),F_3(t))$	Dynamic Gumbel copula	-0.0291	-0.0881	-0.1374
	Static Gumbel copula	-21.2691	-42.5481	-42.5646
	Dynamic Normal Copula	-287.1480	-574.3259	-574.3752
	Static Normal Copula	-3.6206e+04	-7.2412e+04	-7.2412e+04
$C_2(F_1(t),F_3(t))$	Dynamic Gumbel copula	164.2200	328.4102	328.3609
	Static Gumbel copula	143.6518	287.2937	287.2772
	Dynamic Normal Copula	-198.9406	-397.9110	-397.9603
	Static Normal Copula	-2.3250e+04	-4.6500e+04	-4.6500e+04

Reliability degree of suspension system is 0.5208 which is obtained by time-varying Gumbel Copula function, it is just in the interval between complete independence $R_S(t)=0.425$ and perfect correlation $R_S(t)=0.535$. It is conformed that time-varying Gumbel Copula function is more accurate to describe the correlation between each unit or each failure mode. Moreover, the actual engineering data provided in literature [1] show that failure probability of the suspension system at 6000km is 48%, which is equivalent to reliability degree of the suspension system at 6000km is 0.520, and it is slightly lower than the calculation value 0.5208. The results prove the correctness of time-varying Copula model used in this paper to describe the correlation between units or multi-failure modes of suspension system.

6. Conclusion

Based on time-varying copula model (Patton model) and vine Copula function, reliability modeling and solution procedure of multi-failure modes correlation system is presented in this paper. Reliability degree of suspension system is 0.5208 at 6000km which is obtained by the above reliability modeling and solution procedure. Moreover, the actual engineering data show that reliability degree of the suspension system is 0.520. Therefore, the results prove the correctness of the reliability modeling and solution procedure.

References

[1] He Cheng-ming, Wu Wei, Meng Qingjun.A Copula-based Mechanical System Reliability Model and Its Application[J]. *Acta ArmamentarII*, 2012, 33(3): 379-382.

[2] Zhihua Wang, Zhonglai Wang, Shui Yu, Kewang Zhang. Time-dependent mechanism reliability analysis based on envelope function and vine-copula function[J]. *Mechanism and Machine Theory*, 2019 (134) : 667-684.

[3] Wang Zhonglai, Liu Jing, Yu Shui. Time-variant reliability prediction for dynamic systems using partial information[J]. *Reliability Engineering and System Safety*, 2020 (195): 106756.

[4] Zhao-Xia Xua, Xiao-Ping Zhou. Three-dimensional reliability analysis of seismic slopes using the copulabased sampling method[J]. *Engineering Geology*, 2018 (242) : 81-91.

[5] Árpád Rózsás a, Zsuzsa Mogyorósi. The effect of copulas on time-variant reliability involving time-continuous stochastic processes[J]. *Structural Safety*, 2017 (66) : 94-105.

[6] Y.T. Sun, L.F. Luo, Q. Zhang, X.R. Qin. Reliability analysis of stochastic structure with multi-failure modes based on mixed Copula[J]. *Engineering Failure Analysis*, 2019(105): 930-944.

[7] Pan Yue, Ou Shenwei, Zhang Limao, Zhang Wenjing, Wu Xianguo, Li Heng. Modeling risks in dependent systems: A Copula-Bayesian approach[J]. *Reliability Engineering and System Safety*, 2019(188): 416-431.

[8] Dan Xu, Mengli Xing, Qidong Wei, Yong Qin, Jie Xu, Yunxia Chen, Rui Kang. Failure behavior modeling and reliability estimation of product based on vine-copula and accelerated degradation data[J]. *Mechanical Systems and Signal Processing*, 2018(113): 50–64.

[9] Hu Qiguo, Zhou Song. Reliability Analysis of Failure Dynamic Mechanical System using Vine Copula Model[J]. *Mechanical Science and Technology for Aerospace Engineering*, 2018, 37(8): 1149-1155.

[10] Y. Wang, H. Pham. Modeling the dependent competing risks with multiple degradation processes and random shock using time-varying copulas[J]. *IEEE Transactions on Reliability*, 2012, 61(1): 13-22.

[11] Patton A J. Estimation of Multivariate Models for Time Series of Possibly Different Lengths[J]. *Journal of Applied Econometrics*, 2006, 21(2): 147-173.

[12] Daniel Kraus, Claudia Czado. D-Vine copula based quantile regression[J]. *Computational Statistics and Data Analysis*, 2017, 110: 1-18.

Visualization Analysis of Top Papers in the Subject Category of Agricultural Engineering Based on ESI

Baozhong Yuan[1,*] and Jie Sun[2]

[1]College of Plant Science and Technology, Huazhong Agricultural University, Wuhan city, Hubei province, PR China, 430070
[2]Library of Huazhong Agricultural University, Wuhan city, Hubei province, PR China, 430070

*Corresponding author email: yuanbz@mail.hzau.edu.cn

Abstract. All of 404 top papers for Agricultural Engineering were analyzed from 2009 to 2019, which included 396 highly cited papers and 13 hot papers. All papers were written in English, from 56 countries/territories. The 7 core journals were *Bioresource Technology, Industrial Crops and Products, Biomass Bioenergy, Transactions of the ASABE, Biosystems Engineering, Aquacultural Engineering, International Journal of Agricultural and Biological Engineering.* Top 3 countries and regions were Peoples R China, USA, India. The research were separated 6 clusters.

1. Introduction
Bibliometrics technique has been adopted in the subject category of Agricultural Engineering, such as production of bioenergy from biomass [1], Italian agricultural engineers [2], food waste anaerobic digestion [3], analysis of journal metrics of Agricultural Engineering categories [4], solid biofuels [5]. This paper analyzes papers in the subject category of Agricultural Engineering during 11 years period from 2009 to 2019, to understand research status.

2. Data and Methodology

2.1. Essential Science Indicators (ESI)
ESI database has been updated as of March 12, 2020, to cover an 11 year period, January 1, 2009 – December 31, 2019.

2.2. Data Analysis
It was completed on the single day on April 17, 2020. Four processes were separated to get the top papers in the field (Figure 1).

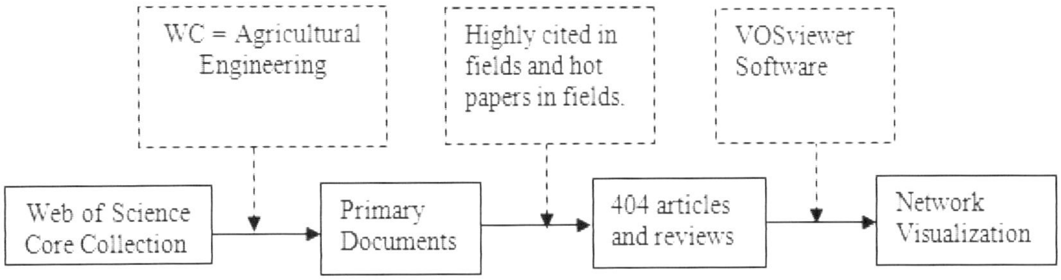

Figure 1. A framework of Agricultural Engineering.

2.3. VOSviewer

VOSviewer (version 1.6.15, 2020) was used to show the international countries collaboration and the research trends by all keywords [6].

3. Results and Discussion

3.1. Document Type and Language of Publication

Based on WoS Index, the 404 papers were SCIE (404), Conference Proceedings Citation Index-Science (10), Social Science Citation Index (3).

Among 404 papers, articles were 292 (72.277%), reviews (112, 27.723 %), proceedings papers (10, 2.475%), published in English. There are 396 highly cited paper and 13 hot papers.

3.2. Publication Output

Figure 2 shows the research trend. The average publication was 36.73, and the highest value is 70 in 2019. Total citations each year were power increased from 2009 to 2019. For 404 papers, the h-index is 165, and the average citation per item is 195.18.

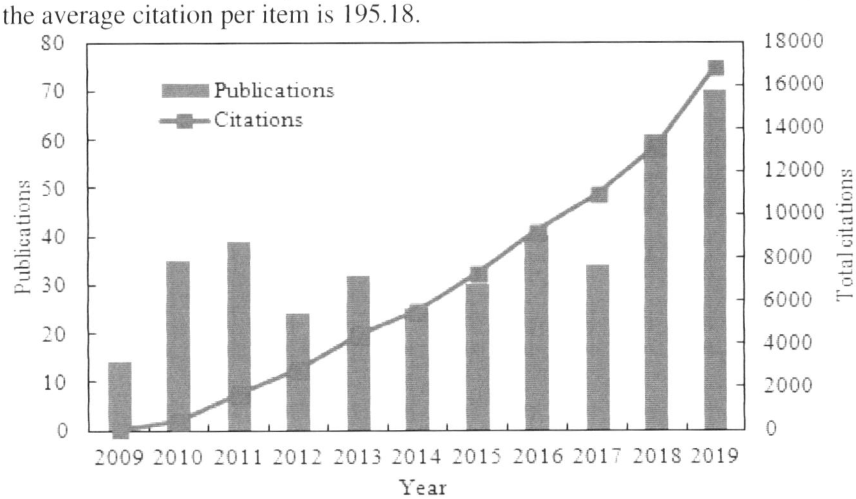

Figure 2. Number of top papers of Agricultural Engineering per year from 2009 to 2019.

3.3. WoS Categories and Research Areas

There are total 6 WoSC of Agricultural Engineering and 4 research areas (Table 1). The WoSC include Agricultural Engineering (404 papers, 100% of 404 papers), Biotechnology Applied Microbiology (276, 68.317 %), Energy Fuels (276, 68.317 %), Agronomy (116, 28.713 %), Agriculture Multidisciplinary (3, 0.743 %), Fisheries (2, 0.495 %). The research areas include Agriculture (404 papers, 100% of 404 papers), Biotechnology Applied Microbiology (276, 68.317 %), Energy Fuels (276, 68.317 %), Fisheries (2, 0.495 %).

Table 1. WoS Categories and Research Area for Agricultural Engineering.

Rank	WoS Categories	Total publications	% of 404	Research Areas	Records	% of 404
1	Agricultural Engineering	404	100	Agriculture	404	100
2	Biotechnology Applied Microbiology	276	68.317	Biotechnology Applied Microbiology	276	68.317
3	Energy Fuels	276	68.317	Energy Fuels	276	68.317
4	Agronomy	116	28.713	Fisheries	2	0.495
5	Agriculture Multidisciplinary	3	0.743			
6	Fisheries	2	0.495			

3.4. Core Journals

There are 13 Journals for Agricultural Engineering. All the 404 publications were published in 7 Journals. Table 2 displayed the IF2018 and IF5 year, JIF Quartile in Agricultural Engineering.

The 7 core journals were *Bioresource Technology* 254 top papers (62.871 %), *Industrial Crops and Products* (116, 28.713 %), *Biomass Bioenergy* (22, 5.446 %), *Transactions of the ASABE* (6, 1.485 %), *Biosystems Engineering* (3, 0.743 %), *Aquacultural Engineering* (2, 0.495 %), *International Journal of Agricultural and Biological Engineering* (1, 0.248 %). Based on Table 2, there are 3, 3, 1 journals ranked in Q1, Q2 and Q3.

Table 2. Journal indexed in the web of science during period from 2009 to 2019.

Rank	Journal	TP	Ratio of 404 (%)	IF 2018	IF5 year	QC
1	*Bioresource Technology*	254	62.871	6.669	6.589	Q1
2	*Industrial Crops and Products*	116	28.713	4.191	4.419	Q1
3	*Biomass Bioenergy*	22	5.446	3.537	4.062	Q1
4	*Transactions of the ASABE*	6	1.485	1.153	1.349	Q3
5	*Biosystems Engineering*	3	0.743	2.983	3.112	Q2
6	*Aquacultural Engineering*	2	0.495	2.143	2.553	Q2
7	*International Journal of Agricultural and Biological Engineering*	1	0.248	1.349	1.465	Q2

TP: total publication; QC: Quartile in Category of Agricultural Engineering.

3.5. Countries/Regions Co-authorship Analysis

All top papers were contributed by 56 countries/regions. In Figure 3, the threshold was set as 5, and 27 countries/regions were divided into 6 clusters. The first cluster consisted of 8 (red color), Spain, Canada, France, Italy, Iran, Czech Republic, Poland, Thailand. The second cluster consisted of 6 (green color), Germany, Sweden, Portugal, Netherlands, Belgium, Finland. The third cluster consisted of 5 (blue color), Australia, Taiwan, Malaysia, Egypt, Japan. The fourth cluster consisted of 4 (yellow color), India, South Korea, England, New Zealand. The fifth cluster consisted of 2 (violet color) of Peoples R China, Singapore. The sixth cluster consisted of 2 (shallow blue color) of USA, Brazil.

Figure 3. The country co-authorship network of Agricultural Engineering.

3.6. Organizations Co-authorship Analysis

The 583 organizations contribute 404 top papers. Table 3 lists the top 7 organizations above 6 papers were ranked by the number of total publications, and also show the total link strength, citations and average citations.

Table 3. Top 7 Organizations indexed using the web of science.

Rank	Organizations	Cluster	Records	Total link strength	Citations	Avg. citations
1	Hunan Univ	3	21	14	1691	80.5
2	Chinese Acad Sci	2	14	12	2570	183.6
3	Univ Technol Sydney	1	14	13	1730	123.6
4	Northeast Agr Univ	2	10	4	262	26.2
5	Tsinghua Univ	2	9	5	1791	199.0
6	Beijing Forestry Univ	2	8	2	1049	131.1
7	Korea Univ	4	6	4	246	41.0

3.7. All Keywords Co-occurrence Analysis

Of the all 2,901 keywords, there were only 199 keywords meet the threshold more than 5 times. There are 6 different viewpoints on Agricultural Engineering (Figure 4), they main focused on food waste disposal (red), pretreatment of lignocellulosic biomass (green), biomass and bio-oil (blue), adsorption of biochar (yellow), nanocomposites (violet), biodiesel and biodiesel production (shallow blue), respectively.

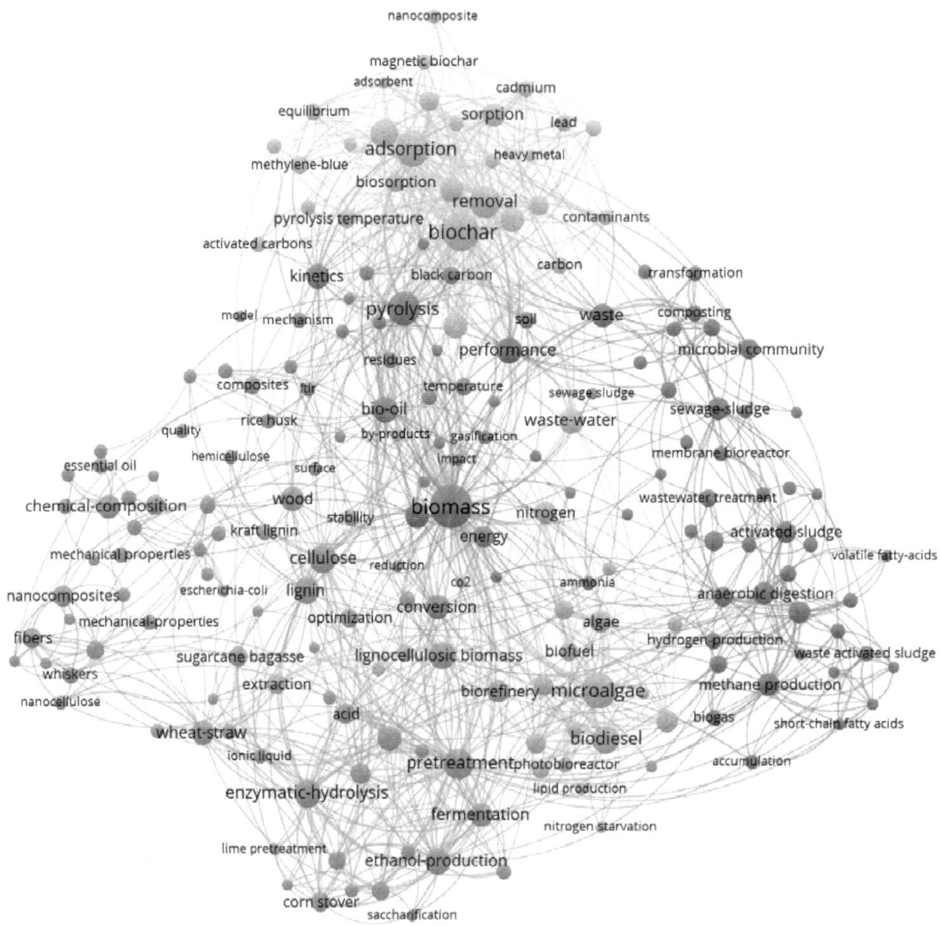

Figure 4. Mapping of most frequent all keywords in Agricultural Engineering.

3.8. The Most Frequently Cited Articles

Table 4 shows the top 4 papers of total citations more than 1000 times, which they are all from the Bioresource Technology.

Table 4. Top 4 highly cited papers with total citations more than 1000 times.

Rank	Title	Source Title	PY	Vol.	TC	APY
1	Pretreatment technologies for an efficient bioethanol production process based on enzymatic hydrolysis: A review	*Bioresource Technology*	2010	101(13)	2121	192.8
2	Pretreatments to enhance the digestibility of lignocellulosic biomass	*Bioresource Technology*	2009	100(1)	2111	175.9
3	An overview of the recent developments in polylactide (PLA) research	*Bioresource Technology*	2010	101(22)	1197	108.8
4	Influence of fatty acid composition of raw materials on biodiesel properties	*Bioresource Technology*	2009	100(1)	1012	84.3

4. Conclusions

The 404 top papers include 396 highly cited papers and 13 hot papers in the field. Results showed that these papers, all written in English, were 56 countries/territories, listed in 7 journals in the field of Agricultural Engineering. All keywords were separated into six clusters.

Acknowledgments
Funding: This research was funded grant number 2017YFD0301400.

References

[1] Konur Ozcan, 2012. The scientometric evaluation of the research on the production of bioenergy from biomass. Biomass and Bioenergy, 47, 504-515.
[2] Montis Andrea De, Amedeo Ganciu, Fabio Recanatesi, Antonio Ledda, Vittorio Serra, Mario Barra, Stefano De Montis, 2017. The scientific production of Italian agricultural engineers: a bibliometric network analysis concerning the scientific sector AGR/10 Rural buildings and agro-forestry territory. Journal of Agricultural Engineering, XLVIII(s1):635. 37-47.
[3] Ren Yuanyuan, Yu Miao, Wu Chuanfu, Wang Qunhui, Gao Ming, Huang Qiqi, Liu Yu. 2018. A comprehensive review on food waste anaerobic digestion: Research updates and tendencies. Bioresource Technology, 247, 1069-1076.
[4] Wang Yuanjie, Gassman Philip W., Wang Yingkuan, Pu Yingjun, 2018. Analysis of journal content characteristics and metrics reported in the Clarivate Analytics Journal Citation Reports and Web of Science Core Collection Agricultural Engineering categories. Int J Agric & Biol Eng, 11(5), 1-26.
[5] Knapczyk Adrian, Francik Slawomir, Fraczek Jaroslaw, Slipek Zbigniew, 2019. Analysis of research trends in production of solid biofuels. Engineering for Rural Development, 1503-1509. DOI: 10.22616/ERDev2019.18.N415
[6] Van Eck, N.J., Waltman, L. 2010. Software survey: VOSviewer, a computer program for bibliometric mapping. *Scientometrics* 84: 523-538.

AUTHOR INDEX